The Immunoglobulins

The Immunoglobulins

STRUCTURE AND FUNCTION

ROALD NEZLIN

Department of Immunology
The Weizmann Institute of Science
Rehovot, Israel

ACADEMIC PRESS
San Diego London Boston New York Sydney Tokyo Toronto

Front Cover Photograph: Courtesy of Dr. Alexander McPherson and Dr. Lisa Harris. Image based on coordinates by Lisa J. Harris, Steven B. Larson, Karl W. Hasel, and Alexander McPherson (1997). Refined structure of an intact IgG2a monoclonal antibody. *Biochemistry* **36,** 1581–1597.

This book is printed on acid-free paper.

Academic Press
15 East 26th St.,15th Floor, New York, New York 10010, USA
http://www.academicpress.com

Academic Press Limited
24-28 Oval Road, London NW1 7DX, UK
http://www.hbuk.co.uk/ap/

Library of Congress Cataloging-in-Publication Data

Nezlin, R. S. (Roal 'd Solomonovich)
The immunoglobulins : structure and function / Roald Nezlin.
p. cm.
Includes bibliographical references and index.
ISBN 0–12–517970–7 (hc : alk. paper)
1. Immunoglobulins. I. Title
QR186.7.N489 1998
616.07'98--dc21 98-4750
CIP

PRINTED IN THE UNITED STATES OF AMERICA
98 99 00 01 02 03 MM 9 8 7 6 5 4 3 2 1

Contents

Preface

Modern immunology evolved from revolutionary discoveries that occurred a hundred years ago, during the last decade of the 19th century. In 1890, Emil Behring and Shibasaburo Kitasato observed that sera obtained from rabbits previously injected with bacterial toxins could neutralize those toxins. Furthermore, they found that the injection of these sera into other animals conferred a long-term resistance to the toxins, and thus could be used as an effective therapeutic agent. These findings provided solid evidence that blood-born substances can mediate immune responses to toxins. Soon afterward, Paul Ehrlich showed that sera obtained from rabbits injected with plant toxins also had toxin-neutralization activity. The neutralizing activity of antitoxin sera was correlated by Ehrlich with a distinct group of macromolecules. Ehrlich hypothesized that cells have surface receptors with an affinity for a particular antigen. Upon detachment of the surface receptors from cells, the receptors could neutralize toxins. The high specificity of antitoxins was attributed by Ehrlich to structural complementarity between chemical groupings on the antitoxins and toxins. Based on the chemical concepts, he developed a quantitative method for the detection of antitoxins and toxins as chemical reactants. All these pioneering studies and novel concepts led to the birth of a new discipline, termed "immunochemistry" by Svante Arrhenius.

During the following decades, these concepts were corroborated and expanded. Karl Landsteiner, Michael Heidelberger, and other immunochemists showed that not only toxins from plants or bacteria can elicit specific immunological responses but also polysaccharides and a seemingly unlimited number of chemical substances (haptens). In the 1930s, the antigen-specific macromolecules in serum were described as a particular class of serum proteins, which were named antibodies. Elvin Kabat during his work in the laboratories of Arne Tiselius and Theodor Svedberg showed that the antibody activity of rabbit antiserum was due to 7S gammaglobulins. This was achieved by subjecting the antiserum to adsorption on the immunizing antigen, which depleted a significant portion of the 7S gammaglobulins from the immune serum. These classic experiments of Kabat initiated studies of antibodies at the molecular level.

A turning point occurred in 1958–1959 with the studies of Rodney Porter and Gerald Edelman. Porter developed a mild procedure for proteolysis of rabbit immunoglobulin G (IgG) using papain and succeeded in separating two large proteolytic fragments. One, the Fab fragment, can interact with antigens and contains one antigen-binding site. The other, a crystallizable Fc fragment, cannot combine with antigen, but has other important biological activities. Concurrently, Edelman separated the human IgG1 molecule into constituent subunits, the large (heavy) and small (light) peptide chains, by reduction of interchain disulfide linkages. In 1962, Porter suggested his famous structural model, according to which each IgG molecule is composed of four polypeptide chains (two identical heavy and two identical light chains) joined by disulfide bonds.

Due to the rapid development of molecular immunology in the 1960s and 1970s, the main features of immunoglobulin structure and biosynthesis were elucidated, as were the genetic mechanisms responsible for the enormous variability of antibodies. This tremendous body of knowledge about antibodies was presented in several books of that period (Kabat, 1968; Pressman and Grossberg, 1968; Nisonoff *et al.*, 1975; Nezlin,1977).

During the 1980s and 1990s, the accumulation of knowledge about immunoglobulins has continued at an accelerated rate. Precise structural data on how the antibody-combining site interacts with antigens has come from x-ray crystallographic studies and has advanced modeling of the antibody-combining sites. Intense effort has been directed toward furthering our understanding of the genetic mechanisms of antibody formation, which has facilitated the generation of new types of antibody molecules *in vitro*. In fact, some of these new antibodies have already been effectively used in biotechnology and medicine.

Achievements in molecular immunology have and continue to be published in a continous flow of original papers and reviews. This monograph provides one source in which recent concepts regarding both immunoglobulin structure and function have been summarized. Structural and functional properties of major immunoglobulin classes of human and animal imunoglobulins are treated at length. The localization and the structure of different binding sites of immunoglobulin molecules, including first of all the antigen-binding site, are discussed on the basis of latest x-ray crystallography studies. Emphasis has been placed on recently solved problems and their practical applications. Recent reviews are cited in each section where additional information and extensive relevant citations can be obtained for topics not extensively covered herein. Even with the tremendous advances in understanding the structure and function of immunoglobulins, achieved during the last century, there are still enough mysteries concerning the imunoglobulin molecules remaining to be solved, which presents an exciting challenge to us all.

REFERENCES

Kabat E.A. (1968). *Structural Concepts in Immunology and Immunochemistry.* Holt, Rinchart and Winston, New York.

Pressman D., and Grossberg A.L. (1968). *The Structural Basis of Antibody Specificity.* W.A. Benjamin, New York.

Nisonoff A., Hopper J.E., and Spring S.B. (1975). *The Antibody Molecule.* Academic Press, New York.

Nezlin R. (1977). *Structure and Biosynthesis of Antibodies.* Plenum Press, New York.

PART I

Structural Aspects

CHAPTER 1

General Characteristics of Immunoglobulin Molecules

Immunoglobulin molecules of all classes are heterodimers and composed from four polypeptide chains—two identical large or heavy (H) chains of about 50–60 kDa and two small or light (L) chains of about 23 kDa, linked by interchain disulfide bridges (Fig. 1). This prototype structure, first suggested by Rodney Porter in 1962, is common for all monomeric immunoglobulin molecules. Polymeric immunoglobulins of higher molecular weights are formed by 2–6 four-chain subunits, similar to the monomeric immunoglobulin molecules. They possess one or two additional peptide chains, which are important for formation and stabilization of immunoglobulin polymers. All immunoglobulin molecules are glycoproteins and contain two or more carbohydrate chains, usually linked to heavy chains. Membrane forms of immunoglobulins serve as a specific part of the B-cell antigen receptors. They have additional peptide stretches on the COOH-terminal ends of their heavy chains, with which they are embedded in cell membranes, as well as short portions of chains inside the cell (intracellular segments of different sizes). Membrane immunoglobulins are associated noncovalently with two accessory peptides forming the B-cell antigen receptor complex in the B lymphocyte membrane.

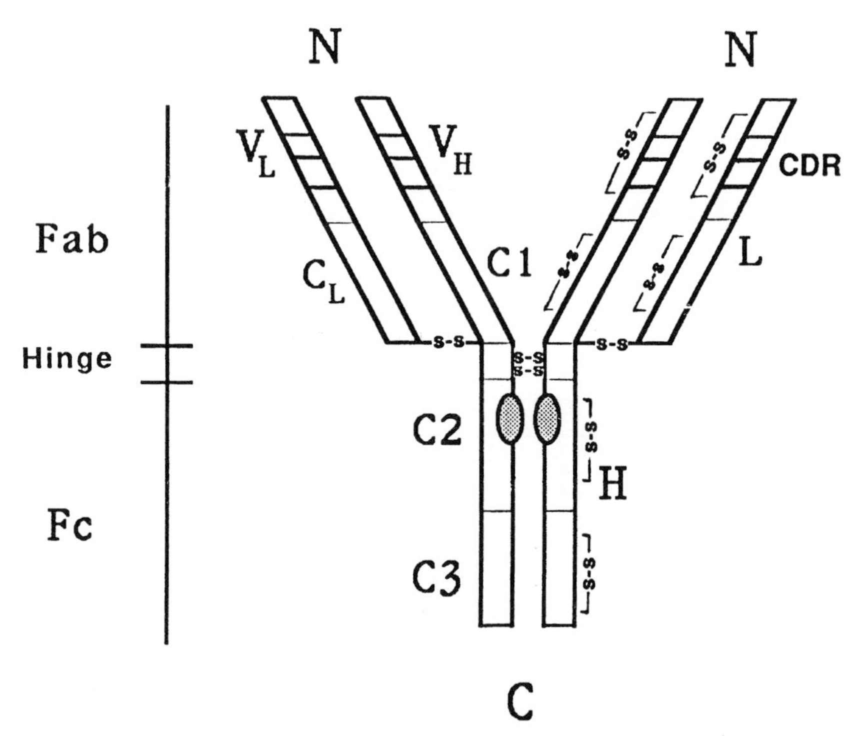

FIGURE 1 The basic four-chain structure of immunoglobulin molecules. H and L, the heavy and light peptide chains; N and C, NH_2- and COOH-terminal ends of the peptide chains; V_L and V_H, variable regions of the light and heavy chains; C1, C2, and C3, homology regions (domains) of the heavy chains; CDR, complementarity-determining region. The distribution of interchain and intrachain disulfide bonds are shown. Shadowed ovals point to localization of oligosaccharides in C_H2 domains. (Nezlin, 1994. Reprinted with permission from Marcel Dekker, NY.)

I. IMMUNOGLOBULIN CLASSES

In humans and rodents, there are five immunoglobulin classes or isotypes, which differ in the primary structure, carbohydrate content, and antigenic properties of their heavy chains (Table 1). By contrast, the light chain types are the same for all immunoglobulin classes. Each immunoglobulin molecule contains light chains of one of two types, either lambda (λ) or kappa ($\varkappa$). The λ and $\varkappa$ light chains have different primary structures and antigenic properties. They are usually free of carbohydrate components. The ratio of $\varkappa/\lambda$ chains in human and swine immunoglobulins is about 60:40, whereas in mouse, rat and rabbit immunoglobulins it is about 95:5. Some other mammals such as dog, cat, and farm animals (ox, sheep, and horse) have mainly λ chains and chicken immunoglobulins contain only λ chains.

TABLE 1 Properties of Major Immunoglobulin Classes

	IgG	IgA	IgM	IgD	IgE
Molecular formula	$\kappa_2\gamma_2$ $\lambda_2\gamma_2$ Monomers	$\kappa_2\alpha_2$ $\lambda_2\alpha_2$ Monomers, dimers (+SC,+ J chain)	$\kappa_2\mu_2$ $\lambda_2\mu_2$ Pentamers (+ J chain), hexamers	$\kappa_2\delta_2$ $\lambda_2\delta_2$ Monomers	$\kappa_2\varepsilon_2$ $\lambda_2\varepsilon_2$ Monomers
Subclasses	γ1–4	α1–2	—	—	—
Molecular weight	150,000	160,000	950,000	180,000	190,000
s_{20}, (w)	6.6S	7S, 9S, 11S	19S	7S	8S
Electrophoretic mobility	γ	Fast γ to β	Fast γ to β	Fast γ	Fast γ
Carbohydrates (%)	3	7.5	12	12	12
No. of oligosachharide chains per constant region	(1)	(5) (α1–2, +5 in hinge; α2–4 or 5)		(3, +4 in hinge)	(6)
Serum level (mg/ml)	~10	~2	~1.2	~0.04	$\sim 3 \times 10^{-4}$
Half life (days)	23 (IgG3: 8)	5–6	5	2.8	2.5
Complement fixation					
Classical	+	–	+	–	–
Alternative	–	+	–	–	–
Transplacental transfer	+	–	–	–	–
Binding to cells	Macrophages, polymorphs	—	—	—	Mast cells, basophils

(Nezlin, 1994. Reprinted with permission from Marcel Dekker, NY.)

A. Immunoglobulin G

Immunoglobulin G (IgG) is the major class of immunoglobulins. About three-quarters of all serum immunoglobulins belong to this class. IgG molecules consist of two heavy γ and two light chains ($2\gamma + 2L$). Normally each molecule of IgG has two identical antigen combining sites. Upon electrophoresis at alkaline pH, IgG migrates slower than almost all other serum proteins. Each IgG molecule contains about 3% carbohydrates, whereas that of other immunoglobulins is usually much higher (8–12%). After a secondary immunization, B lymphocytes secrete predominantly IgG molecules. IgG, unlike other immunoglobulins, can cross the placenta barrier and can penetrate into extravascular areas. IgG molecules are able to react with Fc_{γ} receptors present on the surface of macrophages and some other cells. The interaction with Fc receptors initiates various effector reactions and particularly facilitates the destruction of potentially harmful germs recognized by IgG antibodies. IgG molecules can also activate complement by the classical pathway. In humans and mice there are four IgG subclasses, which differ in the structure of their heavy chains (Table 2) (Shakib, 1990). Other animals have from one (rabbit) and two (sheep) to four (cattle) and five (horse and swine) IgG subclasses.

A cysteine residue near the middle of the human and mouse γ 1 chains (position 220) is involved in the formation of the heavy–light disulfide bridge with the penultimate cysteine residue of the light chains. In other IgG subclasses and other immunoglobulin classes, cysteine residues located between the V_H and C_H1 domains of the heavy chains (position 131) are involved in the formation of the heavy–light disulfide bridges with the penultimate cysteine residue of light chains. According to the x-ray crystallographic studies, both residues 220 and 131 are in close proximity in the three-dimensional globule, and the heavy chains with both variants of the heavy–light disulfide linkage have the same general structure.

B. Immunoglobulin M

Immunoglobulin M (IgM) is a high molecular weight protein (macroglobulin), consisting of five or rarely of six subunits (IgM monomers). Like IgG molecules,

TABLE 2 Properties of Subclasses of Human Immunoglobulin G

Properties	IgG1	IgG2	IgG3	IgG4
Percentage in serum (to total IgG)	66	23	7	4
Number of interheavy S–S linkages	2	4	11	2
Segmental flexibility	Good	Restricted	Very good	Restricted
Complement fixation	+	+	+	–
Placental transfer	+	±	+	+
Protein A binding	+	+	–	+
Protein G binding	+	+	+	+
Reaction with Fc receptors				
FcγRI	+	–	+	+
FcγRII	+	–	+	–
FcγRIII	+	–	+	–
Attachment to guinea pig skin	+	–	+	+

(Nezlin, 1994. Reprinted with permission from Marcel Dekker, NY.)

the IgM monomers are composed of two heavy and two light chains, which are linked together by disulfide bridges. The IgM monomers are found at a low concentration in human serum. Each pentameric IgM molecule is composed of 10 heavy (μ) chains, 10 light chains and usually one joining (J) chain (Metzger, 1970). The B cells can also secrete functionally active IgM hexamers lacking J chains but the amount of hexamers in serum is no more than 5% of total IgM (Brewer *et al.*, 1994). The carbohydrate content of IgM is high, about 12%. A pentameric IgM molecule has 10-antigen combining sites and can bind 10 small antigens (haptens). However, due to steric restrictions, only five large antigen molecules can be bound by one IgM molecule. The antibody activity of IgM is destroyed upon reduction of the intersubunit disulfide linkages. Such a reduction can easily be achieved with very low concentrations of reducing agents, such as dithiothreitol or mercaptoethanol.

IgM appeared early in evolution and primitive vertebrates have macroglobulins whose structure resembles that of mammalian IgM molecules. IgM are the first immunoglobulins synthesized by neonates and are the preponderant class of immunoglobulin molecules appearing during early phases of immune responses. In the monomeric form, IgM functions as an antigen-specific part of the B-cell antigen receptor on the surface of unstimulated B lymphocytes. The antigen receptors with the participation of the μ chains are very important for the normal development of B cells.

The polymeric IgM molecules are very efficient activators of the classical complement cascade. A single IgM molecule can activate complement component C1, whereas for that several IgG molecules are needed. The hexamer IgM molecules are much more efficient in activation of complement than pentameric IgM but they have nearly the same avidity to antigen as pentamers (Davis *et al.*, 1988; Randall *et al.*, 1990). Due to the multiplicity of their combining sites, IgM are very efficient in agglutination and cytolytic reactions. The so-called natural antibodies are mainly IgM and provide the first protection against invading microorganisms. Some "natural" IgM antibody molecules exhibit antiself activity (Potter and Smith-Gill, 1990; Avrameas, 1991; Mouthon *et al.*, 1996).

The B lymphocytes do not secrete pentameric IgM molecules but express IgM monomers on their surface. The plasma cells secrete only IgM polymers. The mechanism of the retention of monomeric IgM inside cells is linked at least partly with the carboxyl terminus of the μ chains and particularly with the Cys-575 residue: if this residue is mutated or removed, the monomeric IgM molecules are secreted at a high rate (Sitia *et al.*, 1990).

C. Immunoglobulin A

Immunoglobulin A (IgA), the third major class of immunoglobulins, plays the most important role in mucosal immunity. More IgA is produced than all other immunoglobulin isotypes combined. IgA molecules are present in serum, in the gut, and in exocrine secretions, such as saliva, colostrum, breast milk, and tears. The IgA molecules of higher vertebrates are synthesized mainly in gastro-intestinal lymphoid tissue. The gut and other mucosal surfaces, which are together 20 times larger than the surface of the skin, are the main sources of the pathogen invasion. On mucosal surfaces, IgA molecules inhibit the binding of microorganisms that try to penetrate through the mucosa, and thus prevent their invasion into the body (Brandtzaeg, 1985; Mestecky and McGhee, 1987; Kerr, 1990; Underdown and Mestecky, 1994; Hexam *et al.*, 1997).

The IgA isotype has been generated during evolution before segregation of birds and mammals. Like the γ chains, the heavy alpha (α) chains of mammalian IgA have three constant regions, whereas the chicken α chain has four (like the μ and delta [δ] heavy chains). The serum IgA molecules are usually monomeric (about 80%). The IgA level in serum is lower than that of IgG, but the synthetic rates of both immunoglobulins are nearly the same. The difference in the serum levels between IgG and IgA is due to the higher rate of IgA catabolism. The charge of the IgA molecules is more negative than that of most IgG molecules and IgA has β mobility on electrophoresis. In secretions IgA are found as dimers and, to a lesser extent, as trimers and tetramers. In addition to the heavy α and light chains, polymeric IgA contains two other polypeptides: a J chain and a secretory component (SC). The polymerization of IgA can occur in the absence of the J chain, which is, however, required for the efficient transportation of polymeric IgA into secretions (Brandtzaeg *et al.*, 1994). IgA dimers can be effectively bound by the poly(Ig) transport receptors and are more resistant to proteolytic digestion.

There are two IgA subclasses in humans, IgA1 and IgA2, with two allelic variants [A2m(1) and A2m(2)]. IgA1 comprises most of the IgA in the serum (up to 90%) but 60% of IgA molecules in the gastrointestinal tract belongs to the IgA2 subclass. The constant region of IgA1 differs from that of IgA2 at 22 residue and there is a 13–residue deletion in the hinge region of IgA2 as compared with the hinge of IgA1. Two IgA subclasses were also found in hominoid primates. The "novel" human IgA2 probably represents a third IgA2 allotype (Chintalacharuvu *et al.*, 1994). It is a product of recombination or gene conversion between two IgA2 alleles. Human IgA2m(1) molecules usually lack disulfide bridges between the heavy and light chains and instead the light chains are linked covalently by a disulfide bond. However, from supernatants of a myeloma line two types of IgA2m(1) molecules have been isolated: one of them lacks the heavy–light disulfide bridge and another, less abundant, has such a bond (Chintalacharuvu and Morrison, 1996).

In mice only one isotype of the alpha chain was found, which has two allelic variants. One of the alleles has no disulfide linkage between heavy and light chains, similar to human IgA2m(1). In rabbits 13 IgA isotypes were found, which differ mainly in C_H1 domains and hinge regions (Burnett *et al.*, 1989). IgA molecules are incapable as a rule of activating the classical complement pathway or binding the C1q complement component. The alternative complement pathway can be activated by aggregated IgA molecules.

D. Immunoglobulin D

Immunoglobulin D (IgD) is a minor class of immunoglobulins. These molecules are monomers composed of two heavy and two light chains ($2\delta + 2L$), without any additional chains. They are extremely sensitive to proteolysis. In membrane form, IgD molecules together with IgM monomers are present on the surface membranes of human and murine mature B lymphocytes and serve as an antigen-specific part of the B-cell antigen receptors. The quantity of IgD molecules on the surface of B cells is an order of magnitude higher than that of IgM. Since the discovery of this class of immunoglobulins (Rowe and Fahey, 1965) a number of studies were performed aimed at understanding the biological role of this protein. However, still the precise functions of membrane IgD are still not known. In IgD deficient mice obtained by gene targeting, the number of mature B cells is the same as in normal mice and antibody response is not different from that in wild type animals (Roes and Raewsky, 1993). IgD molecules can anchor on the cell membrane via a glycosyl phosphatiylinositol linkage.

E. Immunoglobulin E

Immunoglobulin (IgE) is another minor class of immunoglobulins, molecules of which are present in serum at very low concentrations. IgE molecules exist in a monomeric form consisting of two heavy and two light chains ($2\varepsilon + 2L$) and are the most important participants of allergic reactions. Through their Fc portions, IgE molecules bind to the Fc_ε receptors that are present on the surface of mast cells and basophils. The crosslinking of such membrane bound IgE antibodies by multivalent antigens, triggers the release of chemically active substances, such as histamine, leukotrienes, prostaglandins, and chemotactic factors, from the cells. These substances initiate allergic and inflammatory reactions and serve as chemoattractants for other cells. The serum level of IgE antibodies in patients with allergic conditions (hay fever, for example) or with chronic parasitic infections are usually elevated several hundredfold and can be used for diagnostic purposes.

The second secreted isoform of the epsilon chains that differs from the "classical" one in the C-terminal end was described. This epsilon isoform has an additional stretch of six amino acid residues with C-terminal cysteine (Cys-554), which is characteristic for μ and α chains. Cys-554 residue forms an interchain disulfide bond between epsilon chains of the same molecule but does not induced polimerization of the molecule. The IgE molecules with the second epsilon isoform are secreted from B cells and their C-terminal cysteine does not function as a signal for intracellular retention and degradation. There are differences in extent of glycosylation between both epsilon isoforms (Batista *et al.*, 1996b). Two different human epsilon chain isoforms are found as a part of the membrane IgE antigen receptor. In the "long" IgE variant the extracellular membrane-proximal domain has an extra 52-residue stretch, which is absent in the "short" variant. These isoforms have different functional properties. Cross-linking of the IgE molecule with the short epsilon isoform located on cell membrane leads to growth inhibition of the B cells transfected with the corresponding ε gene. Such effect was not found in similar experiments with the second IgE isoform (Batista *et al.*, 1996a).

II. HETEROGENEITY OF IMMUNOGLOBULINS

Immunoglobulin preparations obtained from normal sera are very heterogeneous and usually contain molecules that differ in most of their physical, chemical, and immunological properties (e.g., electrophoretic mobility, antigenic properties, and affinity to antigens). This heterogeneity is the result of the amino acid sequence diversity of immunoglobulin peptide chains, first of all of their N-terminal variable parts, and in the past it hampered the study of the structure and functional properties of immunoglobulins.

In the 1960s, proteins contained in large amount in the sera of patients with multiple myeloma and Waldenström macroglobulinemia (so-called myeloma proteins) were shown to be homogeneous immunoglobulins. Each of them is the product of only one clone of transformed B lymphocytes. All the classes and subclasses of human immunoglobulins were found among homogeneous myeloma proteins. About the same time, Bence Jones proteins, discovered a century ago in the urine of patients with myeloma and used for diagnosis of this neoplastic disease, were identified as the homogeneous immunoglobulin light chains secreted by these same transformed B lymphocytes (Edelman and Gally, 1962; Stevens *et al.*, 1991).

Of equal importance was the discovery that artificially induced mouse plasma cell tumors produce homogeneous immunoglobulins, analogous to human myeloma proteins (Potter, 1972). Later the plasma cell tumors were obtained in rats (Bazin *et al.*, 1988). Almost all the major advances of the early studies of immunoglobulin structure (peptide chain sequencing, disulfide bond arrangement, antigenic properties, and so on) were achieved using human and rodent myeloma proteins.

The development of the hybridoma technique to produce homogeneous rodent monoclonal antibodies with the desired specificity to antigen (Köhler and Milstein, 1975) revolutionized nearly all areas of immunology and further eliminated most difficulties encountered when studying heterogeneous immunoglobulins. New methods have been developed that permit the generation of antibodies, including antibodies of human origin, that cannot be prepared by conventional hybridoma techniques (Winter and Milstein, 1991). They include protein engineering methods to produce human-like antibodies (chimeric and humanized antibody molecules) and the creation of transgenic mice expressing human antibodies. With antibody phage display, homogeneous antibodies can be produced with desired affinity and specificity completely *in vitro* bypassing immunization (see Chapter 3).

III. DETECTION AND ISOLATION OF IMMUNOGLOBULINS

A. Detection of Immunoglobulins

In biological fluids and in cell culture media, immunoglobulins are present with many other proteins. To detect immunoglobulins, various immunological methods that are based on their specific antigenic properties or antibody activity, are applied. Different commercial anti-immunoglobulin antisera are used in several variants of semiquantitative precipitation assays (Johnstone and Thorpe, 1982) and the popular enzyme-linked immunoadsorbent assay (ELISA) technique (Kemeny, 1991). To determine the exact quantity of immunoglobulins in a sample, anti-immunoglobulin antibodies can be immobilized on one or another insoluble support, such as cyanogen bromide-activated Sepharose or small particles of modified cellulose. The resulting immunoadsorbents are used to specifically and quantitatively adsorb immunoglobulins from the sample. The amount of adsorbed immunoglobulins can be measured by one of the assays for protein detection. If the immunoglobulin preparation to be studied is radiolabeled, the sensitivity of the immunoadsorbent technique is enhanced greatly. The immunoadsorbent pellet with adsorbed immunoglobulins are put on filter paper disks and the radioactivity is measured in a scintillation counter. To ensure that all immunoglobulins are adsorbed, an additional portion of the immunoadsorbent can be added to the sample after the sedimentation of the first portion.

To detect and quantitate immunoglobulins with specific antibody activity, several methods were devised, based on their interaction with antigens (Harlow and Lane, 1988; Coligan *et. al.*, 1991; Herzenberg *et al.*, 1996). Such quantitative assays are essential to immunochemists. The first method for antibody quantitation, developed by Michael Heidelberger, is based on the determination of the amount of protein in the precipitate formed after addition of a slight excess of the relevant antigen to a sample of an antiserum. This analytical test was an important step for the progress of immunochemistry and was later used to calibrate many semi-quantitative assays (Kabat and Mayer, 1961). However, Heidelberger's method cannot detect nonprecipitable antibodies. In this respect, the immunoadsorbent technique is advantageous, since antigens immobilized on insoluble supports (Sepharose, cellulose, plastics, etc.) bind all kinds of antibodies. If a large number of samples must be studied regularly, as in clinical laboratories, modern automated immunoassay systems are recommended (Chan, 1996).

Immunoadsorbents generated from small cellulose particles are well suited for analytical as well for preparative purposes (Gurvich, 1964). They are especially effective for studying small samples, because cellulose particles pack well upon centrifugation, even at low speed, the nonspecific adsorption to cellulose is negligible, and the specific capacity is very high. One of the best and perhaps the cheapest matrix for immunoadsorbents is dialdehyde cellulose (Gurvich and Lechtzind, 1982). The small particles of dialdehyde cellulose are very easily prepared even in a small laboratory. They are very stable and can be stored in cold as a water suspension for years. For the column technique, the dialdehyde cellulose can be prepared in the form of beads. The immunoadsorbent can be applied for quantitative detection or preparation of large quantities of either antigens or antibodies. The small particles of dialdehyde cellulose can be effectively used also for immunization in the form of antigen–cellulose complexes (Gurvich and Korukova, 1986). The preparation and usage of this immunoadsorbent was recently described (Nezlin, 1997).

If an antibody specifically recognizes a low-molecular weight substance (hapten), the equilibrium dialysis method can be used to quantitate the antibody and to assess its parameters, such as heterogeneity, valence, and affinity (Eisen and Karush, 1949; Eisen, 1990; Day, 1990; van Oss, 1994).

B. Isolation of Immunoglobulins

Several methods to isolate immunoglobulins based on their physical and chemical differences from other serum proteins have been devised. One of the most useful techniques to purify IgG involves precipitation with sulfate ions and subsequent chromatography on DEAE-cellulose (Manson, 1992). The high affinity of the staphylococcal proteins A and G for IgG has been widely used to obtain highly purified preparations of IgG of most mammalian species (Godfrey, 1997). The capacity of these proteins immobilized on Sepharose is very high, especially for human IgG (Kerr and Thorpe, 1994). However, the efficiency of such adsorbents for rodent IgGs is lower.

IgG subclasses can be isolated by several methods. For example, IgG3 is the only human IgG subclass that does not react with protein A. Therefore, application of an IgG pool to a protein A–Sepharose column will result in an effluent containing only IgG3. Various immunoglobulin subclasses can be separated on the basis of their differential susceptibilities to proteolysis. For instance, the papain resistance of human IgG2 and trypsin resistance of rat IgG2a facilitates their separation from other IgG subclasses.

To isolate large quantities of IgA for passive protection on mucosal membranes is not an easy task. However, a procedure based on classical chromatographic methods was described to isolate IgA from different sources, such as milk, bile, hybridomas and transfected cells, on a laboratory scale as well as to separate different forms of these immunoglobulin molecules (Lüllau *et al.*, 1996). For isolation of human IgA1, a lectin, Jacalin, can be used. It has a high specificity for *O*-linked oligosaccharide located in the IgA1 hinge and only few other serum glycoproteins are bound by Jacalin (see Chapter 6).

IV. PROTEOLYTIC AND CHEMICAL FRAGMENTS OF IMMUNOGLOBULINS

A. Fragmentation by Proteinases

Immunoglobulin molecules can be digested under mild conditions by proteolytic enzymes of different origin into large fragments. After the classical publication by Porter (1959) on the limited papain hydrolysis of rabbit IgG into Fab and Fc fragments, a number of other enzymes, including pepsin (Nisonoff *et al.*, 1975), was used for fragmentation of immunoglobulins from many species. These studies not only furthered understanding of immunoglobulin structure on different levels and localization of various antibody activities, but also development in many other areas of immunology.

The immunoglobulin domains are compact structural units and without denaturation they cannot be degraded by proteolysis. Therefore, proteolytic enzymes digest immunoglobulin chains in the regions connecting domains (Fig. 2). The middle of the heavy chains (the hinge region) has an unusual configuration accessible for different proteolytic enzymes. Usually, all immunoglobulins are digested in mild conditions by proteases from different sources at the hinge with formation of large proteolytic fragments that are stable to further proteolysis (Fig. 2).

Papain and trypsin cleave gamma chains at the NH_2-terminal side of disulfide bridges between heavy chains, to generate two univalent Fabs with antigen binding capacity and Fc. A short exposure of human IgG2, IgG1 and IgG3 to papain results in the formation of $F(ab)_2$, Fab/c (one Fab + Fc), and Fch

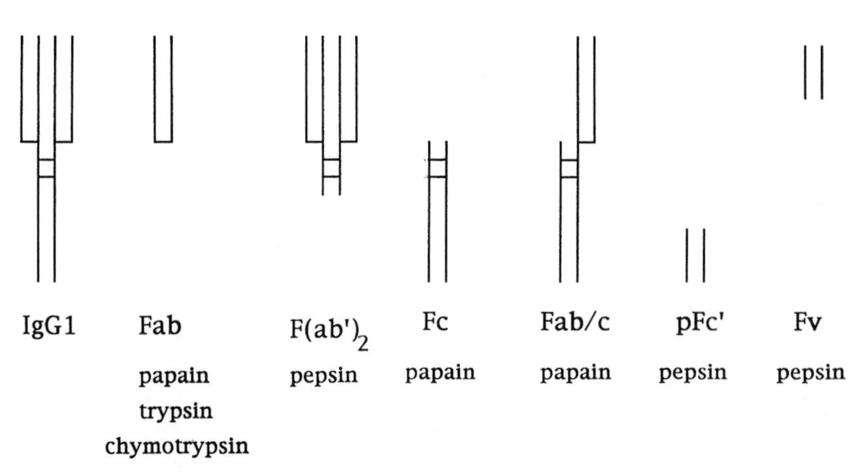

FIGURE 2 Proteolytic fragmentation of the IgG1 molecule. Different types of proteolytic fragments are shown (Nezlin, 1994. Reprinted with permission from Marcel Dekker, NY.)

(Fc + hinge) fragments, respectively. Digestion by papain is more rapid in the presence of reducing agents such as cysteine or mercaptoethanol. Fab/c fragments can also be obtained by trypsin digestion of human IgG1.

The sensitivity of human IgGs to proteolysis varies according to their subclasses and reflects the length and the primary structure of their hinge regions (Michaelsen, 1990). Human IgG3 and IgG1 are more sensitive than IgG2 and IgG4 to proteolytic cleavage in the presence of reducing agents. If activated papain is used without reducing agents, only IgG1 and IgG3 are digested. For digestion of IgG2 and IgG4, the reduction of the interchain disulfide is necessary. The hierarchy of the sensitivity of mouse IgGs to proteolytic cleavage is IgG2b > IgG3 > IgG2a > IgG1 (Parham, 1983). These variations in proteolytic fragmentation can be used to isolate fragments from different IgG subclasses. The native Fab fragments are resistant to the action of proteolytic enzymes and cannot be digested without denaturation, whereas Fc, which has less compact structure, is degraded by pepsin and papain at acidic pH.

Pepsin cleaves the hinge region of IgG on the C-terminal side of the inter-heavy disulfide bridges. After peptic digestion at pH 4–4.5, both bivalent $F(ab')_2$ and the C-terminal domain of heavy chains (pFc′) can be isolated. This pepsin treatment hydrolyses the C_H2 domain into smaller fragments. That the native structure of Fc is destroyed below pH 5.1 (Abaturov *et al.*, 1969) is explained by the sensitivity of Fc to pepsin hydrolysis under acidic conditions. At acidic pH, papain also cleaves Fc fragments into smaller pieces. After reduction of its disulfide bridges, the pepsin-derived $F(ab')_2$ dissociates into univalent Fab′. Upon removal of the reducing agent, spontaneous reoxidation of the bridges can occur, resulting in the reappearance of bivalent $F(ab')_2$. Hybrid $F(ab')_2$ fragments can be obtained if Fab′ fragments from different antibodies are present in the mixture being reoxidized (Nisonoff *et al.*, 1975). Pepsin treatment at pH 3.5–4 of mouse IgG1 can generate $F(ab')_2$ from this subclass even in nonpurified hybridoma ascitic fluids or cell culture supernatants, whereas IgG2 molecules are degraded in such conditions (Parham, 1983).

Mouse IgG2a has two protease susceptible sites, one on each side of the interheavy disulfide bridges (Fig. 3). Proteases first cleave the IgG2a heavy chain on the C-terminal side of the disulfide bridges to form $F(ab')_2$, and then cleave it on the N-terminal side of the hinge region to yield monovalent Fab. Exposure of murine IgG2b, whose major protease susceptible site is on the N-terminal side of the hinge region, to proteases results mainly in the formation of Fab. Significant amounts of Fab/c fragments are also generated by this treatment, apparently due to the glycosylation of a threonine residue on only one of the two heavy chains. This glycosylated threonine residue is on the N-terminal side of the interheavy chain disulfide bridges and prevents the action of proteases. In the absence of reducing agents, mouse IgG1 molecules are digested by different proteolytic enzymes to $F(ab')_2$ (Parham, 1983).

Classes of immunoglobulins other than IgG can also be cleaved by proteolytic enzymes. Proteases can lead to formation of Fab, $F(ab')_2$ and Fc_5 from IgM; Fab from IgA; and Fab and Fc from IgE. The Fc portion of IgA is hydrolysed into smaller pieces by papain. IgD is extremely sensitive to proteolysis and is rapidly cleaved to yield Fab and Fc fragments.

There are several sites in the hinge region of human IgA1 molecules that are susceptible to highly specialized proteinases of several very pathogenic bacteria that can colonize or infect human mucosal membranes. They include pathogens, which cause meningitis, respiratory and urogenital diseases, and

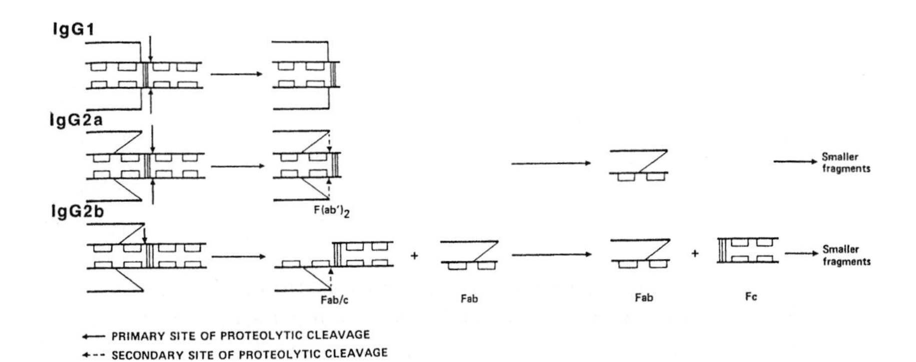

FIGURE 3 The proteolytic fragmentation of IgG1, IgG2a, and IgG2b of BALB/c mice (Parham, 1983. Reprinted with permission from The American Association of Immunologists.)

plaque formation on teeth. The bacterial proteolytic enzymes attack the Pro–Ser or Pro–Thr peptide bonds of the IgA1 hinge segment absent in the IgA2 hinge region (Fig. 4). The proteolysis yield the intact monomeric Fab and Fc fragments of IgA1 (Kerr, 1990; Kilian and Russell, 1994).

A minimal antigen-binding fragment, Fv, was obtained by pepsin digestion of a mouse myeloma IgA MOPC 315 (Givol, 1991). Fv consists of only the variable regions of the heavy and light chains and retains the complete antigen-binding capacity of the parent IgA. Different Fv proteins are prepared by genetic engineering methods.

Trypsin digestion of human and rat kappa light chains yields variable and constant halves without significantly changing their general structure. Isoelectrofocusing of such digest of rat κ light chains results in the concentration of the constant domain in one main band and the distribution of the variable domains in many thin bands with different pI ("spectrum of variability") (Fig. 5).

B. Chemical Fragmentation

Treatment of rabbit, horse, and human IgGs with cyanogen bromide under mild conditions can generate an active bivalent $F(ab'')_2$ fragment (Lahav *et al*., 1967). This chemical $F(ab'')_2$ fragment is similar to the pepsin-generated $F(ab')_2$ fragment. The chemically generated monovalent Fab fragment is slightly larger than the Fab′ fragment obtained by digestion with pepsin.

The Fab/c fragments of rabbit, human, and mice IgG are obtained by cleavage at cyanocysteine residues (Wines and Easterbrook-Smith, 1991). At first IgG molecules were reduced mildly by either dithiotreitol or sulfite and S-cyanocysteines (SCN) generated by 2-nitro-5-thiocyanobenzoic acid or KCN, respectively. The Fab/c fragment is produced when base-catalyzed cleavage of

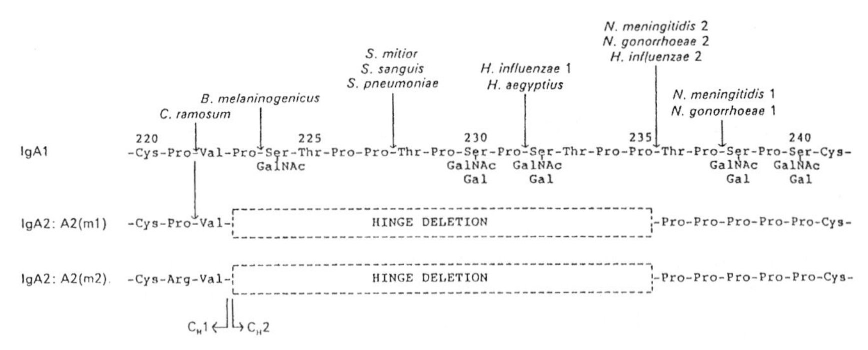

FIGURE 4 The hinge region of human IgA1 and IgA2 [allotypes A2(m1) and A2(m2)] showing the cleavage sites of various bacterial IgA-specific proteinases. (Kerr, 1990. Reprinted with permission.)

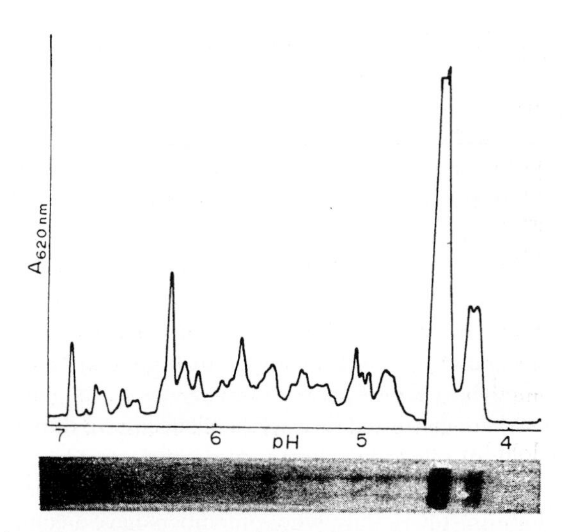

FIGURE 5 Electrofocusing patterns of polyclonal rat $\varkappa$ chains digested by trypsin into halves. Stained gel isoeletrophoregramm (below) and its densitometric scanning is shown. The large peak at pI 4.5 is represented in the constant part of the chains and the variable parts are distributed in the range of pI from 4.7 to 7. (Nezlin, R., *et al.*, (1975). Reprinted with permission from Elsevier Science, Ltd.)

the peptide bond occurs at an SCN residue in the hinge region of one heavy chain. By a similar procedure the Fv fragment was obtained from a human macroglobulin (Rodwell and Karush, 1978). After reduction of the disulfide linkage between the light and heavy chains, the resulting sulfhydryl groups are converted to the S-cyanoform by treatment with 2-nitro-5-thiocyanobenzoic acid. Cleavage of the mu chain is effected by incubation at pH 9 and V_H fragment, the only one-domain subunit after cleavage, is separated by gel filtration. This method is probably generally applicable for isolation of V_H from IgM of all species.

V. IMMUNOGLOBULIN PEPTIDE CHAINS

A. General Features

Antibodies are multifunctional molecules. They are able to react specifically with antigens and with many other types of ligands such as complement components, specific cell receptors, bacterial proteins, and so on. Different kind of binding sites located at various parts of the molecule participate in these reactions. It is not surprising, therefore, that the primary structure of the heavy and light chains have characteristic features that distinguish them from the peptide chains of other proteins.

1. *The main structural elements of immunoglobulin chains are homology units.*
   According to the amino acid sequence studies, immunoglobulin chains are composed of homology units of about 110 amino acid residues long. Each unit has one intrachain disulfide bond forming a loop of about 60 amino acid residues. The homology units fold in an identical manner to form compact globules (immunoglobulin folds or domains), each of them plays a definite functional role. The light chains have two homology units, whereas the γ, δ and α heavy chains have four homology units and the μ and ε heavy chains have five. Apparently, the genes coding for the heavy and light chains were generated by a series of successive duplications of a primordial gene that coded for a short peptide of about 110 residues long (Hill *et al.*, 1966). Similar units are also present in proteins belonging to the immunoglobulin superfamily, whose members have a common evolutionary origin (Williams and Barclay, 1988).
2. *Disulfide bridges are arranged in a specific way.*
   Usually each homology unit has one disulfide bond, which are distributed in a characteristic, symmetrical order along the immunoglobulin chain (Fig. 1). The exceptions are the C_H1 domains of human IgE and amphibian, chicken, rabbit, and ruminant IgGs, which have additional intrachain disulfide bridges (Fig. 6). Some rabbit $\varkappa$ light chains have an additional disulfide bridge linking their constant and variable domains. Nearly all disulfide bonds between the heavy chains occur in the hinge regions of these chains.
3. *Immunoglobulin chains are composed of variable and constant regions.*
   Early studies using peptide mapping indicated that various light chains contain some common peptides while the structure of others is varied (Putnam *et al.*, 1967). After the first sequencing experiments (Hilschman and Craig, 1965), it became clear that immunoglobulin chains are composed of two distinct regions, a variable (V) and a constant (C) region. The structure of the variable region determines the ability of antibodies to combine with antigens, while the constant region is characteristic for a given immunoglobulin class or subclass. Immunoglobulin functions, other than interaction with antigens (so-called effector functions), depend mainly on the structure of the constant region.

   The amino acid sequences of the variable and constant regions are coded by separate sets of genes. Various constant regions (e.g., different isotypes of heavy chains) can be united in a single chain with one of the variable regions, and various the variable regions (e.g., with different allotypic specificities) can form a single chain with one definite constant region.

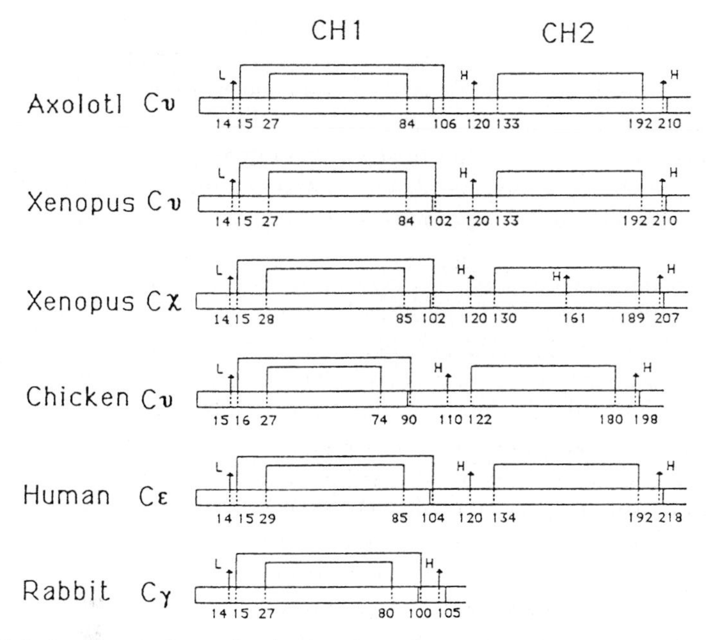

FIGURE 6 Intradomain (square brackets) and interchain disulfide bridges to heavy and light chains (arrows) in C_H1 and C_H2 domains of heavy chains of different species. Cysteine residues are enumerated from the beginning of the C_H1 domain (Fellah *et al.*, 1993. Reprinted with permission.)

4. *A specific hinge region is locate in the middle of the heavy chains.*
   A segment located in the middle of γ, α and δ chains (the hinge region) has no homology with the other sections of the heavy chains. There are significant variations in the sequence of these regions of heavy chains of different classes and subclasses. The heavy chains of IgM and IgE lack the typical hinge region and in its place there are additional homology units. Due to their peculiar secondary structure, hinge regions are very sensitive to different proteolytic enzymes. Segmental flexibility, an important functional property of immunoglobulin molecules, also depends on the hinge structure.

B. Variable Regions

A large number of heavy and light chains from different species have been sequenced during the past decades. These studies culminated in the description of the structure of the entire immunoglobulin molecule (IgG1 Eu) (Edelman *et al.*, 1969). The results of sequencing performed before 1991 are summarized in a special publication (Kabat *et al.*, 1991). Analysis of the primary structures of the variable regions helps in understanding the basic principles of their organization as well as in elucidating the genetic mechanisms that determine the biosynthesis of these regions.

The two types of light chains ($\varkappa$ and λ) have different groups of variable regions ($V_\varkappa$ and V_λ), whereas all heavy chains have common variable regions (V_H). Each of these variable region groups can be divided into families according to their homology. Members of each family are more similar to each other (about 70% or greater homology) than to members of different families (less than 60% homology). Human V_H sequences are grouped into 6 families (Matsuda and Honjo, 1996), each with an intrafamily similarity of more than 80%, and murine V_H sequences are grouped into 14 families (Honjo and Matsuda, 1995).

For analysis of variability at different positions throughout the immunoglobulin chains a parameter (v, variability) was suggested. It is defined as the number of different amino acids that occur in a given position divided by the frequency of the most common amino acid found at that position (Wu and Kabat, 1970). The variability plots for human heavy and light chains are presented in Figure 7. The variability distribution along variable regions of both chains is not random, but is concentrated in three clusters (called the hypervariable sequences), located around position 30, between positions 50 and 60, and between positions 90 and 100. The residues of these hypervariable sequences, the complementarity-determining regions (CDRs), are responsible for contacts with antigens. In a three-dimensional fold, they form the antigen-combining site, which has been confirmed by x-ray crystallography studies of antibody molecules. The differences between variable regions are due to amino acid exchanges, insertions, and deletions. The most variable is the third CDR of the heavy chains. This CDR3 varies in composition as well as in length, which accounts for some V_H regions being longer than others. The variable region residues outside the CDRs are more conserved and are organized in four framework regions (FR). These regions are involved mainly in maintaining the three-dimensional structure of the variable regions and the contacts between the V_H and V_L regions. However, in many antibodies some residues of the framework regions participate in direct contacts with antigen.

The variable regions are coded by two (V_L) or three (V_H) variable gene segments, which rearrange at early stages of B-lymphocyte development. The variable gene segments for the light chains code amino acid residues to position 95 (from FR1 to CDR3) and J gene segments from positions 97 to 107 for $V_\varkappa$ and 108 for V_λ (FR4). The variable gene segments for the heavy chains code to position 94 and the rest of V_H region is coded by two other gene segments, D and J_H (Kabat *et al.*, 1991). The boundaries between sections coded by these three gene segments are variable.

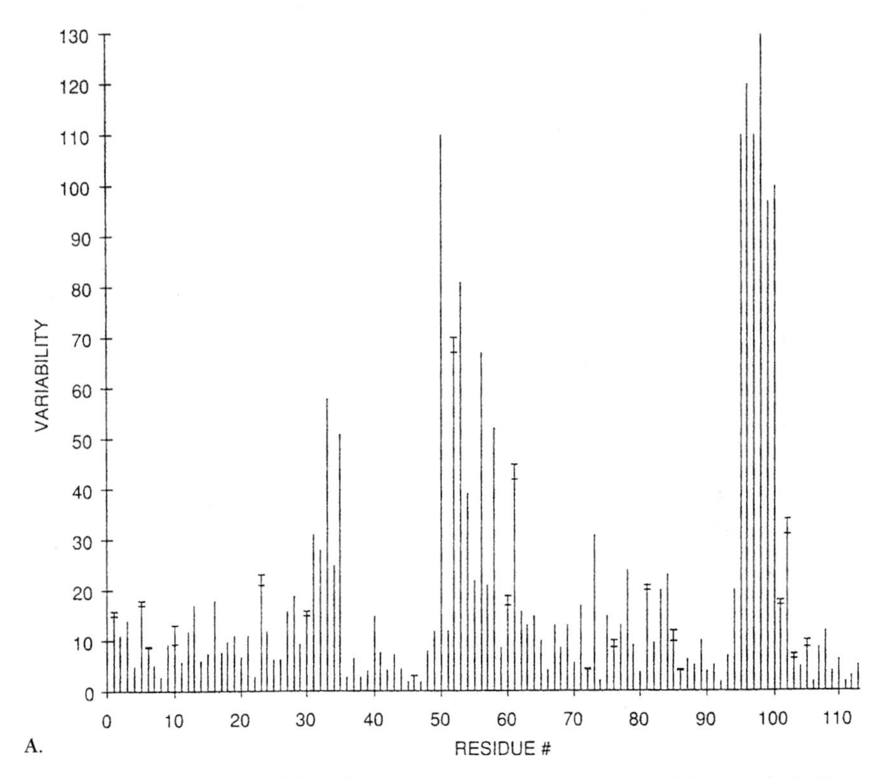

FIGURE 7 Amino acid variability plots according to Wu and Kabat. Variability is remarkable in three areas of variable regions of (A) heavy and (B) light chains. (Kabat, E.A., *et al.*, (1991). Reprinted with permission from NIH.)

In V_H there are conserved motifs, Trp/Phe–Gly–X–Gly (residues 103–106) and Gly–Leu–Glu–Trp–Leu/Ile (residues 44–48), which provide formation of β-bulges in strands G and C′, respectively. In V_L there are conserved sequences (residues 98–101 and 44–48) that participate in similar β-bulges (Colman, 1988; Carayannopoulos and Capra, 1993).

C. Constant Regions

The constant region of all light chains has only one homology unit, whereas the same regions of heavy chains consist of three or four homology units. Despite their general similarity and the presence of some conserved residues, especially cysteines and tryptophans, the homology between variable and constant regions

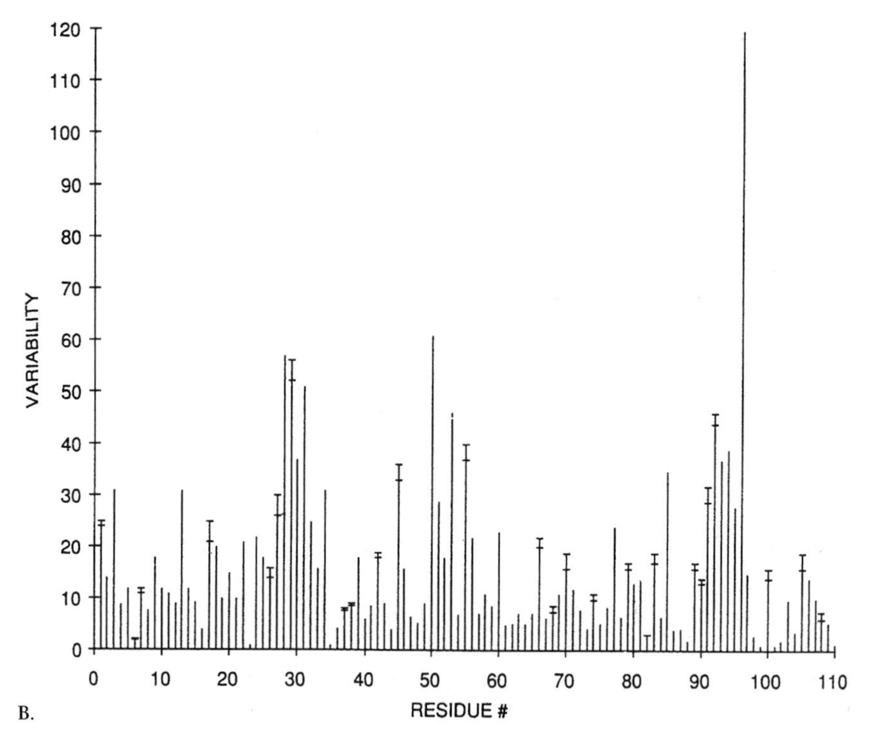

FIGURE 7 *(Continued)*

is small, (about 15%). The homology between the homology units of one constant region and between constant regions of different classes of immunoglobulins is significant (about 30%). The homology of the constant regions of human IgG subclasses is much higher, up to 90%. The three-dimensional structure of constant regions is also similar among different IgG subclasses. For example, the murine IgG C_H1 domain structures are similar overall with some local conformational variations. However, the C_H1 domain of IgA has several significant structural differences in comparison with the C_H1 domain of IgG (Sheriff *et al.*, 1996).

There is a high homology between the constant regions of rodent and human chains of the same class and type. For example, sequence homology between the C_γ of rabbit and human IgG is about 70% and between murine and human C_κ is about 60%, which is higher than the overall homology between classes of the same species. The lowest homology, about 25%, is between the C_H1 domains of human and mouse delta chains. The latter chains lack the second constant domain and the C-terminal half of the hinge.

1. Hinge Region

The hinge region is a stretch of heavy chains between the Fab and Fc portions. Its unique structure and position provide segmental flexibility, which is essential for normal functioning of antibodies (e.g., for crosslinking two antigens or binding two antigenic determinants on the same antigen molecule). A region from residues 216 to 230 of γ chains (according to the IgG1 Eu protein numbering) is encoded by its own exon and is designated as the genetic hinge. Some myeloma IgG proteins, such as Dob and Mcg IgG1 proteins, due to a deletion, are hingeless (Steiner and Lopes, 1979). The hinge of the δ heavy chains is also encoded by a separate exon, whereas the hinge of the α chains is encoded at the beginning of $C_{\alpha}2$ exon. The structural hinge (residues 221–237), a definition based on crystallographic data, can be divided on two short segments, which flank the central polyproline core (residues 226–229): upper hinge, the section between the end of Fab up to the first interheavy chain disulfide bridge (residues 221–225), and (γ) low hinge, the stretch from the last disulfide bridge to the beginning of Fc (residues 230–237) (Fig. 8; Table 3).

The upper hinge forms a one-turn helix, with little inherent stability, and is exposed to the solvent. The low hinge, which is similar in different subclasses but appears absent in IgA, has an extended conformation. This segment provides segmental flexibility of immunoglobulin molecules. The rigid core forms two parallel polyproline double helices linked by disulfide bridges, the number of which varied between different immunoglobulin classes and subclasses. The length of the central core (or the middle hinge), is different in immunoglobulin molecules of various classes and subclasses and also is variable between species. That the heavy chains of the human IgG3 have four repetitions in the

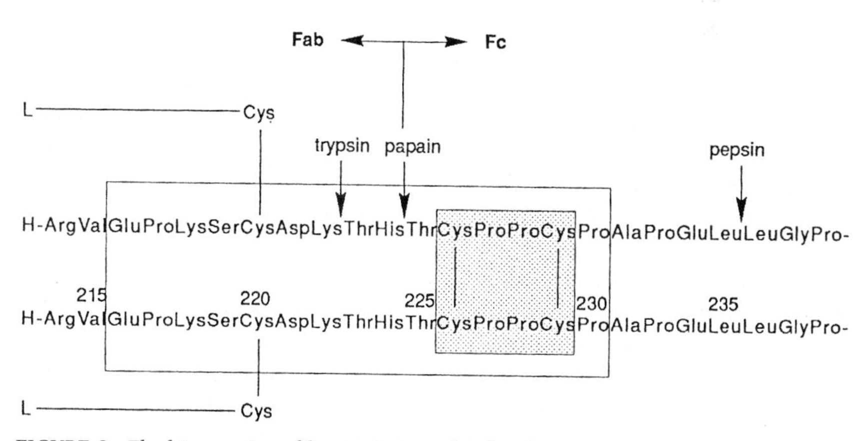

FIGURE 8 The hinge region of human IgG1 molecule. The residues in the box (216–230) are deleted in Dob and Mcg myeloma proteins. The central part of the hinge (core) is shadowed.

TABLE 3. Structure of Hinge Region of Immunoglobulin G[a]

Protein	216 Upper hinge	Core	Low hinge 238
Human IgG1	EPKSCDKTHT	CPPC	PAPELLGGP
Human IgG2	ERK	CCVECPPC	PAPPVAGP
Human IgG3	ELKTPLGDTTHT	CPRCP(EPKSCDTPPPCPRCP)3	APELLGGP
Human IgG4	ESKYGPP	CPSC	PAPEFLGGP
Mouse IgG1	VPRDCG	CKPCIC	TVPSEVS
Mouse IgG2a	EPRGPTIKP	CPPCKC	PAPNLLGGP
Mouse IgG2b	EPSGPISTINP	CPPCKECHKC	PAPNLEGGP
Mouse IgG3	EPRIPKPSTPPGSS	C	PPGNILGGP
Rabbit IgG	APSTCSKPT	C	PPPELLGGP
Guinea pig IgG1	QSWGHT	CPPCIPC	GAPZLLGGP
Guinea pig IgG2	EPIRTPZBPBP	CTCPKC	PPPENLGGP
Rat IgG1	VPRNCGGD	CKPCIC	TGSEVSS
Rat IgG2a	VPRECNP	CGC	TGSEVSS
Rat IgG2b	ERRNGGIGHK	CPTCPTCHKC	PVPELLGGP
Rat IgG2c	EPRRPKPRPPTDI	CSC	DDNLGRP
Bovine IgG1	DPTP	CKPSPCDCC	PPPELPGGP
Bovine IgG2A1	GVSSD	CSKPNNQH C	VREP
Bovine IgG2A2	GVSID	CSKCHNQPC	VRER
Ovine IgG1	EPG	CPDPCKHCRC	PPPELPGGP
Ovine IgG2	GISSDVSK	CSKPPC	VSRP

[a]The amino acid sequences are aligned from residues 216 to 238 (human IgG1 Eu numbering).

hinge accounts for the high sensitivity of this immunoglobulin to poteolysis. The hinge region, owing to its noncompact, extended conformation is highly susceptible to proteolysis, and proteolytic enzymes of different specificities (papain, pepsin, and trypsin) attack immunoglobulin molecules in this section of heavy chains (Fig. 9).

The information on the dynamic structure of the hinge region of mouse IgG2a in solution was obtained by nuclear magnetic resonance (NMR) spectroscopy (Kim *et al.*, 1994). The hinge of this molecule has a mosaic structure with rigid and more or less flexible sections. The upper hinge may be functionally divided on two parts. The motion of the tripeptide Thr 221–Ile-222–Lys 223 in the C-terminal side of the upper hinge is much faster than that of the N-terminal side of this hinge. Several contacts with the C_H1 domain of the N-terminal side are responsible for the restrictions in mobility. The other section with high flexibility is the large part of the lower hinge. The middle part of the

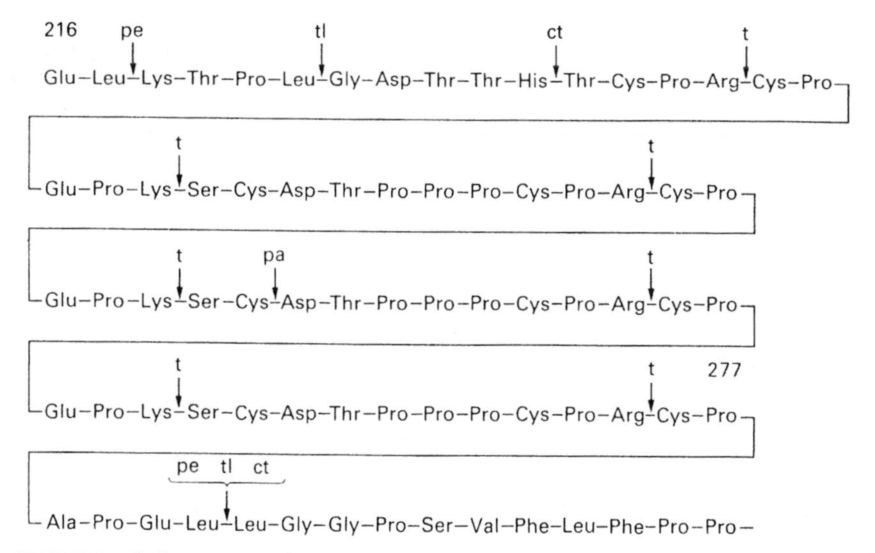

FIGURE 9 The hinge region of human IgG3 (residues Glu-216—Pro-277). The sites of proteolysis by different proteolytic enzymes are shown. ct, α-chymotrypsin; pa, papain; pe, pepsin; t, trypsin, and tl, thermolysin. (Michaelsen, 1990. Reprinted with permission from Pergamon Press, Oxford.)

hinge (or core hinge) is most rigid and comparable in flexibility to the other parts of the IgG molecule. After cleavage of the interheavy chain disulfides, the hinge, except for the N-terminal part of the upper hinge, becomes uniformly mobile.

The human IgD hinge region is unusually long and is composed of 64 amino acid residues. Its N-terminal portion is proteolytic-resistant with five *O*-type oligosaccharide units in close proximity, whereas the C-terminal portion is highly charged and protease-sensitive. The mouse IgD molecules lack a long segment of 135 amino acid residues, including half of the hinge region and the whole $C_{\delta}2$ domain (Takahashi *et al.*, 1982).

In the human IgA1 hinge, there is a duplication of the octapeptide Pro–Ser–Thr–Pro–Pro–Thr–Pro–Ser. This sequence is absent in the IgA2 hinge, which has instead a pentaproline sequence (Toraño and Putnam, 1978). Five oligosaccharides are present in a segment 17 residues long of IgA1 hinge (Toraño *et al.*, 1977). The unusual presence of so many carbohydrate residues in this segment provides the segment's resistance to proteolysis by enzymes produced by different microorganisms colonized on mucosal membranes. By contrast, the human IgA2m(2) has no oligosaccharides in the hinge region, due to a deletion of 13 residues, and it is resistant to some bacterial proteases.

IgM and IgE molecules have no hinge regions such as those in other immunoglobulin classes. Their heavy chain extra domains ($C_{\mu}2$ and $C_{\varepsilon}2$) are exposed and can be cleaved by proteases.

There are two (IgG1 and IgG4) or more disulfide bridges between the human heavy chains, except in IgD molecules, where there is only one. In IgE molecules, the two interheavy disulfide bridges are situated on both sides of the additional domain $C_\varepsilon 2$, and are parallel, like those in IgG. IgM molecules have a cysteine residue at the C-end of the mu chains that participates in the disulfide linkage between these chains. Murine IgG1 and IgG2 molecules have three interheavy chain disulfides, whereas IgG3 possess two disulfide bridges. There are no such linkages between heavy chains in murine IgD molecules, which may exist also as half-molecules. Heavy chains of rabbit and goat IgGs are linked by only one disulfide bridge (Table 4). The amount of interheavy chain disulfide bridges in rat immunoglobulins varies from two (IgG2c) to three (IgG2a) and four (IgG1 and IgG2b).

2. Secretory Tailpiece

The μ and α heavy chains of the secreted polymeric IgM and IgA molecules have an 18-amino acid long extension at the C-terminal called the secretory tailpiece. Its primary structure has no homology to other parts of mu and alpha chains. A significant homology between the tailpieces of both chains was found: 11 residues are identical in the human IgM and IgA molecules. A cysteine residue of the tailpiece (Cys–575 in the μ chain and Cys–495 in the alpha chain) forms a disulfide bond between heavy chains of subunits in the polymeric IgM and IgA molecules. Substitution of this cysteine residue with serine or the deletion of the entire tailpiece prevents the formation of IgA dimers and a mixture of IgA monomers and half-molecules (αL) are secreted (Atkin *et al.*, 1996).

In both IgM and IgA the tailpiece interacts with the J chain. However, the presence of the tailpiece is not the only factor for incorporation of the J chains into polymeric immunoglobulin molecules. IgG molecules, which have gamma chains fused with the mu chain tailpiece, are synthesized as oligomers of different sizes but they do not associate with the J chains (Smith *et al.*, 1995).

The IgM tailpiece contains a single Asp-563 glycosylation site that is conserved in the IgM and IgA of all studied species, including mammals and fish (the one exception is the mouse IgA). Oligosaccharides isolated from this tailpiece are a mixture of three main glycans of oligomannose type, either unprocessed or only partially processed, with the general formula $(\text{Man})_n(\text{GlcNac})_2$, where $n = 6, 8$, or 7/9 (Wormald *et al.*, 1991). The glycosylation site at Asp-563 is important for J chain incorporation. After replacement of Asp-563 by tyrosine (or Ser-565 by phenylalanine), the 563 residue is not glycosylated and the J chain incorporation diminishes (Wiersma *et al.*, 1997). These mutant IgM molecules are predominantly secreted as hexamers.

The mu tailpiece probably participates in the retention of unpolymerized IgM monomers in the endoplasmic reticulum of B cells by binding to the im-

TABLE 4 Disulphide Bridges between Immunoglobulin Heavy Chains

Immunoglobulin	No. of interheavy disulfide bridges
Human	
IgG1	2
IgG2	4
IgG3	11
IgG4	2
IgM	3
IgA	2
IgD	1
IgE	2
Mouse	
IgG1, 2a and 2b	3
IgG3	2
Guinea pig	
IgG1 and IgG2	3
Rabbit	
IgG	1
Goat	
IgG2	1
Bovine	
IgG1	3
IgG2A2	2
IgG2A1	1
Swine	
IgG	3
Ovine	
IgG2	1
IgG1	3

munoglobulin-binding protein (BiP chaperonin) (Sitia *et al.*, 1990). However, chimeric IgG molecules with the μ tailpiece are secreted by cells not only as polymers but also as monomers and dimers (Sensel *et al.*, 1997). Perhaps, some other features of IgM are necessary for association with BiP.

3. Membrane Segments

Instead the secretory tailpieces, membrane-anchored forms of immunoglobulins have membrane segments. Immunoglobulin molecules localized on cell membranes serve as the antigen recognition part of the B-lymphocytes antigen receptors. These membrane-bound forms have additional segments at the

C-end of their heavy chains: the extracellular spacer, which varied among the different isotypes in sequence and length, the hydrophobic transmembrane segment, and the intracellular ones. Membrane μ chains are more basic than secretory μ chains (Vassalli *et al.*, 1980).

The transmembrane segments are composed of 25–26 amino acid residues and are very similar in all heavy chains—13 residues are identical for all isotypes. The same residues are conserved in evolution. Probably, these residues of the transmembrane segments associate with a single integral membrane protein. As it is now known, this protein is an accessory peptide chain of the B-cell antigen receptor. In contrast to other membrane proteins, the transmembrane segments of immunoglobulins have many hydrophilic residues. In the transmembrane segment of mu chains there are 10 residues with hydroxyl group (serine, threonine, and tyrosine residues). Most likely the transmembrane segments of the α, μ, and δ chains, as for similar portions of other membrane proteins, are in α-helix forms.

Membrane murine IgG and IgE molecules have an intracellular segment 25 residues long and in IgA it is 14 residues long. However, the cytoplasmic tail of membrane-bound IgM and IgD molecules is composed from only three residues (Lys–Val–Lys). Membrane-bound IgG, IgE, and IgA serve as specific parts of the antigen receptors on mature B lymphocytes, whereas membrane-bound IgM and IgD comprise the receptors on unstimulated B lymphocytes. Different mechanisms are probably responsible for signal transduction in these two types of B cells upon reaction of antigens with the B-cell antibody receptors. The long cytoplasmic tail of IgG permits mediation of antigen endocytosis and presentation even in the absence of the Igα/Igβ accessory chain dimer (Knight *et al.*, 1997).

The levels of IgG and IgE in serum are reduced significantly in mice lacking most of the cytoplasmic portions of gamma or epsilon chains. This could be explained by the inefficiency of B cells with tailless heavy chains of the antigen receptor to present antigen to T lymphocytes (Wieser *et al.*, 1997; Achatz *et al.*, 1997).

VI. ADDITIONAL PEPTIDE CHAINS

A. J Chain

The J chain is a part of the polymeric immunoglobulin molecules IgM and IgA. It is generally assumed that the J chain participates somehow in the polymerization of these molecules (Koshland, 1985). If B lymphocytes express high levels of J chain, the secreted IgM molecules are mostly pentamers. In the absence of J chain, B cells secrete mainly hexameric IgM molecules that are much more efficient in activation of complement (Randall *et al.*, 1992; Niles *et al.*, 1995). The J chain is linked to IgM on the late steps of the polymerization when monomers have already formed a pentameric structure (Brewer and Corley, 1997). The J chains are not necessary for formation of IgA dimers. However, they are essential for binding polymeric IgA to secretory component (SC) and are required for the efficient transportation of serum IgA into bile (Hendrikson *et al.*, 1995).

The J chains are synthesized in plasma cells as polypeptides 137 amino acid residues long. They have no significant homology with any other types of proteins or immunoglobulin peptide chains. There is a striking similarity, 77% homology, between the human and mouse J chains and both also display more or less homology to the J chains of other mammals, birds, reptiles, and fishes.

The J chains are glycoproteins and have one *N*-linked oligosaccharide chain composed of five different sugar residues. The human and mouse J chains contain eight cysteine residues and an abundance of acidic amino acid residues, which accounts for the low isoelectric point of the J chains. In the J chains, glycine, serine, and phenylalanine occur rarely and no tryptophan residues are present. The J chain molecule is composed of two domain-like structures (Frutiger *et al.*, 1992). One of them consists of two parallel β-sheets held together by two S–S bridges located in opposite loops connecting β-strands. The second domain is built from a mixture of α-helix and β-strands (Fig. 10). Each domain has a cysteine residue involved in the linkage to IgM subunits. In IgM molecules, the J chain is probably present within the ring composed of five Fc_{μ} fragments. In IgA dimers the monomers are most probably linked end-to-end with the J chain interposed (Fig. 11).

The J chain arose early in evolution independently from immunoglobulins and its structure is highly conserved. A peptide with high degree of homology to the J chain of mammals was found in different species of invertebrates, which do not synthesize any immunoglobulins (Takahashi *et al.*, 1996). The function of the J chain-like protein in invertebrates is unknown.

B. Secretory Component

In polymeric IgA molecules presented in external secretions, there is another polypeptide, the secretory component (SC). In human IgA, it consists of 558 amino acid residues. The secretory component sequence composed of five segments of about 110 residues that are homologous to each other and to the variable regions of immunoglobulins. The secretory component contains 20 half-cystines and oligosaccharides in seven positions. In rabbit milk and some other secretions the secretory component occurs in a free form, as a heterogeneous mixture of glycosylated molecules with molecular masses of about 80 kDa and 55 kDa. In the 55-kDa secretory component, a portion of about 25

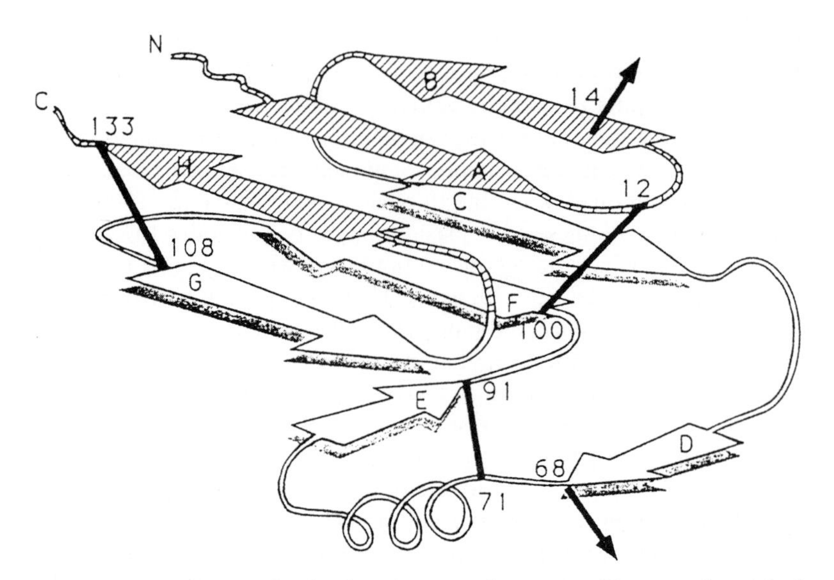

FIGURE 10 Predicted model for the three-dimensional structure of human J chain. The large arrows designated A–H indicate β strands and the thick black lines represent the intradisulfide pairings. Black arrows indicate the interdisulfide bonds with Cys-515 of the mu chain. (Frutiger, *et al.*, 1992. Reprinted with permission from American Chemical Society.)

kDa in the middle of the chain is absent (Frutiger *et al.*, 1987). Both 55 kDa and 80 kDa forms of the secretory component bind dimeric IgA molecules via the $C_\alpha 2$ domains with equal efficiency, since their N-terminal portion (CDR loops) are responsible for the interaction (Bakos *et al.*, 1991, 1993; Coyne *et al.*, 1994). In human and rat IgA–SC complexes, the secretory components are

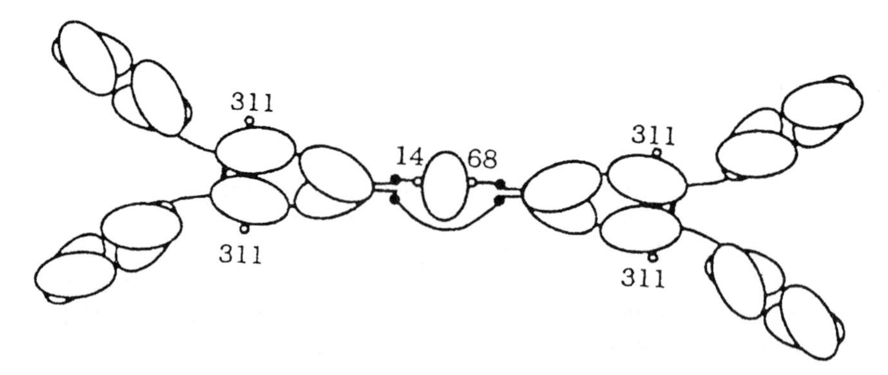

FIGURE 11 End-to-end arrangement of IgA monomers and J chain in dimeric IgA. The J chain is represented as an oval between two IgA molecules. Cys-14 and Cys-68 of the J chain form disulfide linkages with Cys-471 of the IgA monomers. (Krugman *et al.*, 1997. Reprinted with permission from The American Association of Immunologists.)

covalently linked to one of the monomeric subunit of the IgA dimers. Human IgA is linked to the secretory component by a single disulfide bond between Cys-311 of one alpha chain and Cys-467 of the secretory component (Fallgreen-Gebauer *et al.*, 1993). However, in rabbits, both covalent and noncovalent SC–IgA complexes can be found, and their ratio depends on the relative proportions of IgA subclasses. Noncovalent interactions between pentameric IgM and the secretory component are much stronger than between the latter and IgA (Brandtzaeg, 1985).

The secretory component is an ectoplasmic portion of a transport receptor protein that is specifically responsible for the translocation of polymeric immunoglobulins (IgA and IgM) through glandular epithelial cells into external secretions (Mostov *et al.*, 1984; Underdown and Schiff, 1986; Apodaca *et al.*, 1991). The synthesis of this protein takes place in the epithelial cells. It is transferred to their basolateral surface, where it serves as a receptor for polymeric immunoglobulins (pIgR). The complexes of immunoglobulin molecule with the receptor are endocytosed by cells and transferred to the cell apical surface. There the exterior portion of the receptor (i.e., the secretory component) is proteolytically cleaved and the complex of the secretory component with immunoglobulin is released into the secretions. The formation of the stable high affinity complex ($K_a = 10^8$–10^9 M^{-1}) of polymeric immunoglobulins and the secretory component is mediated first of all by a highly conserved portion (residues 15–37) of the first domain of the secretory component (Bakos *et al.*, 1991)

C. Surrogate Chains on the B-Cell Precursors

Precursors of B lymphocytes (pre-B cells), which are present in fetal liver and adult bone marrow, express only the mu heavy chains and no conventional light chains. It was found that the pre-B cells as well as the progenotor B cells (pro-B) of humans and mice synthesize other polypeptides homologous to immunoglobulins. One of them, a 18-kDa protein (Pillai and Baltimor, 1987) has equal homology to the variable regions of the kappa, lambda, and heavy chains (approximately 45%) and the second one, a 22-kDa protein has strong homology to the constant regions of the λ light chains (70%). They together form a light chainlike structure, so-called surrogate light chains (ΨL). These polypeptides are the products of two genes, $V_{pre\text{-}B}$ and λ_5, which are expressed specifically in pre-B cells. Their activity is turned off after differentiation of pre-B cells with the surface mu chains into immature B cells with the surface IgM (Melchers *et al.*, 1993, 1994; Melchers, 1995; Karasuyama *et al.*, 1996). The 18-kDa ($V_{pre\text{-}B}$) and 22-kDa (λ_5) polypeptides form a complex with the mu chains on the surface of pre-BII cells. In the complex, the λ_5 peptide is linked to the μ chain by a disulfide bond through the μ chain's penultimate cysteine residue but the $V_{pre\text{-}B}$ peptide is associated noncovalently (Karasuyama *et al.*, 1990).

The surrogate light chains facilitate the expression of μ chains on the surface of pre-BII cells, like the conventional light chains facilitate the expression of immunoglobulins in mature B lymphocytes. The ΨL and μ chains together with Ig-α and Ig-β signal production polypeptides of the B-cell-receptor complex form a receptor-like complex that promotes differentiation of pre-B cells. Knockouts of either μ chain genes or ΨL genes (or signaling Ig-α and Ig-β polypeptide genes, see the following) block the normal pathway of B cell development. The ligand for the ΨL-μ receptors is unknown but they probably recognize a protein on the surface of the neighboring stromal cells in spleen and lymph nodes. The contacts of pre-B cells with the stromal cells are necessary for the normal differentiation pathway of pre-B cells.

The surrogate light chains also cannot be expressed on the cell surface alone. On the B-cell progenitors (pro-B) and on the early pre-B cells (pre-BI) they associate noncovalently with a complex of 130 kDa / 35–65 kDa glycoproteins (surrogate heavy chains, ΨH) forming surrogate receptors (Karasuyama *et al.*, 1993, 1996; Shinjo *et al.*, 1994). The third partner of ΨL in the formation of membrane receptors is a short μ chain that is a product of the $D_HJ_HC_\mu$ gene and lacks the amino-terminal portion of the V_H region. $D_HJ_HC_\mu$ protein can be expressed on membranes of pre-B cells in a form of disulfide-bonded complexes with ΨL. Not only the complexes ΨL-μ but also L-μ and ΨL/L-μ can be found on the surface membranes of immature B cells. The surrogate light chains are secreted by most of the studied pre-B cell line (Melchers *et al.*, 1993).

These data were obtained in experiments with mouse B cells. In general similar results were found with the human B cells (Lassoued *et al.*, 1996; Sanz and de la Hera, 1996). On early steps of the ΨL-μ receptor assembly, the nascent μ chains bind to molecular chaperones including BiP, calnexin, GR P94 and a 17–kDa protein. The complex of the μ chains with 17-kDa protein is later substituted by the ΨL-μ complex and Ig-α and Ig-β signal transduction chains are added to the μ chains. The recombinant $V_{pre\text{-}B}$ protein exists as a homodimer in solution (an apparent K_d of dimer formation is 5×10^7 M^{-1}). The $V_{pre\text{-}B}$ protein is able to bind to the human and mouse V_H domains and to the human V_L domain as well as to the Fab fragment of human IgG with an apparent $K_d = 6 \times 10^7$ M^{-1} (Hirabayashi *et al.*, 1995). Hence, the Fab binding sites for ΨL are different from that involved in the V_H–V_L interactions.

D. Accessory Peptide Chains of the B-Cell Antigen Receptor Complex

Two polypeptide chains, the α and β (Ig-α and Ig-β), which form a heterodimer, comprise the B-cell antigen receptor together with immunoglobulin molecules (Fig. 12) (Cambier, 1992; DeFranco, 1993; Gold and DeFranco, 1994; Reth,

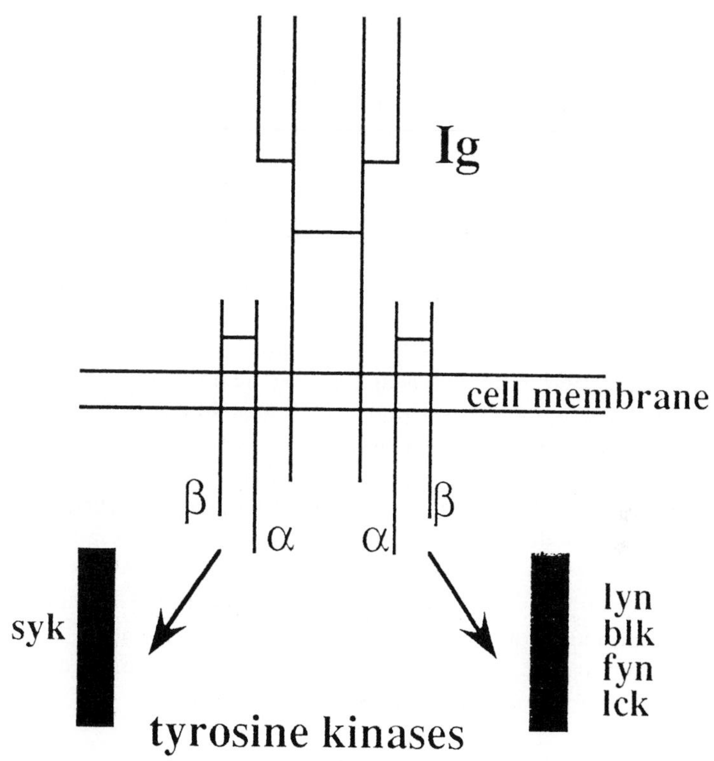

FIGURE 12 The B-cell antigen receptor complex. The antigen-binding part is a membrane-bound immunoglobulin molecule noncovalently associated with two disulfide-connected Ig-α and Ig-β accessory peptide chains. The signal transduction functions are mediated by Ig-α–Ig-β complex.

1995; Burrows *et al.*, 1995). Two such heterodimers associate probably noncovalently with one IgM or IgD molecule in the B-cell membrane. Other immunoglobulin isotypes also form multimeric complexes with the same heterodimer (Venkitaraman *et al.*, 1991).

Both α and β chains, protein products of the *mb-1* (Ig-α) and *B29* (Ig-β) genes, have a extracellular part, a transmembrane segment and a cytoplasmic tail (Table 5). The size of mouse Ig-α and Ig-β chains is 34 and 39 kDa and of human α and β chains is 47 and 37 kDa. The sequence of the N-terminal extracellular part is homologous to a typical immunoglobulin domain—it has a intradomain disulfide bridge (the Ig-β chain probably has two such bridges) and several conserved amino acid residues characteristic to proteins of the immunoglobulin superfamily. The accessory chains of the B-cell receptor are glyco-proteins and their oligosaccharides link to the extracellular parts of the chains. After deglycosylation the size of mouse α and β chains is reduced to 23–26 kDa

TABLE 5 Characteristics of the Murine B-Cell Receptor's α and β chains

		Amino acid residues				N-linked glycosylation sites
Chain	kDa	Total	Extracellular part	Transmembrane region	Cytoplasmic tail	
Ig-α	34	220	109	22	61	2
Ig-β	39	228	129	22	48	3

Reth, 1995. Reprinted with permission.

(Campbell *et al.*, 1991; Friedrich *et al.*, 1993). The human Ig-α chain is larger than murine due to more glycosylation. Both Ig-α and Ig-β chains have cysteine residues that form a disulfide bridge linking them together near the B-cell membrane. A C-terminal truncated form of the Ig-β chain (γ chain) was found on splenocytes and B-cell blasts but not in small lymphocytes (Friedrich *et al.*, 1993). This chain is shorter by about 30–36 residues than the Ig-β chain.

The sequence of the cytoplasmic tail is nearly similar for the mouse and human Ig-α and Ig-β chains and includes the so-called immunoreceptor tyrosine-based activation motif (ITAM) (Reth, 1989; Cambier, 1995). The motif has two tyrosines and two leucines (or isoleucines), which are precisely separated by other amino acid residues:

<u>residue</u> 149 160 168

<u>murine α</u> Glu Asp Glu Asn Leu <u>Tyr</u> Thr Gly <u>Leu</u> Asn Leu Asp Asp Cys Ser Met <u>Tyr</u> Glu Asp <u>Ile</u>

<u>residue</u> 161 170 179

<u>murine β</u> Glu Glu Asp His Thr <u>Tyr</u> Thr Gly <u>Leu</u> Asn Ile Asp Gln Thr Ala Thr <u>Tyr</u> Glu Asp <u>Ile</u>

The same motif xxAsp(or Glu)xxx**Tyr**xx**Leu**(or Ile) xxxxxxx**Tyr**xx**Leu**(or Ile) was also found in components of the T-cell receptor, $Fc_{\varepsilon}RI$ receptor, and in two viral transmembrane proteins. The ITAM has a specific binding site for the ZAP-70 and syk-tyrosine kinases (Chan and Shaw, 1995).

The α–β heterodimer is responsible for several important functions, including signal transduction and expression of IgM on the cell surface. The contact of antigen with the membrane immunoglobulins results in the aggregation of the whole B-cell receptor. The first response to the aggregation is tyrosine phosphorylation of several intracellular proteins by tyrosine kinases. The cytoplasmic tail of the membrane immunoglobulins is short, especially in IgM and IgD, and the Ig-α and Ig-β chains participate in contacts with two types of tyrosine kinases (Fig. 12). The signal transduction function is mediated by the

ITAM of both chains. The development of B cells is blocked at an early step in the absence of the Ig-β chain. In the cells that lack this chain, V_H to D_HJ_H recombination is not performed and therefore the μ chains are not expressed (Gong and Nussenzweig, 1996). Probably for these development steps, a functional complex of the surrogate receptor ΨH–ΨL and the α–β heterodimer signal transducer is necessary. The elimination of the C-terminal part of Ig-α chain containing the ITAM by gene targeting results in a sharp decrease of B cells in the peripheral lymphoid organs. However, mice with the truncated Ig-α chain can generate some B cells in the bone marrow (Torres *et al.*, 1996). The exact cause of this defect is not known but seemingly it is linked with an absence of a signal generated by the B-cell receptors.

The other important function of the α–β heterodimer is linked with the surface expression of the IgM molecules. The IgM monomers are retained in cytoplasm by chaperone protein(s) localized in the endoplasmic reticulum, such as immunoglobulin binding protein (BiP), a member of HSP70 protein family. The IgM monomers are expressed on the B-cell surface only after complexation with the α–β heterodimer. The IgD, IgG2a, and IgG2b molecules are expressed in the absence of the α–β heterodimer. In the case of IgD molecules, they anchor to cell membranes through glycosylphosphatidylnositol (Wienands and Reth, 1992). After activation of B cells, for example, by crosslinking of membrane IgM with anti-mu chain antibodies, the Ig-α and Ig-β chains are phosphorylated on tyrosine residues by tyrosine kinases (Gold *et al.*, 1991).

The B-cell receptor participates also in the first steps of antigen presentation. After capture of the antigen by immunoglobulin, the receptor complex is internalized and transported to the intracellular compartments, where antigen is digested by intracellular proteolytic enzymes. The resulting antigen peptides are presented to the T cells in the context of MHC II molecules. The antigen internalization with the B-cell receptor is much more effective than direct antigen pinocytosis by the B cells.

The presence of a functional B-cell receptor is required for the survival of mature B cells. After ablation of the receptor, the numbers of mature B cells in spleen fall rapidly (Lam *et al.*, 1997). The destruction of the receptor leads to the decreased expression of MHC molecules and increased level of Fas, a cell-surface molecule that participates in programmed cell death. Such changes make B cells susceptible to the T cell and NK cell—mediating killing. The survival of mature B lymphocytes is probably dependent on an essential weak signal given by the B-cell receptor, which is not related to an activation signal after antigen binding (Nueberger, 1997).

Some other peptide chains have been discovered, which specifically associate only with IgM or only with IgD (BAP; B-cell-receptor associated proteins). Three proteins of 32, 37, and 41 kDa were found in association with the IgM membrane molecules (Terashima *et al.*, 1994). The 32-kDa protein (BAP32) has homology with prohibitin, a highly conserved protein, that is an inhibitor of cell proliferation and a candidate for a tumor suppresser gene product. The 37-kDa protein is structurally related to BAP32. Probably, both BAPs form a heterodimer that noncovalently combines with the transmembrane region of the mu chain. Two other proteins bind preferentially with IgD and in much lesser degree with IgM (Adachi *et al.*, 1996). These proteins, BAP29 and BAP31, contain 240 and 245 amino acids, respectively, and have a 43% sequence identity. All four BAPs has some common properties: they are nonglycosylated proteins of a similar size, form heterodimers, and are composed of transmembrane hydrophobic domains and hydrophylic tails. They cannot be labeled on the cell surface. The functional role of BAPs is unknown.

VII. IMMUNOGLOBULIN FOLD (DOMAIN)

The independent folding of each homology unit of the heavy or light chains results in formation of an immunoglobulin fold or domain. Our knowledge of its structure is based primarily on x-ray crystallographic studies of immunoglobulins. In 1970s, three research groups published the first detailed descriptions of the Fab fragments (R. Poljak and coworkers and D. Davies and coworkers) and the immunoglobulin domains of a human light chain dimer (A. Edmundson and coworkers). Since then the high-resolution structures of many immunoglobulin fragments have been obtained. The same characteristic features, specific for domains, were found for all immunoglobulins and other members of immunoglobulin superfamily (Alzari *et al.*, 1988; Poljak, 1991; Davies and Chacko, 1993, Padlan, 1994, 1996).

A. Structural Characteristics

An immunoglobulin fold is a compact globule with dimensions of approximately 40 × 25 × 25 Å. It is formed by antiparallel strands (seven in C domains or nine in V domains) arranged in two β-pleated sheets (Poljak *et al.*, 1973; Edmundson *et al.*, 1975). The sheets are tightly packed face-to-face and are connected by a conserved disulfide bond, which maintains the stability of the domain (Fig. 13). The side chains of hydrophobic amino acid residues are located between sheets and form a hydrophobic core. According to NMR studies, the structure of immunoglobulin domains in solution is nearly the same as in crystals (Constantine *et al.*, 1994). After formation of V_L–V_H dimers the general structure of each of the constituent domains does not change considerably.

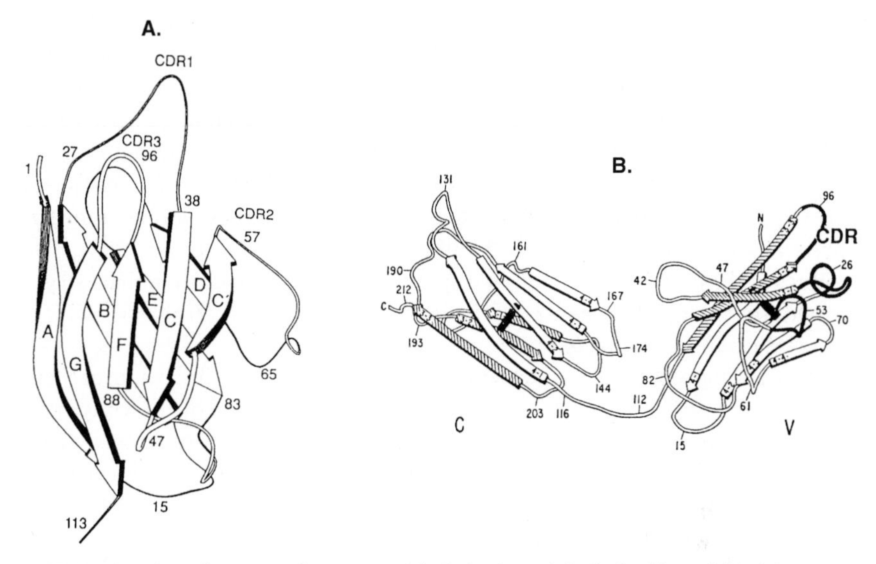

FIGURE 13 Three-dimensional structure of the light chain. (A) The backbone fold of the murine V_L domain. The β-strands (indicated by white arrows) are labeled by capital letters. β-strand C″ is not indicated by an arrow, since it is composed of only two residues (Gly-62 and Val-63). (Constantine *et al.*, 1994.) (B) Constant (C) and variable (V) domains of a human lambda chain, separated by a switch peptide. White directional arrows belong to three-chain β-sheets and striated directional arrows belong to four-chain β-sheets. CDR residues of three loops are concentrated at one end of the molecule (black lines). Intradomain disulfide bonds linking both sheets of each domain are indicated by black bars. (Edmundson *et al.*, 1975. Reprinted with permission from American Chemical Society.)

1. Constant Domains

In the constant domains of the heavy and light chains, one sheet is formed by three strands and the other by four strands. In C_H2 domains both sheets have four strands. The strands are connected by short sequences, or loops, which have no organized secondary structure. Most amino acid residues belong to the strands of β-pleated sheets. Usually, conserved residues are located in the β-strands that are highly homologous between domains of different classes. Only short α-helical segments could be found in domains. For example, in the C_L domain of antibodies against *p*-azophenylarsonate the residues 121–128 form two α-helical turns and the residues from 184 to 187 form one such turn (Strong *et al.*, 1991).

2. Variable Domains

The structure of the variable domains is different from that of the consant domains. Two connected strands in the form of a hairpin or one single strand are inserted in a three-strand sheet. The loops of the variable domains are longer. The four-chain layers located outside and the five-strand sheets interact with each other stabilizing the V–V dimer. The six extended loops of the V_H and V_L domains (CDR loops) (Fig. 13 right, black lines) are clustered together in space and their residues form the antigen-combining site. There is a limited number of conformations, so called canonical structures for each of the five CDR loops (L1, L2, L3, H1, and H2) (Chotia *et al.*, 1989). Many of these loop conformations were found in other protein families (Tramontano and Lesk, 1992). The residues of the β-strands are mainly responsible for the domain's compact structure and the interdomain contacts. However, the CDR loops can also participate in the interdomain contacts.

The CDR loops are the most mobile parts of the V domains and some of them do not contribute significantly to the diffraction data obtained by the x-ray crystallography (Rose *et al.*, 1990). The antigen-combining site formed by a given set of the CDR loops can be found not in one but in two or more conformations (Foote and Milstein, 1994). There is equilibrium between such isomers and a ligand can preferentially bind to one of them.

The stability of the domain structure is maintained by the internal disulfide bond and by hydrogen bonds as well. The arrangement of the hydrogen bonds connecting strands of the Fab domains is shown in Figure 14. As a rule, the hydrogen bonds link antiparallel strands but there are also short parallel stretches connected by these bonds. The tertiary structure of variable domains from different immunoglobulin molecules is very similar. Being superimposed, variable domains are overlapped satisfactorily, with the exception of the CDR1 and CDR3 loops of the V_H domains (Fan *et al.*, 1992).

B. Domains of Members of Immunoglobulin Superfamily

Members of the large immunoglobulin superfamily of proteins are built from immunoglobulin-related domains, which can be described as being V-like or C-like according to their type of β-strand patterns (Williams and Barclay, 1989). The V-like sequences usually have 65–75 residues between two cysteine residues that form an intrachain disulfide bridge. The C-type domains can be subdivided into C1 and C2 variants or sets (Williams and Barclay, 1988; Hunkapiller and Hood, 1989). The C1 domains are found in immune recognition molecules of vertebrate origin (immunoglobulins, T-cell receptors and MHC molecules), whereas C2 domains are common for secreted and membrane-bound molecules of invertebrates, as well. C2 domains have only 55–60 residues between the intrachain disulfide bridge and some C2 domains have

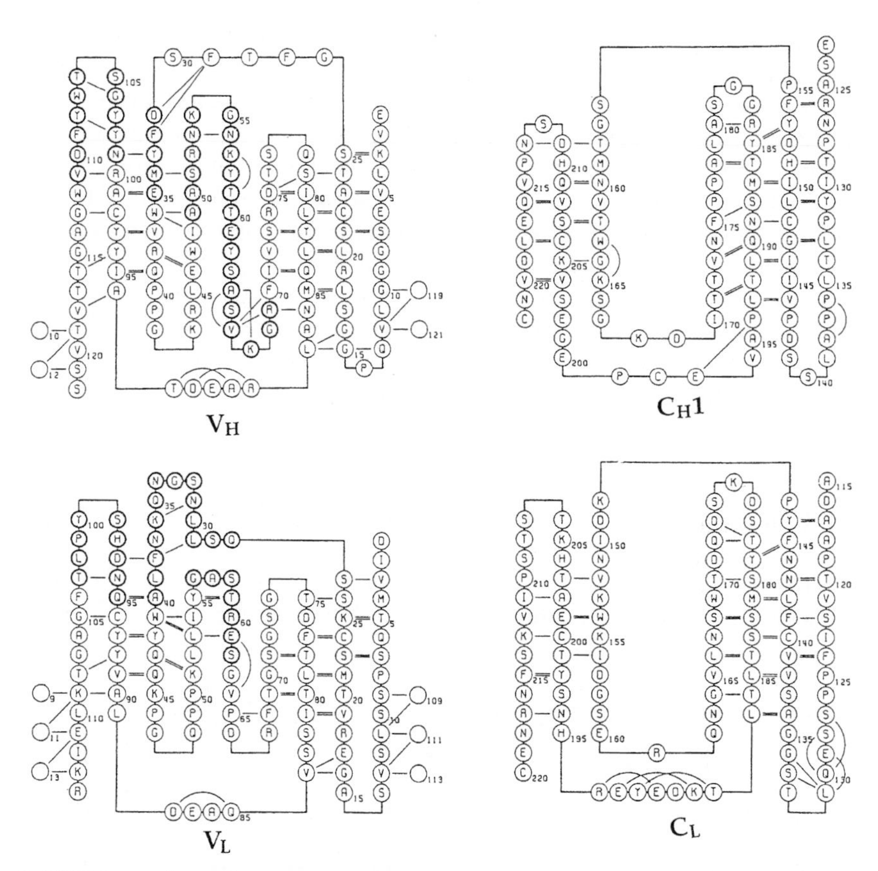

FIGURE 14 Hydrogen bonds between the main-chain amide residues of the mouse Fab domain. Hypervariable domain residues in V_H and V_L are drawn with a double line. Amino acid residues are designated by one letter code. (Satow *et al.*, 1986. Reprinted with permission.)

even less (as few as 40 residues). Their structure is in between V and C1 structures because C2 has a C-type fold, but some of its stretches (links between E and F β-strands) look like that of the variable domains, with a conserved sequence Asp–X–Gly/Ala (Williams and Barclay, 1988). The V domains can also be subdivided into V1 and V2 types (Du Pasquier and Chrétien, 1996). The first of them is found in vertebrates and has "the J-feature," a characteristic conserved sequence Gly–X–Gly–X–X–X–X–(Thr, Val, Leu, Val) (β-bulge at G strand). In V2-type domains, such a J-like sequence is absent. Another type of domain, "I set," is closely related to variable domains but at the same time has some features that are observed only in C1 domains. The I domains were found in cell adhesion and receptor proteins (Harpaz and Chothia, 1994).

The domain classification which was developed is based on the number of strands and the location of strand *c′/d* (Fig. 15) (Bork *et al.*, 1994). Four distinct types of domains are defined. The first of them is the classical seven-strand type of the constant domains of immunoglobulins (c-type). The second one is a nine-strand type that occurs in the variable domains of immunoglobulins (v-type). The position of strand *a* varies in this case: it could be a part of one or another β-pleated sheet. The s-type domain has seven strands but it is different from the c-type in that the fourth strand switches sheets (therefore its name on Fig. 15 is changed from *d* to *c′*). The last type or h-type is composed from eight strands and is a hybrid between c- and s-types. The spacing between *c′* and *e* strands is varied. In the c-type domain, it is 21 residues, in the v-type domain, 25 residues; and in the s-type domain, 19 residues.

C. Contacts between Domains

Domains of V_H–V_L and C_H–C_L pairs have different types of contacts. The constant domains interact by four-strand sheets. However, the variable domains

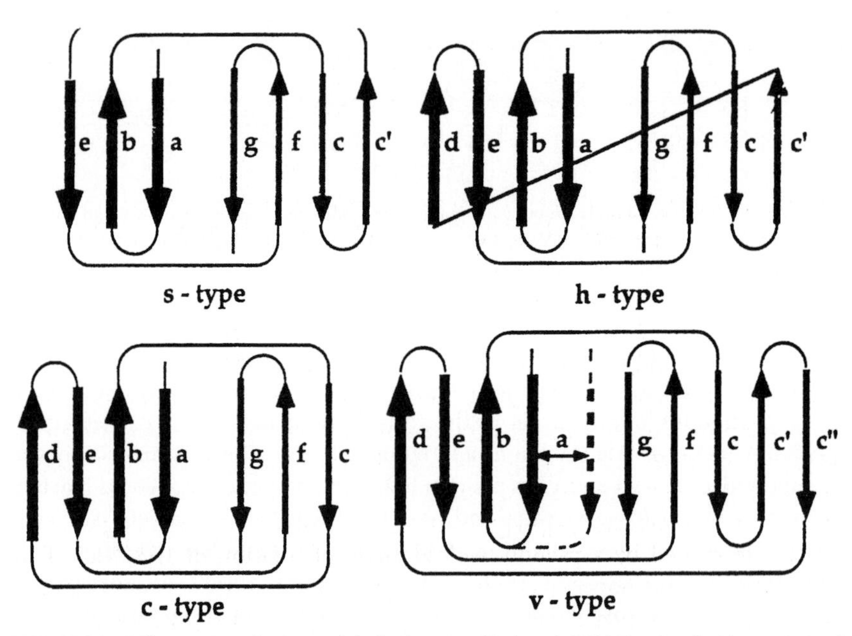

FIGURE 15 Different types of immunoglobulin domains. (Bork *et al.*, 1994. Reprinted with permission.)

contact by five-strand sheets. In C_H1–C_L pairs, the interface area is about 850 $Å^2$ for each domain. Most of the highly interacting residues of the constant domains are conserved (six of C_H and seven of C_L) but three residues from each domain are varied in the Fab fragments studied at high resolution. A cavity is found, between the C_H1 and C_L domains the volume of which is varied in different Fabs (about 50–150 $Å^3$). The cavity is probably filled by solvent and it can be occupied by amino acid side chains without distorting interdomain contacts (Padlan *et al.*, 1987).

The contact areas of the $C_\gamma3$ domains are large and about 2200 $Å^2$ of the interacting surfaces are buried in the interface. These contacts involve more than 20 residues of each domain (Deisenhofer, 1981). Many of the amino acid residues involved in the $C_\gamma3$-$C_\gamma3$ contacts are highly conserved and are present in corresponding domains of other immunoglobulin classes. Carbohydrate chains cover about 520 $Å^2$ of the four-strand sheets of the $C_\gamma2$ domains and oligosaccharides prevent lateral contacts between them. As a consequence a cavity exists between two $C_\gamma2$ domains, which can be seen in structural models based on x-ray crystallographic studies and also in electronmicrographs of IgG molecules. The structural differences between the $C_\gamma2$–$C_\gamma2$ and $C_\gamma3$–$C_\gamma3$ domain pairs is also illustrated by the distance between the centers of the domains, 35 Å in the former pair and 20 Å in the latter. The $C_\gamma2$ and $C_\gamma3$ domains also has longitudinal interactions. They involved about 780 $Å^2$ of surface area and 17 amino acid residues of the both domains. The angle between both domain pairs is small or absent (Padlan, 1990, 1996).

The framework residues of the β-sheets are mainly mediated interactions between the V_H and V_L domains. However, 26–57% of the V–V contacts are mediated by the residues of CDRs involving the interactions between framework–CDR and CDR–CDR residues (Novotny and Haber, 1985; Padlan, 1994). The interface area between the V_H and V_L domains involves about 700 $Å^2$ of each. It can be considered as a three-layered structure with a central part, formed by several aromatic side chains between two "backbone" side layers. Twelve residues of the center part are conserved in all immunoglobulin chains (Chothia *et al.*, 1985). This is a unique type of β-sheet packings and is required to form the antigen-combining site. The V_H–V_L interface involves many aromatic residues. For example, in FabR19.9 six aromatic residues of V_L and also six aromatic residues of V_H participate in the contacts (Lascombe *et al.*, 1992). The interrelationship between variable domains in different antibody molecules varies. The V domains of V_H–V_L pairs in Fabs or of V_L–V_L pairs in light chain dimers are related by a pseudo-twofold angle of rotation of 165–180° (Fan *et al.*, 1992). The noncovalent interactions between the $C_\mu2$ domains are stronger that the similar interactions between the C-terminal domains of the mu chain, $C_\mu3$ and $C_\mu4$ (Wiersma and Shulman, 1995).

VIII. Fab PORTION

In all immunoglobulin classes, the Fab portions are formed by two pairs of domains, V_H–V_L and C_H1–C_L. They are connected by short sections of the heavy and light chains built from several residues (switch peptides). In solution the proteolytic Fab fragments behave in nanosecond range as rigid globules, as determined by steady-state fluorescence polarization and spin-label measurements (Nezlin, 1990). According to x-ray crystallographic studies, there are no extensive contacts between the variable and constant pairs of the Fab fragments and a large cavity between them is filled with tightly bound water molecules. The large portion of V–C interdomain contacts is mediated through these water molecules. In the center of the cavity there is no water and significant contacts are absent in this region (Strong *et al.*, 1991).

The angle between the pseudo-axis of the V and C pairs ("elbow bend") varies in different immunoglobulins between 127 degrees (Tormo *et al.*, 1992) and 179 degrees (Strong *et al.*, 1991). Structurally identical Fab fragments or light chain dimers in different crystal forms can also have varied elbow angles (Abola *et al.*, 1980; Mol *et al.*, 1994). The Fab fragments from the antibody R19.9 have the same elbow angle in two different crystal packings and they behave like rigid units (Lascombe *et al.*, 1992). This type of flexibility was also observed in a light chain dimer (Edmundson *et al.*, 1978), where despite the identity of both dimer chains, their elbow angles cab be different, 70 and 110 degrees.

The evidence of the Fab bending was obtained also by electron microscopy of antibody molecules in complexes with protein antigens (Wrigley *et al.*, 1983; Roux, 1984). The angle between the variable and constant portions of the antibody Fab fragments decreased to 90 degrees in closed complexes of two molecules of influenza hemagglutinin with two anti-hemagglutinin antibody molecules. By time-resolved fluorescence spectroscopy two different rotational components for Fab were found. A longer correlation corresponds to the global motion of Fab and a shorter correlation time is probably due to local and/or segmental movement of the pair of constant domains. The value of the shorter correlation time increased by more than 50% upon antigen binding. Probably, the conformation of the Fab fragment is more rigid in the liganded form (Lim *et al.*, 1995). In solution, the Fab portions, either in free form or as a part of an intact IgG molecule, rotate as a compact structure, but their elbow angle can be changed substantially by external forces, such as restraints due to antigen–antibody complexes or crystal lattice.

The majority of contacts between V_H and C_H1 domains in the Fabγ fragments are mediated by interactions of several conserved residues. These conserved residues, three and two of which are in the V_H and C_H1 domains, respectively, were identical in all Fabs studied so far by x-ray crystallography. The other contacts observed between the V_H and C_H1 domains in the Fabg fragments varied in other known immunoglobulin structures.

IX. Fc PORTION

The Fc fragment of IgG has no compact structure similar to its Fab fragment. According to fluorescent polarization and spin-label experiments, the rotational correlation time, τ, a parameter that provides information about the rotational volume of a macromolecule, is about twofold lower for Fcγ than for the Fab fragment (about 12 and 25 nsec, correspondingly), even though both fragments are of almost equal dimensions (Timofeev *et al.*,1978; Nezlin, 1990). These data point out that the Fc portion, in free form and as a part of the IgG molecule, is less compact than the Fab portion. The flexibility of Fcγ is due mainly to the lack of lateral contact between the $C_\gamma 2$ domains (Huber *et al.*, 1976) and probably between homologous domains in other immunoglobulin classes as well. The Fcγ portion reacts with many ligands, such as the Cq1 complement component and the Fc receptors of different cells. Thus, for a multispecific globule, as the Fcγ portion is, a flexible structure facilitating interaction with different types of ligands simultaneously would be optimal.

By tryptic digestion of the IgM molecule at 60°C the polymeric $Fc_{5\mu}$ fragment can be isolated. After reduction of disulfide bonds the $Fc_{5\mu}$ fragment dissociated into the monomeric Fc_μ fragments, which are unable to form noncovalent dimers or higher polymers. So, noncovalent interactions between the Fc regions of the μ chains are practically absent (Hester *et al.*, 1975). By contrast, there are strong noncovalent interactions in the C-terminal regions of the IgG molecules (between the $C_\gamma 3$ domains of Fc).

X. IMMUNOGLOBULIN MOLECULES

A. Dissociation and Reassociation of Immunoglobulin Peptide Chains

Peptide chains of immunoglobulin molecules are linked covalently by disulfide bonds, one for each heavy–light chain pair and from 1 to 11 between heavy chains, and by strong noncovalent interactions. To separate the immunoglobulin chains, all the interchain disulfide bridges must be cleaved. In the first experiments β-mercaptoethanol (Edelman, 1959) or oxidative sulfitolysis (Franék, 1961) were used for the cleavage of interheavy disulfide bridges. Mild reducing conditions are achieved with thiol reagent such as 0.01 M dithiothreitol or 0.1 M β-mercaptoethanol solutions. Reduced samples are then subjected to gel filtration under conditions that dissociate noncovalent bonds, such as solutions containing organic acids or high concentrations of urea or guanidinium chloride. However, the interchain bridges of some IgG molecules are not cleaved by mild reducing conditions and the nonreduced molecules must be separated from the heavy and light chains by column gel filtration. A milder reducing treatment of 0.01 *M* β-mercaptoethanol splits rabbit IgG molecules into halves containing one light and one heavy chain because the interheavy chain disulfide bonds are more susceptible to reduction than the heavy–light chain ones. The disulfide bonds between IgM monomers are the most labile and exposure of IgM to 0.000125 M dithiothreitol or 0.015 M mercaptoethylamine can yield subunits with the intact L–μ chain disulfide bonds, which at higher concentration of reducing agents, such as 0.1 M β-mercaptoethanol, dissociate into halves (L + μ).

The light chains are usually soluble in water and neutral buffers, whereas the isolated heavy chains precipitate gradually. The addition of the light chains to such precipitating heavy chains results in their solubilization with formation of immunoglobulin molecules. The heavy chains are also soluble in 0.01 M acetate buffer, pH 5.5. After complete reduction under strong denaturation conditions, the native structure of the heavy and light immunoglobulin chains is lost. However, gradual removal of the denaturant, by prolonged dialysis, can restore the original structure of the immunoglobulin molecules and can even result in the recovery of the part of antibody activity (Freedman and Sela, 1966; Huston *et al.*, 1996).

If the IgM molecules are reduced without subsequent alkylation and their sulfhydril groups are free, separated chains, half-molecules (L–μ), or IgM monomers can reassociate into full molecules with reoxydized cysteine residues upon removal of the reducing agent. An equilibrium constant of the formation of IgM subunit from half subunits (halfmers) is equal to $K = 2.3 \times 10^6 M^{-1}$ half subunits per liter (Solheim and Harboe, 1972). Hybrid immunoglobulin molecules can be formed if chains, half-molecules, or the IgM monomers from different antibodies are mixed prior to renaturation. Polymerization of IgM subunits can occur under such conditions without the J chains.

Even if interheavy SH-groups are blocked, the heavy and light chains of IgG can reassemble, due to noncovalent interactions, into molecules that resemble native immunoglobulin molecules and express antibody activity. However, the mu chains with blocked interheavy disulfide bonds can form with light chains only halfmers of IgM (μ + L) and for the assembly of monomers or polymers intact inter-μ chain disulfide bonds are necessary. For formation of IgM polymeric molecules, the penultimate cysteine residue of mouse mu chains (Cys-575) is the most important and after its elimination very little IgM polymer molecules are seen. The cysteine residue located in mouse $C_\mu 2$ domain (Cys-337) is essential for formation of monomers from halfmers (Wiersma and Shulman, 1995).

After the initial observations on recombination of immunoglobulin chains (Franěk and Nezlin, 1963; Edelman *et al.*, 1963), many other similar studies were performed (for review of early works see Nisonoff *et al.*, 1975; Nezlin, 1977). For the heavy–light chain interactions several conserved residues of the variable and constant domains of both chains are responsible. Therefore, there is practically no preferences for pairing of immunoglobulin chains and the light chains associate on average in random with heavy chains both *in vitro* and *in vivo*. The affinity between different heavy and light chains is usually high—on average about 10^{10} M^{-1}. In contacts between the variable domains, a number of variable residues are also involved. That is why the variations in the affinity between different pairs of the heavy and light chains were found (Hamel *et al.*, 1987). The light chains are able to form dimers. The dimerization constant ranged from 10^3 to 10^6 M^{-1}. For these variations the CDR3 residues are mainly responsible, particularly residues at position 96. The presence of an aromatic or hydrophobic residue at that position enhances dimer formation, but a charged residue prevents this process (Stevens *et al.*, 1980).

During the past two decades, many experiments were performed to study the ability of recombinant antibody molecules built from heterologous peptide chains to react with antigens. The results obtained in these studies vary from clearly positive to negative. To understand the apparent contradictions, it is necessary to take into account the following considerations. In these studies, the antibody molecules against different kind of antigens were used as a source of the heavy and light chains for recombination experiments. The structure and dimensions of the antigen combining sites and the relative participation of both chains in the construction of the site vary widely. Usually, the heavy chains donate more contact residues than light chains, although there are some exceptions from the rule. The light chains can not only directly contribute its residues for the combining site but also modify the conformation of the site by contacting the variable loops and framework residues of the heavy chains. Therefore, the more similar is a light chain used for recombination to the autologous heavy chain, the higher chance of finding the antigen-binding activity of the recombined antibody molecule.

A good illustration is experiments with transfection of the V_H gene of the anti-DNA antibody 3H9 into hybridoma cells synthesizing different light chains (Radic *et al.*, 1991). The 3H9 heavy chain contributes the main part of residues that determine the anti-DNA activity. However, the light chains can modulate or prevent the binding to DNA or even expand the specificity of recombinant antibodies. The difference in only one residue (L-97) between the light chains originated from the anti-DNA antibodies is enough to change the binding properties of the recombinant antibodies. The specific activity of cold agglutinins (CA), which bind to the carbohydrate antigen I of erythrocytes and of some other cells, also depends on residues of the heavy chains. It was shown, however, that the binding to the I antigen depends greatly on the light chains used for recombination (Li *et al.*, 1996). In another experiment, eight hybrid hybridoma were used, which expressed chain alleles of both parental origin (De Lau *et al.*, 1991). The synthesized antibody molecules with specificity against different cell membrane proteins were isolated and their activity and chain composition was tested. All three possible types of chain association in the IgG2a$\varkappa$ and IgG1$\varkappa$ molecules were found—with the preference for homologous pairing, with the preference for heterologous pairing, and with random pairing.

A large number of recombinant molecules were studied in experiments with antibodies against hapten nitrophenyl phosphonamidate (Kang *et al.*, 1991). When 22 heavy or light chains from these antibodies were recombined with 2×10^6 heavy or light chains from nonimmunized NZB mice, no antigen-binding molecules were found. After recombination of the heavy and light chains from 22 hapten-binding antibody molecules (total of 484 possible combinations), two types of recombinant molecules were found, with the antigen-binding activity and without it. The frequency of the antigen-binding molecules was fivefold greater than expected if unique heavy–light chain combinations are obligatory for creating active molecules. In another experiment, chains from the nonactive recombinant molecules were used. After repeated recombinations, functional molecules were found among new recombinants (11%).

All these data confirm the ability of the light chains to recombine with different heavy chains to form immunoglobulin molecules. The functional antigen-binding activity of the newly formed molecules depends on the properties of chains used for recombination and their ability to reconstruct the antigen-binding site.

B. Intact Immunoglobulin Molecules

Our concepts of the general arrangement of immunoglobulin molecules are based on several ideas formulated about three decades ago. They are Porter's scheme of the IgG molecule as a complex of two light and two heavy peptide chains, Edelman's description of immunoglobulin chains as a tandem sequence of several homologous basic units or domains, and a model of Tanford's group representing IgG molecules as tripartite system from 2 Fabs and Fc tethered by a flexible hinge. Electron microscopy and x-ray diffraction studies confirm these general concepts and present the detailed pictures of various immunoglobulin molecules.

In electron micrographs the IgG molecules are observed in Y configuration with different angles between their Fab arms up to a T configuration (Fig. 16). Such variations in the angle up to 180 degrees was the first convincing evidence that the antibody molecules possess segmental flexibility, an important func-

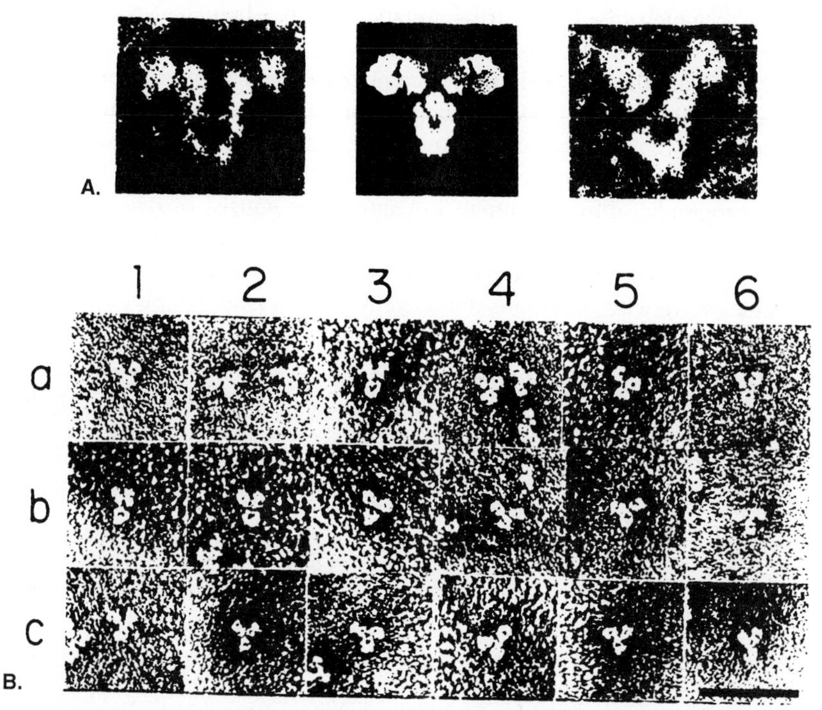

FIGURE 16 (A) Electron micrographs of rabbit IgG molecules (left and right) and a space-filling model of the Dob IgG molecule with inserted hinge (middle).(Roux and Metzger, 1982. Reprinted with permission from The American Association of Immunologists.) (B) Electron micrographs of human monoclonal IgG1 molecule. Bar = 50 nm. (Roux, *et al.*, 1994. Reprinted with permission from Elsevier Science, Ltd.)

tional property common to these proteins (see Chapter 7). The best electron microphotographs have shown that the Fab arms consist of two entities, a variable domain dimer V_L–V_H (or Fv) and a constant domain dimer C_H1–C_L that are attached. The structure of the Fc portion is different. It is composed of two C_H3 domains, closely attached to each other by noncovalent forces and two C_H2 domains that are separated from each other by a cavity where oligosaccharide chains are located. The general shape of human IgG subclass molecules depends on the length of the hinge. The IgG3 molecules with a long hinge are more extended than other human IgGs that are more compact. The IgG1 molecules are slightly more extended than IgG2 and IgG4 molecules.

Early electron microscopic examination of IgM molecules gave important information not only on the structure of these molecules but also on their functional activity. Free IgM molecules reveals a stellar conformation with a central

Fc_5 disk to which five $F(ab)_2$ arms are attached in a branched configuration, similar to a Y (Fig. 17A). Only after binding to a particle with multiple identical epitopes, did the IgM molecules become five-legged staple-like or table-like structures (Feinstein *et al.*, 1971). Such transformation can be explained by the flexible nature of the hinge peptide linked Fabμ and Fcμ units. The isolated $Fc_{\mu5}$ fragment is seen in micrographs as a ring-shaped structure (Fig. 17B). In electron micrographs, dimeric human IgA molecules consist of two identical Y-shaped units, linked by their Fcα units end-to-end to a rigid structure (Fig.18).

All reported findings concerning the three-dimensional structure of IgG molecules agree with the results obtained by x-ray crystallography. To date, five

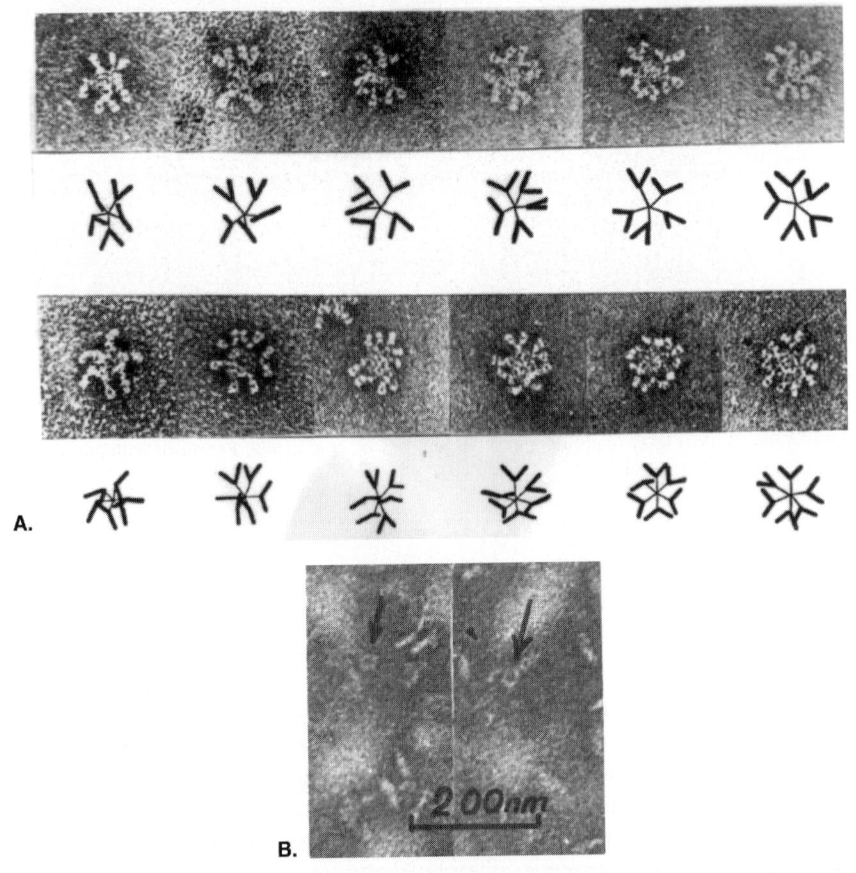

FIGURE 17 (A) Electron micrographs of mouse IgM pentamers (top row) and IgM hexamers (bottom row). Magnification × 340,000. (Davis *et al.*, 1988. Reprinted with permission.) (B) Electron micrographs of human $Fc_{\mu5}$ fragment. Ring-shaped $Fc_{\mu5}$ is shown by arrows. (Sykulev and Nezlin, 1982. Reprinted with permission.)

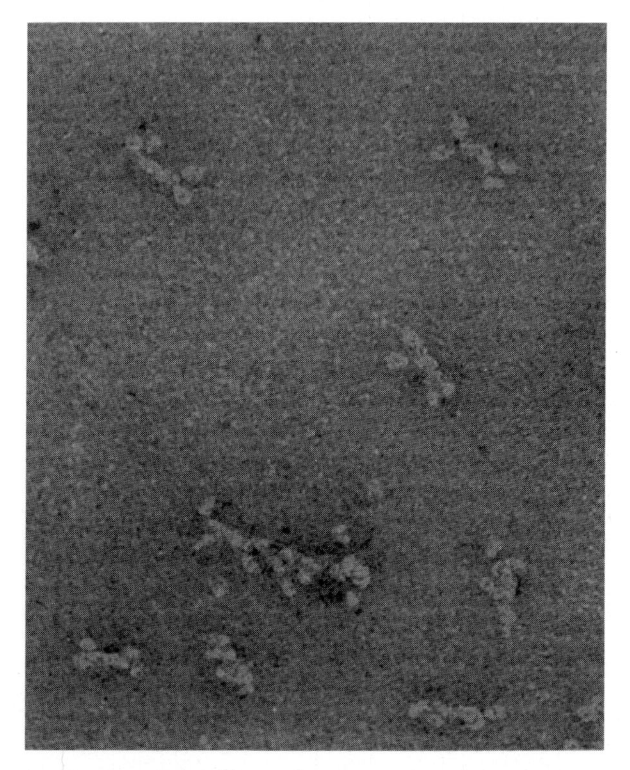

FIGURE 18 Electron micrographs of dimeric human IgA. Magnification × 450,000. (Courtesy of K.H. Roux, Florida State University, Tallahassee.)

immunoglobulin molecules were studied by this method (Table 6) (Edmundson *et al.*, 1993). Two of them are human IgG1 monoclonal myeloma proteins (Dob and Mcg) with a similar 15-amino acid residue deletion in the hinge region (Silverton *et al.*, 1977; Guddat *et al.*, 1993). At low resolution, both molecules have a compact T-type configuration and consist of three globular parts representing two Fab arms and the Fc portion. Two halves of the molecules (H–L) are related by a twofold axis of rotation. The upper part of the Fc fragment is pulled tightly upward between the lower portions of the Fab arms, thus limiting Fc mobility and shielding the potential C1q binding site. In the Mcg molecule there is no interheavy disulfide linkage. By contrast, the light chains are linked covalently by an interchain disulfide bridge. The antigen-combining sites of the Mcg molecule can be described as large irregular cavities at the tips of the Fab portions.

Two other studied human IgG1 (Kol and Zie) molecules have an intact hinge (Marquart *et al.*, 1980; Ely *et al.*, 1978). The most striking finding was that, in

TABLE 6 IgG Molecules Studied by X-Ray Crystallography

Protein	Resolution	Hinge	Fc	Reference
Human myeloma				
Dob IgG1ϰ	3.0 Å	Deleted	Well-defined	Silverton *et al.*, 1977
Mcg IgG1λ	3.2 Å	Deleted	Well-defined	Guddat *et al.*, 1993
Kol IgG1λ	3.5 Å	Intact	Disordered	Marquart *et al.*, 1980
Zie IgG2λ	6.5 Å	Intact	Disordered	Ely *et al.*, 1978
Murine monoclonal				
Mab231 IgG2a	2.8 Å	Intact, long	Well-defined	Harris *et al.*, 1997

crystals of these intact immunoglobulin molecules, the Fab portions were well ordered, while no density related to the Fc portion was found. This Fc disorder is probably due to the Fc distribution at several different sites of the crystal lattice, which would explain why no density related to Fc was seen. Fc is able to adopt at least four different conformations.

The murine monoclonal antibody Mab231 with a specificity to malignant canine lymphocytes is the only intact antibody molecule with known structure (Harris *et al.*, 1992; 1997). All parts of the molecule, including the hinge and oligosaccharide chains, are visible in electron density maps (color figure on the book cover). The Fc portion of Mab231 is located asymetrically, in an oblique orientation to both Fab subunits. The angles between the Fc and two Fabs are about 65 and 115 degrees. One dyad relates two halves in the Fc and the other the constant domains of the Fab subunits. Mab231 has a long hinge region and, consequently, pronounced segmental flexibility. Therefore, the Fc subunit can assume without restraint its single orientation, which depends on the lattice interactions and gives a well-defined electron density for the Fc. The C_H2 domains of Mab231 diverge further apart from one another compared with corresponding human free Fcs despite the presence of an intact hinge. Both C_H2 domains incline toward a pair of C_H3 domains. The elbow angles of the Fabs are different, for one Fab it is 159 degrees and for the second, 143 degrees.

The angle between two Fab arms varies in different immunoglobulin molecules studied by x-ray crystallography. In the human Kol IgG1 it is equal to 125 degrees (Y-shape model) and in the murine Mab 231 IgG2a it is 170 degrees (nearly T-shape model). According to electron microscopic studies the shape of a human IgG1 can vary from Y to T due to the various angles between the Fabs (Fig. 16) (Roux *et al.*, 1994). The variation of the angle is more pronounced in IgG3 molecules (from 45 to 180 degrees), which could be ascribed to the longer hinge region and the higher flexibility of this human IgG subclass (Ryazantsev *et al.*, 1990). The most probable models of pig IgG studied in solution by neutron and x-ray small-angle scattering have the Fab arms either fully extended or slightly bent down toward the Fc (shirt-shape model) (Cser *et al.*, 1981).

There are no x-ray crystallographic studies of immunoglobulin molecules belonging to classes other than IgG because their crystals have not yet been obtained. Thus, our knowledge of their structure is based on other types of experiments and molecular models rather than on x-ray diffraction studies. According to the electron microscopic and synchrotron x-ray scattering experiments (Perkins *et al.*, 1991), the IgM molecule has a planar structure with a central ring of five Fc monomers and side-to-side displacement of the $F(ab')_2$ arms in the plane of the molecule. Diameters of the IgM molecule and the Fc_5 ring are approximately equal to 36 and 17 nm, respectively, and the thickness of the Fc_5 ring is 4 nm. The $C_\mu 2$ domains link the Fab arms at an angle of about 90 degrees. For free IgM molecules in solution, out-of-plane displacements of $F(ab')_2$ are probably rare. In the presence of antigen excess, IgM antibodies crosslink two particulate antigens in a planar complex. However, in antibody excess IgM antibody molecules are able to form a "staple" conformation with $F(ab')_2$ portions out of the plane (Feinstein *et al.*, 1971).

The information on the general structure of IgE was obtained in the experiments, in which the head-to-tail distances of IgE molecules were studied (Zheng *et al.*, 1992; Baird *et al.*, 1993). For these experiments, a mouse chimeric IgE molecule with anti-dansyl antibody activity was constructed that has a fluorescein donor molecule at the C-end of the epsilon chain. The energy transfer distance between coupled fluorecsein and dansyl acceptor located in the combining site was determined. According to these measurements, the average distance between the tips of Fab and the end of the Fc was equal to 71 Å (within range of 65–77 Å), whereas the calculated distance for a planar Y-shaped IgE molecule is much larger (175 Å). These results are consisted with the bent structure of the IgE molecule in solution differing from the planar configuration. They are also in agreement with the findings obtained in polarization fluorescence experiments, which pointed to a rigid, compact structure of the human IgE molecule (Nezlin *et al.*, 1973). The bent structure of IgE was substantiated by hydrodynamic measurements (Beavil *et al.*, 1995). The head-to-tail distance found for the mouse anti-dansyl IgG1 antibody molecule was on average 75 Å (within a range of 57–143 Å). Hence, IgG1 molecules can also exist in solution as a bent structure. However, the IgG1 molecule is much more flexible than the IgE molecule and therefore the IgG1 head-to-tail distances are varied to a greater extent (Zheng *et al.*, 1992).

In serum from an individual normal donor, IgG molecules present mainly as monomers, but in IgG preparations from human plasma pooled from many donors, many IgG dimers are present. In plasma pooled from 10,000 or 100,000 donors up to 25% and even 30–40%, respectively, of the IgG molecules are present as dimers (Gronski *et al.*, 1991). The dimerization does not involve Fc and can be related to idiotype–anti-idiotype interactions between Fab portions of IgG. In pooled preparations, idiotype diversity is expected to be higher, which would explain the high percentage of IgG dimers in pooled plasma. The flexibility of IgG molecules is important for the formation of dimers. More molecules of the most flexible subclass of human IgG, namely IgG3, than of the less flexible subclass IgG4, are present as dimers. Solutions of commercial IgG preparations can be freed from dimers and higher complexes by gel-filtration on a column with Sephadex G-200 or other suitable gels. However, after storage at 4°C for 2–3 or more days many large IgG aggregates can be found by using sensitive assays, like fluorescence polarization.

The isolated human light chains or their homogeneous counterpart, Bence Jones proteins, can form dimers in solution. The λ light chains exist primarily as disulfide-linked dimers, but the κ light chains usually are found as noncovalent dimers and monomers or as mixtures of both (Stevens *et al.*, 1991). Some light chains are precursors of amyloid, fibrillar tissue deposits (Klafki *et al.*, 1992). The formation of such stable noncovalent assemblies of the light chains is dependent on the structure of the variable domains and first of all on their CDR residues. The amyloid-associated residues are located mainly along the surface and are not responsible for the internal packing of the variable domain. It was suggested that at the first step of the polymerization two variable domains form a noncovalent dimer, which then associate in a proamyloid filament by head-to-tail interaction. The last step is the formation of amyloid fibrils by lateral interactions of the β-sheet surfaces of proamyloid filaments (Stevens *et al.*, 1995).

XI. CARBOHYDRATE COMPONENTS OF IMMUNOGLOBULINS

After formation of peptide chains, they are modified by the attachment of sugar residues at some definite sites. Such covalent modification diversifies proteins to a great extent and each glycoprotein molecule usually can be found in many variants, or glycoforms (Lis and Sharon, 1993; Dwek, 1995; Sharon and Lis, 1997). The oligosaccharide units play a significant structural and functional role and may be important for proper protein folding, solubility, supporting a local structure of folded peptide chains, recognition, and ligand binding.

All immunoglobulin molecules, as are other serum proteins, are glycosylated and contain carbohydrates (glycans), the amount of which varies between classes. IgG molecules contain about 3% carbohydrates, whereas the carbohydrate contents of other classes is higher, usually 8–12%. The carbohydrate chains of immunoglobulins are usually composed from the following sugar residues: D-galactose (Gal), *N*-acetyl-D-galactosamine (GalNAc), *N*-acetyl-D-glucosamine (GlcNAc), L-fucose (Fuc), D-mannose (Man), and *N*-acetylneuraminic (sialic) acid (NeuAc).

Carbohydrate chains of immunoglobulins are highly heterogeneous and their diversity derives from their various chemical composition, size, sequence of residues attached to a common pentasaccharide core, and types of linkage to the peptide chains (*N*- or *O*-linkage).

A. Linkage of Glycans to Immunoglobulin Peptide Chains

Most of the oligosaccharides of immunoglobulin molecules are covalently linked through *N*-acetylglucosamine residues to the amino group of asparagine residues, which are present in the glycosylation sequence Asn–X–Ser/Thr, where X can be any amino acid except proline. The carbohydrates thus linked are called *N*-linked oligosaccharides.

N-acetylglucosaminyl-asparagine (GlcNAc-Asn):

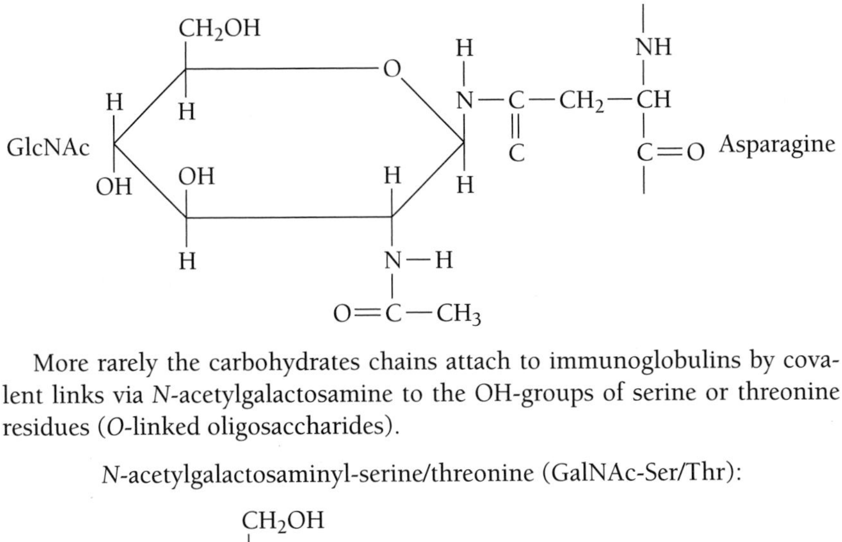

CH_2OH
H NH
O
H H N—C—CH_2—CH
GlcNAc C C=O Asparagine
OH OH H H
H N—H
O=C—CH_3

More rarely the carbohydrates chains attach to immunoglobulins by covalent links via *N*-acetylgalactosamine to the OH-groups of serine or threonine residues (*O*-linked oligosaccharides).

N-acetylgalactosaminyl-serine/threonine (GalNAc-Ser/Thr):

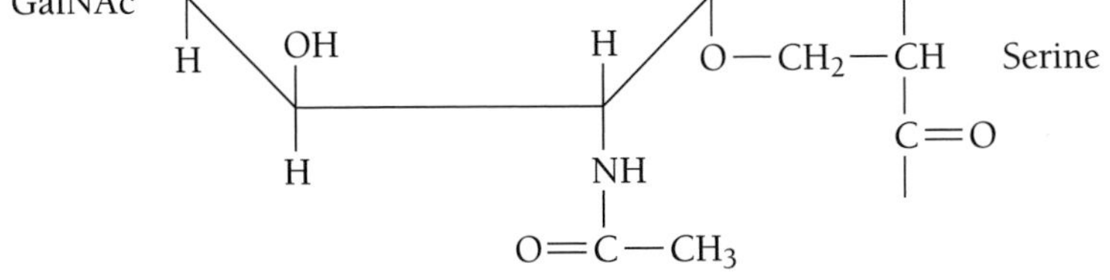

CH_2OH
O
HO H H
GalNAc NH
H OH H O—CH_2—CH Serine
C=O
H NH
O=C—CH_3

B. Chemical Structure of Immunoglobulin Glycans

Immunoglobulin oligosaccharide chains are extremely heterogeneous, and differ in both composition and structure. The extreme heterogeneity of the carbohydrate components, whose biosynthesis is not template-dependent, contrasts with the structure of peptide chains, whose sequence is strictly determined by nucleic acid templates. One source of carbohydrate heterogeneity is caused by differences between the composition and structure of *O*-linked and *N*-linked oligosaccharides.

O-linked glycans are located in the hinge regions of human IgA1 and IgD and a genetic variant of the rabbit IgG, the d12 allotype molecules. Rarely, these are found in the variable domains of myeloma proteins. *O*-linked glycans attached to hydroxyl groups of threonine and serine are small and heterogeneous. For example, normal human serum IgA1 has in the hinge region up to five glycans that consist of one to four of the following sugar residues (number per α chain cited in parentheses): GalNAc (1.0); Galβ1 $\rightarrow$ 3GalNAc (1.0); NeuAcα2 $\rightarrow$ 3Galβ1 $\rightarrow$ 3GalNA (2.5); and NeuAcα2 $\rightarrow$ 3Galβ1 $\rightarrow$ 3(NeuAcα2 $\rightarrow$ 6) GalNAc (0.5) (Field *et al.*, 1989). These sugar residues are not distributed uniformly and human serum IgA1 molecules are microheterogeneous in this respect.

N-**linked glycans** belong to two main groups, complex oligosaccharides or high-mannose oligosaccharides, which have the same core region (Fig. 19, boxed). In the oligomannose oligosaccharides, different numbers of mannose residues are attached to the core. The complex oligosaccharides are the best-studied group of immunoglobulin glycans. They have two arms or antennae that are bound to the core and are composed of GlcNAc and Gal with or without NeuAc units. *N*-linked oligosaccharide chains have been found in all immunoglobulin classes of all studied mammalian species. A total of 36 different

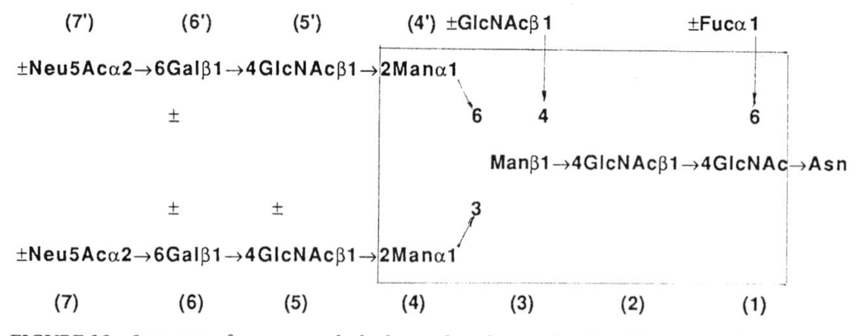

FIGURE 19 Structure of asparagine-linked complex oligosaccharides. The core region is boxed.

variants or glycoforms of the *N*-linked complex glycan of human IgG are known. The glycosylation process can be asymmetric and an IgG molecule may have two different oligosaccharide chains at each Asn-297 residue. Therefore, one B cell can produce as much as 648 structurally different IgG glycoforms (Jefferis and Lund, 1997). The heterogeneity can be even more pronounced because about one fourth of all polyclonal IgG molecules have glycans in their Fabs.

One source of the glycan heterogeneity is the substitution of various sugars in the core. Four core variants have been described without L-fucose and/or *N*-acetylglucosamine residues (marked $\pm$ in Fig. 19). Another source of variation is due to the absence or presence of galactose, sialic acid, and acetylglucosamine residues (marked $\pm$ in Fig. 19) in either one or both antennae. All human IgG subclasses have the same sets of sugars and no subclass specific variations were found. However, there are variations in the structure of carbohydrate chains in IgG molecules of different animal species. For example, the oligosaccharides of mouse IgGs, in contrast to those of human IgGs, are almost all fucosylated and contain no bisecting *N*-acetylglucosamine linked to the C-4 position of β-mannosyl residues. The functional significance of the carbohydrate heterogeneity of immunoglobulins is not known.

C. Distribution of Glycans on Immunoglobulin Molecules

Most of the immunoglobulin glycans are located in the constant regions of their heavy chains. Each immunoglobulin class has specific sites for glycan attachment and the number of oligosaccharide chains is characteristic for a given class. There is only one *N*-glycosylation site in the heavy chain constant region of most IgG classes. In mouse IgG3 molecules, an additional *N*-glycosylation site was found at residues 471-473 (Asp—Leu—Ser) of the C_H3 domain (Wels *et al.*, 1984). The glycan located at this extra site is responsible for the self-associating ability of IgG3 molecules (Panka, 1997). Four *N*-glycosylation sites are in the IgM and IgE heavy chain constant regions. The IgA and IgD heavy chain constant regions have from two to five such sites as well as several *O*-linked sites (Fig. 20).

The IgG molecules contain an average of 2.8 mol of asparagine-linked sugar chains per human and 2.3 mol per mouse and rabbit IgG molecule. In the Fcγ portions, there are two *N*-linked oligosaccharide chains and the rest of the sugar units are located in the Fabγ portions. About 15–20% of all IgG molecules have carbohydrates in the Fab portions. Some of the Fab carbohydrates are attached to the variable domains of the heavy chains, whereas others can be found in the light chains. The light chain glycans are *O*- or *N*-glycosidic linked

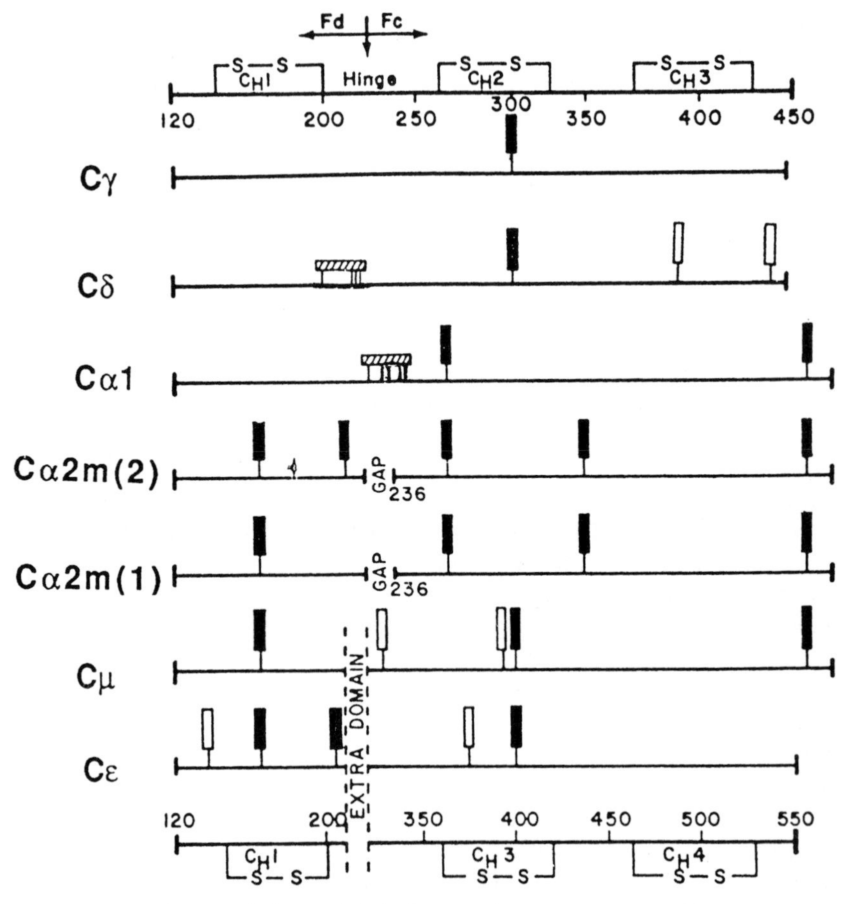

FIGURE 20 Localization of oligosaccharide chains in the constant regions of human immunoglobulins. Complex oligosaccharides represented as vertical bars (black bars—chains found in homologous positions in two or more molecules). Short simple chains in hinge regions of IgD and IgA1 are shown as horizontal bars. The upper scale is for top five C regions and the lower scale is for the bottom two C regions. $C_{\varepsilon}2$ and $C_{\mu}2$ domains (extra domains) are omitted. (Toraño *et al.*, 1977, adapted.)

carbohydrate chains located on V_L domains (Chandrasekaran *et al.*, 1981; Savvidou *et al.*, 1981).

The glycan structure in the Fab and Fc regions of IgG differs. The Fab oligosaccharide chains contain more sialic acid, galactose, and bisecting *N*-acetylglucosamine residues than those of Fc glycans. Disialylated sugars are present only in the Fab fragments (Taniguchi *et al.*, 1985). The Fab and Fc portions of one IgG molecule are exposed to the same set of cell enzymes that participate in the biosynthesis of glycans. Therefore, the structural differences between the Fab and Fc glycans can be ascribed to the influence of different conformational structures of peptide sections surrounding the glycosylation sites. The experiments with three types of mouse (V_H)–human ($C\gamma4$) chimeric IgG molecules point to the importance of the vicinity of glycosylation sites on the glycan structure (Wright *et al.*, 1991; Endo *et al.*, 1995). These chimeric molecules are identical in peptide sequence, except for the introduced carbohydrate sites in the second complementary determining region of V_H (Asn-54, Asn-58 or Asn-60). The glycans attached at Asn-54 and Asn-58 are complex type oligosaccharides but are more highly sialyted than the Fc glycans. Many of the V_H glycans contain an unusual terminal Gal$\alpha1 \rightarrow 3$ residue that is absent in the Fc glycans. The carbohydrate attached at Asn-60 is a high-mannose oligosaccharide. The V_H region carbohydrates can modify antibody affinity. The carbohydrate chain at Asn-60 increases the antibody affinity but less effectively than that positioned at Asn-58. On the contrary, glycosylation at Asn-54 blocks the antigen binding.

In some nonprecipitating antibodies, the presence of a high-mannose content oligosaccharide in the Fd portion of only one of their two Fabs sterically hinders their ability to react with antigens. Such asymmetrical IgG antibody molecules, which are present in antisera to all studied antigens at a level of about 10%, have one antibody-combining site of high affinity and one of low affinity. The association constants of these two antigen-combining sites differ by about 100-fold. The nonprecipitating antibodies can precipitate antigen only in the presence of normal, symmetrical antibodies. The nonprecipitating antibodies are functionally deficient; they do not activate complement by the classical or alternative pathways, are not cytotoxic, and do not induce phagocytosis or opsonization (Margni and Binaghi, 1988). The treatment by *N*-glycanase, which removes *N*-linked glycosaccharides, restores the precipitating activity of asymmetric murine IgG antibodies (Mathov *et al.*, 1995).

D. Three-Dimensional Studies of Immunoglobulin Glycans

Oligosaccharides are flexible and may adopt different conformations. For example, different branches of one complex glycan can accept different conformations due to their differential flexibilities (Wu *et al.*, 1991) However, glycans isolated from glycoproteins exist in solution with regions of well-defined three-dimensional structures and only a few degrees of freedom (Homans *et al.*, 1987). Such regional rigidity minimizes the number of overall glycan conformations and probably favors the formation of glycan contacts with certain parts of the protein surface.

In several high-resolution x-ray crystallography studies, the interrelationship between the residues of *N*-linked glycans and the protein moiety of the Fc fragment was described. In two of them, the isolated Fc fragments of human IgG1 and rabbit IgG were investigated (Deisenhofer, 1981; Sutton and Philipps, 1983). In others, the Fc was studied as a part of human IgG1 or murine monoclonal IgG2a antibody molecules (Guddat *et al.*, 1993; Harris *et al.*, 1997). These studies determined that significant parts of both Fc oligosaccharide chains make contact with the surface of the C_H2 domain (Fig. 21). Carbohydrates cover the area of C_H2, which is usually in interdomain contacts in other pairs of domains. The area covered by glycan is, however, about two times less than the area involved in C_H3–C_H3 contacts. In the human Fc fragment, six side-chain atoms are within hydrogen bonding distance to glycan polar atoms. Numerous contacts are found between the terminal galactose residue and several amino acid residues of human Fc, especially with Lys-246 (14 contacts) and Glu-258 (9 contacts) (Padlan, 1990). The six amino acid residues of the murine Fc are in sugar contacts, Phe-256, Lys-259, Asp-262, Thr-273, and Tyr-313. The latter residue is in contact with the fucose residue of the murine and human Fc glycans. It is near the Fcγ receptor binding site and fucose could influence the activity of the receptor. There are no contacts between two oligosaccarides chains in the murine and human Fc fragments (Fig. 21, left).

In the rabbit Fc fragment, the core pentasaccharides of both carbohydrate chains (residues (1), (2), (3), (4), (4′); Fig. 19) and residues of the 1 → 6 arms

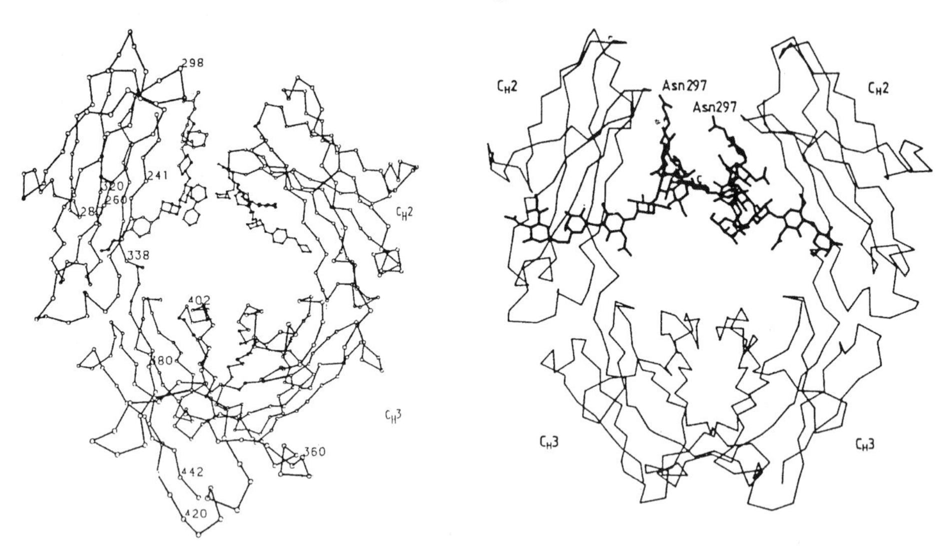

FIGURE 21 The α-carbon backbone of the Fcγ fragment. (left) Fc of human IgG1 Kol. (Deisenhofer, 1981.) (right) Fc of rabbit IgG. (Rademacher *et al.*, 1988; crystal data of Sutton and Philipps, 1983.) Positions of carbohydrate chains are shown between C_H2 domains.

are in a single conformation. They are both in close contact with the amino acid residues of C_H2. The Fc portion of IgG is nonsymmetrical, since the $\alpha 1 \rightarrow 3$ antennae of the two Fc oligosaccharide chains have nonidentical conformations. Furthermore, the two glycan moieties of the rabbit Fc interact with each other, in contrast to the murine and human Fc glycans. The GlcNAc(5) of one rabbit glycan chain is in contact with the GlcNAc(2) and Man(3) of the other (Fig. 21, right). The major portion of the Fc oligosaccharide chain is buried inside the Fc. Only its terminal sialic acid or galactose residues are exposed on the surface and thus are available for interaction with lectins and enzymes. Only after immunoglobulin denaturation can sugar residues, other than sialic acid and galactose, be detected, for example by methods employing lectins.

Most residues of the Fc glycans are relatively immobile, which was confirmed by spin-label studies. Indeed, spin labels, introduced in the sugar units of Fc, rotate in solution together with the neighboring protein units and display no rotational freedom of their own (Timofeev *et al.*, 1978; Sykulev and Nezlin, 1990).

The glycans can influence the conformation of the peptide chain in the region of the glycosylation site. A high-mannose glycan is linked to a single asparagine glycosylation site of the mu chain tailpiece. The tailpiece has no effect on the conformation of the glycan. However, the presence of the mannose glycan causes a decrease on the conformational mobility of the tailpiece peptide at the glycan attachment site. The *N*-glycosidic linkage between the *N*-acetylglucosamine (GlcNAc) and an asparagine residue is rigid and planar. Therefore, the conformational space for the oligosaccharide relative to the protein moiety of a glycoprotein depends on the flexibility of the asparagine side chain and on the local mobility of the peptide chain (Wormald *et al.*, 1991).

E. Functions of Immunoglobulin Carbohydrates

The conservation of the carbohydrate components of immunoglobulins suggests an important structural and functional role for them. In the Fcγ portion they could facilitate the maintenance of the structure of C_H2 domains and/or be directly involved in IgG interactions with ligands. To study the functions of immunoglobulin carbohydrates, aglycosylated immunoglobulin molecules were obtained by treating native immunoglobulins with specific glycosidases or by adding tunicamycin, an inhibitor of glycosylation, to immunoglobulin synthesizing cells (Leatherbarrow *et al.*, 1985; Furukawa and Kobata, 1991).

The replacement of the glycosylation site of IgG molecules, Asn-297 of the γ chains, by site-directed mutagenesis IgG molecules did not influence binding on staphylococcal protein A and antigens (Tao and Morrison, 1989). Since IgG molecules, with and without carbohydrate components, are secreted equally well from B cells, oligosaccharide chains are not necessary for IgG secretion. The catabolism of IgG is also not affected by the absence of its carbohydrate. However, IgG's binding of C1q and its ability to activate complement are diminished significantly or completely lost upon aglycosylation. The absence of glycans is also accompanied by a decrease in the ability of IgG to bind to all three types of the cell Fcγ receptors, and by an increase in IgG's sensitivity to proteolysis by pepsin and chymotrypsin, which cleave C_H2. All these effects are probably due to the influence of carbohydrate components on the structure of the low hinge region of IgG (amino acid residues 234–239). The small localized structural changes in the vicinity of His-268, the residue, which is in spatial proximity to the Fc glycosylation sites, were observed by NMR studies in the absence of the Asp-297 glycan (Lund *et al.*, 1990). The differences between the glycosylated and aglycosylated mouse IgG2b molecules were found in papain cleavage profiles. The glycosylated Fc is cleaved at residue 229, whereas the aglycosylated molecules have nearby three additional cleavage points. The aglycosylated and glycosylated Fc have different thermodynamic parameters, which were revealed by differential scanning microcalorometry (Jefferis and Lund, 1997).

Glycosylation is also important for the effector activity of IgM. Replacement of Asn-402, a residue analogous to Asn-297 of IgG, abrogate the glycosylation on this site. This mutation causes the increased production of monomeric IgM and the decrease in complement-dependent cytolysis in two-thirds (Muraoka and Shulman, 1989). Removal of the IgM and IgA tailpiece oligosaccharides leads to increase the antibody avidity (Bazin *et al.*, 1992; Chuang and Morrison, 1997)

The interactions of some sugar residues with the protein moiety are important for the biological activity of immunoglobulins. Noncovalent contacts of *N*-acetylglucosamine residue that is covalently linked to Asn-297 with Asp-265 is essential for recognition of IgG by Fcγ receptors. The IgG molecules, in which Asp-265 was replaced, lose their ability to interact with FcγR. Replacements of Lys-246, Asp-249 or Glu-258, which make contacts with sugar residues of the Fc glycan outer $\alpha 1 \rightarrow 6$ arm, have no effects on this IgG activity (Lund *et al.*, 1995)

F. Variations of Galactosylation

The most studied variation of the immunoglobulin glycosylation is linked with the reduction in galactosylation associated with several diseases, particularly with rheumatoid arthritis (Parekh *et al.*, 1989). Some IgG molecules in normal serum lack galactose residues in the outer arms of their oligosaccharide chains that terminate in *N*-acetylglucosamine. In the sera of humans, rabbits, mice, and Lewis rats 25, 40, 30, and 70%, respectively, of their IgG is agalactosylated. In patients with rheumatoid arthritis, the quantity of IgG oligosaccharides lacking galactose is linked to their health condition. In acute phases of the disease, the content of agalacto-IgG increases and during remissions it returns to normal or near normal levels. Elevated levels of agalactosyl IgG are also associated with other chronic diseases, such as tuberculosis, Crohn's disease, myositis, scleroderma, sponsdylitis, and lupus erythematosus with Sjögren's syndrome (Parekh *et al.*, 1989). Similar abnormal degree of galactosylation occurs in mice with autoimmune disorders (e.g., lupus-like syndrome and rheumatoid arthritis-like articular changes) (Mizuochi *et al.*, 1990). The percentage of maternal IgG molecules with agalactosyl *N*-glycans decreases during normal human pregnancy and increases following delivery. The amount of IgG molecules bearing agalactosyl *N*-glycans is age related, decreasing from birth to a minimum at 25 years of age and then rising again (Parekh *et al.*, 1988, Rudd *et al.*, 1991).

Variations in the galactose content of IgG can be attributed to changes in the β-galactotransferase activity in B lymphocytes. The significance of such variations is unknown. Rheumatoid arthritis is an autoimmune disorder, whose clinical symptoms correlate with the appearance of IgM or IgG autoantibodies (rheumatoid factors) in serum and synovial liquid, which recognize epitopes located in the Fc portion of IgG molecules. It was proposed (Parekh *et al.*, 1989) that an abnormal galactosylation of IgG glycans in rheumatoid arthritis could cause self-aggregation of IgG molecules or induce some structural changes, which enhance the autoantigenicity of Fc portions and thus the generation of anti-Fc rheumatiod factors .

It was suggested that the terminal galactose residue (Galβ1 $\rightarrow$ 4) is bound on the C_H2 surface at a pocket-like site that immobilized the oligosaccahride chain (Parekh *et al.*, 1989). By NMR studies it was found that in the absence of this residue in the agalacto-IgG molecules, the mobility of the carbohydrate chain is increased. (Wormald *et al.*, 1997). These findings were supported by x-ray crystallographic studies of the Fc fragment from an agalacto-IgG4 molecule (Corper *et al.*, 1997). No electron density was visible either for the carbohydrate or for the part of peptide chain containing Asn-297 glycosylation site, an indication of the increased mobility of this part of C_H2, including the glycan. The carbohydrate component was not involved directly in the Fc recognition by an anti-Fc autoantibody (rheumatoid factor).

However, agalactosyl IgG molecules are indistinguishable from fully galactosylated IgG in some important aspects. For example, binding of IgG to the FcγRII receptor is not dependent on the absence of terminal galactose residues in the Fc portion. Thus, the suppression of antibody production, which is realized through the inhibitory activity of FcgRII, is not impaired by the absence of the terminal galactose. These data do not support the view that defective interactions of agalactosyl antibodies with FcgRII are responsible for uncontrolled excessive autoantibody production (Groeninck *et al.*, 1996). The outer arm galactose residues are also not essential for the FcγRI recognition (Lund *et al.*, 1995).

Glycosylation of monoclonal antibodies depends on several factors. One is the origin of antibody-synthesizing cells that could possess different sets of enzymes, the others are linked with methods of cell cultivation, the supply of nutrients, elimination of metabolics, the longevity of cultivation, and other parameters (Jefferis and Lund, 1997).

REFERENCES

Abaturov L.V., Nezlin R., Vengerova T.I., and Varshavski Y.M. (1969). Conformational studies of immunoglobulin G and its subunits by the methods of hydrogen-deuterium exchange and infrared spectroscopy. *Biochem. Biophys. Acta* **194**, 386–396.

Abola E.E., Ely K. R., and Edmundson A. B. (1980). Marked structural differences of the Mcg Bence-Jones dimer in two crystal systems. *Biochemistry* **19**, 423–439.

Achatz G., Nitschke L., and Lamers M.C. (1997). Effect of transmembrane and cytoplasmic domains of IgE on the IgE response. *Science* **276**, 409–411.

Adachi T., Schamel W.W.A., Kim K.-M., Watanabe T., Becker B., Nielsen P.J., and Reth M. (1996). The specificity of association of the IgD molecule with the accessory proteins BAP31/BAP29 lies in the IgD transmembrane sequence. *EMBO J.* **15**, 1534–1541.

Alzari P.M., Lascombe M.-B., and Poljak, R.J. (1988). Three-dimensional structure of antibodies. *Ann. Rev. Immunol.* **6**, 555–580.

Apodaca G., Bomsel M., Avden J., Breitfeld P.P., Tang K., and Mostov K.E. (1991). The polymeric immunoglobulin receptor. A model protein to study transcytosis. *J. Clin. Invest.* **87**, 1877–1882.

Atkin J.D., Pleass R.J., Owens R.J., and Woof J.M. (1996). Mutagenesis of the human IgA1 heavy chain tailpiece that prevents dimer assembly. *J. Immunol.* **157**, 156–159

Avrameas S. (1991). Natural autoantibodies: from "horror autotoxicus" to "gnothi seauton." *Immunol. Today* **12**, 154–159.

Baird B., Zheng Y., and Holowka D. (1993). Structural mapping of IgE-FceRI, an immunoreceptor complex. *Accounts Chem Res (Published by Am. Chew Soc)* **26**, 428–434.

Bakos M.-A., Kurosky A., and Goldblum R.M. (1991). Characterization of a critical binding site for human polymeric Ig on secretory component. *J. Immunol.* **147**, 3419–3426.

Bakos M.-A., Kurosky A., Czerwinski E.W., and Goldblum R.M. (1993). A conserved binding site on the receptor for polymeric Ig is homologous to CDR1 of Ig $V_\varkappa$ domains. *J. Immumol.* **151**, 1346–1352.

Batista F.D., Anand S., Presani G., Efremov D.G., and Burrone O.R. (1996a). Two membrane isoforms of human IgE assemble into functionally distinct B cell antigen receptors. *J. Exp. Med.* **184**, 2197–2205.

Batista F.D., Efremov D.G., and Burrone O.R. (1996b). Characterization of a second secreted IgE isoform and identification of an asymetric pathway of IgE assembly. *Proc. Natl. Acad. Sci. USA* **93**, 3399–3404.

Bazin H., Darveau A., Martel F., Pelletier A., Piche L., St-Laurent, M., Thibault L., Demers A., Boyer L., Limieux G., and Limieux R. (1992). Increased avidity of mutant IgM antibodies caused by the absence of COOH-terminal glycosylation of the m H chain. *J. Immunol.* **149**, 3889–3893.

Bazin H., Pear W.S. and Sumegi J. (1988). Louvain rat immunocytomas. *Adv. Cancer Res.* **50**, 279–310.

Beavil A.J., Young R.J., Sutton B.J. and Perkins S.J. (1995). Bent domain structure of recombinant human IgE-Fc in solution by x-ray and neutron scattering in conjunction with an automated curve fitting procedure. *Biochemistry* **34**, 14449–14461.

Bork P., Holm L., and Sander C. (1994). The immunoglobulin fold. Structural classification, sequence patterns and common core. *J. Mol. Biol.* **242**, 309–320.

Brandtzaeg P. (1985). Role of J chain and secretory component in receptor-mediated glandular and hepatic transport of immunoglobulins in man. *Scand. J. Immunol.* **22**, 111–146.

Brandtzaeg P., Krajci P., Lamm M.E., and Kaetzel C.S. (1994). Epithelial and hepatobiliary transport of polymeric immunoglobulins In: *Handbook of Mucosal Immunology* (Ed. Ogra P.L. *et al*). pp. 113–126. Academic Press, San Diego.

Brewer J.W., and Corley R.B. (1997). Late events in assembly determine the polymeric structure and biological activity of secretory IgM. *Mol. Immunol.* **34**, 323–331.

Brewer J.W., Randall T.D., Parkhouse R.M.E., and Corley R.B. (1994). IgM hexamers? *Immunol. Today* **15**, 165–168.

Burnett R.B., Hanly W.C., Zhai S.K., and Knight K.L. (1989). The IgA heavy-chain gene family in rabbit: cloning and sequece analysis of 13 Ca genes. *EMBO J.* **8**, 4041–4047.

Burrows P.D., Schroeder H.W., and Cooper M.D. (1995). B-cell differentiation in humans. In: *Immunoglobulin Genes,* 2nd ed. (Eds. Honjo T., and Alt F.W.). pp. 3–30. Academic Press, San Diego.

Cambier J.C. (1992) Signal transduction by T- and B-cell antigen receptors: converging structures and concepts. *Curr. Opin. Immunol.* **4**, 257–264.

Cambier J.C. (1995). Antigen and Fc receptor signaling. The awesome power of the immunoreceptor tyrosine based activation motif (ITAM). *J. Immunol.* **155**, 3281–3285.

Campbell K.S., Hager E.J., and Cambier J.C. (1991). α-chains of IgM and IgD antigen receptor complexes are differentially *N*-glycosylated mb-1–related molecules. *J. Immunol.* **147**, 1575–1588.

Carayannopoulos L., and Capra J.D. (1993). Immunoglobulins: structure and function. In: *Fundamental Immunology,* 3rd ed. (Ed. Paul W.P.). pp. 283–314. Lippincott-Raven, Philadelphia.

Chan D.W. (Ed) (1996). *Immunoassay Automation. An Updated Guide to Systems.* Academic Press, San Diego.

Chan A.C., and Shaw A.S. (1995) Regulation of antigen receptor signal transduction by protein tyrosine kinases. *Curr. Opin. Immunol.* **8**, 394–401.

Chandrasekaran E.V., Mendicino A., Garver F.A., and Mendicino J. (1981). Structures of sialylated *O*-glycosidically and *N*-glycosidically linked oligosaccarides in a monoclonal immunoglobulin light chain. *J. Biol. Chem.* **256**, 1549–1555.

Chintalacharuvu K.R., Raines M., and Morrison S.L. (1994). Divergence of human α-chain constant region gene sequences. A novel recombinant $\alpha 2$ gene. *J. Immunol.* **152**, 5299–5304.

Chintalacharuvu K.R., and Morrison S.L. (1996). Residues critical for H–L disulfide bond formation in human IgA1 and IgA2. *J. Immunol.* **157**, 3443–3449.

Chothia C., Lesk A.M., Tramontano A., Levitt M., Smith-Gill S., Air G., Sheriff S., Padlan E.A., Davies D., Tulip W.R., Colman P.M., Spinelli S., Alzari P.M., and Poljak P.J. (1989). Conformations of immunoglobulin hypervariable regions. *Nature* **342**, 877–883.

Chothia C., Novotny J., Bruccoleri R. and Karplus M. (1985). Domain association in immunoglobulin molecules. The packing of variable domains. *J. Mol. Biol.* **186**, 651–663.

Chuang P.D., and Morrison S.L. (1997). Elimination of *N*-linked glycosylation sites from the human IgA1 constant region. *J. Immunol.* **158**, 724–732.

Coligan J.E., Kruisbeek A.M., Margulies D.M., Shevach E.M., and Strober W., (Eds) (1991). *Current Protocols in Immunology.* Greene Publishing Assoc. and Wiley-Interscience, New York.

Colman P.M. (1988). Structure of antibody-antigen complexes: implications for immune recognition. *Adv. Immunol.* **48**, 100–132.

Constantine K.L., Friedrichs M.S., Metzler W.J., Wittekind M., Hensley P., and Mueller L. (1994). Solution structure of an isolated antibody V_L domain. *J. Mol. Biol.* **236**, 310–327.

Corper A.L., Sohi M.K., Bonagura V.R., Steinitz M., Jefferis R., Feinstein A., Beale D., Tausiig M.J., and Sutton B.J. (1997). Structure of human IgM rheumatoid factor Fab bound to its utoantigen IgG Fc reveals a novel topology of antibody-antigen interaction. *Nature Struct. Biol.* **4**, 374–381.

Coyne R.S., Siebrecht M., Peitsch M.C., and Casanova J.E. (1994). Mutational analysis of polymeric immunoglobulin receptor/ligand interactions. Evidence for the involvement of multiple complementary determining region (CDR)-like loops in receptor domain I. *J. Biol. Chem.* **269**, 31620–31625.

Cser L., Franěk F., Gladkikh I.A., Kunchenko A.B., and Ostanevich Y. (1981). General shape and hapten-induced conformational changes of pig antibody against dinitrophenyl. *Eur. J. Biochem.* **116**, 109–116.

Davis A.C., Roux K.H., and Shulman M.J. (1988). On the structure of polymeric IgM. *Eur. J. Immunol.* **18**, 1001–1008.

Davies D.R., and Chacko S. (1993). Antibody structure. *Acc. Chem. Res.* **26**, 421–427.

Day E.D. (1990). *Advanced Immunochemistry,* 2nd ed. Wiley-Liss, New York.

DeFranco A.L. (1993). Structure and function of the B cell antigen receptor. *Annu. Rev. Cell. Biol.* **9**, 377–410.

Deisenhofer J. (1981). Crystallographic refinement and atomic models of a human Fc fragment and its complex with fragment B of protein A from Staphylococcus aureus at 2.9 and 2.8 Å resolution. *Biochemistry* **20**, 2361–2370.

De Lau W.B.M., Heije K., Neefjes J.J., Oosterwegel M, Rozemuller E., and Bast B.J.E.G. (1991). Absence of preferential homologous H/L chain association in hybrid hybridomas. *J. Immunol.* **146**, 906–914.

Du Pasquier L., and Chrétien I. (1996). CTX, a new lymphocyte receptor in *Xenopus,* and the early evolution of Ig domains. *Res. Immunol.* **147**, 218–226.

Dwek R.A. (1995). Glycobiology: towards understanding the function of sugars. *Biochem. Soc. Trans.* **23**, 1–25.

Edelman G.M. (1959). Dissociation of γ-globulins. *J. Am. Chem. Soc.* **81**, 3155–3156.

Edelman G.M., Cunningham B.A., Gall W.E., Gottlieb P.D., Rutishauser U., and Waxdal M.J. (1969). The covalent strcuture of an entire γG immunoglobulin molecule. *Proc. Natl. Acad. Sci. USA* **63**, 78–85.

Edelman G.M., and Gally J.A. (1962). The nature of Bence-Jones proteins: chemical similarities to polypeptides chains of myeloma globulins and normal γ-globulins. *J. Exp. Med.* **116**, 207–227.

Edelman G.M., Olins D.E., Gally J.A., and Zinder N.D. (1963). Reconstitution of immunologic activity by interaction of polypeptide chains of antibodies. *Proc. Natl. Acad. Sci. USA* **50**, 753–759.

Edmundson A.B., Ely K.R., Abola E.E., Schiffer M., and Panagiotopoulos N. (1975). Rotational allomerism and divergent evolution of domains in immunoglobulin light chains. *Biochemistry* **14**, 3953–3961.

Edmundson A.B., Ely K.R., and Abola E.E. (1978). Conformational flexibilty in immunoglobulins. *Contemp. Top. Mol. Immunol.* **7**, 95–118.

Edmundson A.B., Guddat L.W., and Kim N.A. (1993). Crystal structures of intact IgG antibodies. *Immunomethods* **3**, 197–210.

Eisen H., and Karush F. (1949). The interaction of purified antibody with homologous hapten: antibody valence and binding constant. *J. Am. Chem. Soc.* **71**, 363–364.

Eisen H.N. (1990). *General Immunology.* pp. 11–38. Lippincott-Raven, Philadelphia.

Ely K.R., Coiman P.M., Abola E.E., Hess A.C., Peabody D.S., Parr D.M., Connel G.E., Laschinger C.A., and Edmundson A.B. (1978). Mobile Fc region in the Zie IgG2 cryoglobulin: comparison of crystals of the $F(ab')_2$ fragment and the intact immunoglobulin. *Biochemistry* **17**, 820–823.

Endo T., Wright A., Morrsion S.L., and Kobata A. (1995). Glycosylation of the variable region of immunoglobulin G—site specific maturation of the sugar chains. *Mol. Immunol.* **32**, 931–940.

Fallgreen-Gebauer E., Gebauer W., Bastian A., Kratzin H.D., Eiffert H., Zimmermann B., Karas M., and Hilschmann N. (1993). The covalent linkage of secretory component to IgA. Structure of IgA. *Biol. Chem. Hoppe Seyler* **374**, 1023–1028.

Fan Z., Shan L., Guddat L.W., He X., Gray W.R., Raison R.L., and Edmundson A.B. (1992). Three-dimensional structure of an Fv from a human IgM immunoglobulin. *J. Mol. Biol.* **228**, 188–207.

Feinstein A., Munn E.A., and Richardson N.E. (1971) The three-dimensional conformation of γM and γA globulin molecules. *Annu. Rev. N.Y. Acad. Sci.* **190**, 104–121

Fellah J.S., Kerfourn F., Wiles M.V., Schwager J., and Charlemagne J. (1993). Phylogeny of immunoglobulin heavy chain isotypes: structure of the constant region of *Ambystoma mexicanum* υ chain deduced from cDNA sequence. *Immunogenetics* **38**, 311–317.

Field M.C., Dwek R.A., Edge C.J., and Rademacher T.W. (1989). *O*-Linked oligosaccharides from human serum immunoglobulin A1. *Biochem. Soc. Trans.* **17**, 1034–1035.

Foote J., and Milstein C. (1994). Conformational isomerism and the diversity of antibodies. *Proc. Natl. Acad. Sci. USA* **91**, 10370–10374.

Franěk F. (1961). Dissociation of animal 7S γ-globulin by cleavage of disulfide bonds. *Biochem. Biophys. Res. Commun.* **4**, 28–32.

Franěk F., and Nezlin R. (1963). Recovery of antibody combinig activity by interaction of different peptide chains isolated from purified horse antitoxin. *Folia Microbiol.* **8**, 197–201.

Freedman M.H., and Sela M. (1966). Recovery of specific activity upon reoxidation of completely reduced polyalanyl rabbit antibody. *J. Biol. Chem.* **241**, 5225–5232.

Friedrich R.J., Campbell K.S., and Cambier J.C. (1993). The γ subunit of the B cell antigen-receptor complex is a C-terminally truncated poduct of the B29 gene. *J. Immunol.* **150**, 2814–2822.

Frutiger S., Hughes G.J., Fonck C., and Jaton J.-C. (1987). High and low molecular weight rabbit secretory components. *J. Biol. Chem.* **262**, 1712–1715.

Frutiger S., Hughes G.J., Paquet N., Lüthy R., and Jaton J.-C. (1992). Disulphide bond assignment in human J chain and its covalent pairing with immunoglobulin M. *Biochemistry* **31**, 12643–12647.

Furukawa K., and Kobata A. (1991). IgG galactosylation—its bological significance and pathology. *Mol. Immunol.***28**, 1333–1340.

Givol D. (1991). The minimal antigen-binding fragment of antibodies—Fv fragment. *Mol. Immunol.* 28, 1379–1387

Godfrey M.A.J. (1997). Immunoaffinity and IgG receptor technologies. In: *Affinity Separations.* (Ed. P. Matejtschuk). pp. 141–194. IRL press, Oxford.

Gold M.R., Matsuuchi L., Kelly R.B., and DeFranco A.L. (1991). Tyrosine phosphorylation of components of the B cell antigen receptor following receptor crosslinking. *Proc. Natl. Acad. Sci. USA* **88**, 3436–3440.

Gold M.R., and DeFranco A.L. (1994). Biochemistry of B lymphocyte activation. *Adv. Immunol.* **55**, 221–295.

Gong S., and Nussenzweig M. (1996). Regulation of an early development checkpoint in the B cell pathway by Ig beta. *Science* **272**, 411–414.

Groenink J., Spijker J., van der Herik-Oudijk I.E., Boeije L., Rook G., Aarden L., Smeenk R., van der Winkel J.G.J., and van den Broek M.F. (1996). On the interaction between galactosyl IgG and Fcg receptors. *Eur. J. Immunol.* **26**, 1404–1407.

Gronski P., Seiler F.R., and Schwick H.G. (1991). Discovery of antitoxins and development of antibody preparations for clinical uses from 1890 to 1990. *Mol. Immunol.* **28**, 1321–1332.

Guddat L.W., Herron J.N., and Edmundson A.B. (1993). Three-dimensional structure of a human immunoglobulin with a hinge deletion. *Proc. Natl. Acad. Sci. USA* **90**, 4271–4275.

Gurvich A.E. (1964). The use of antigens on an insolube support. In: *Immunological Methods* (Ed. Ackroyd J.F.), Blackwell, Oxford, 113–136.

Gurvich A.E., and Lechtzind E.V. (1982). High capacity immunoadsorbents based on preparations of reprecipitatetd cellulose. *Mol. Immunol.* **19**, 637–640.

Gurvich A.E., and Korukova A. (1986). Induction of abundant antibody formation with a protein cellulose complex in mice. *J. Immunol. Methods* **87**, 161–167.

Hamel P.A., Klein M.H., Smith-Gill S.J., and Dorrington K.J. (1987). Relative noncovalent association constant between immunoglobulin H and L chains is unrelated to their expression or antigen-binding activity. *J. Immunol.* **139**, 3012–3020.

Harlow E., and Lane D. (1988). *Antibodies: A Laboratory Manual.* Cold Spring Harbor Laboratory, Cold Spring Harbor, New York.

Harpaz Y., and Chothia C. (1994). Many of the immunoglobulin superfamily domains in cell adhesion molecules and surface receptors belong to a new structutal set which is close to that containing variable domains. *J. Mol. Biol.* **238**, 528–539.

Harris L.J., Larson S.B., Hasel K.W., Day J., Greenwood A., and McPherson A. (1992). The three-dimensional structure of an intact monoclonal antibody for canine lymphoma. *Nature* **360**, 369–372.

Harris L.J., Larson S.B., Hasel K.W., and McPherson A. (1997). Refined structure of an intact IgG2a monoclonal antibody. *Biochemistry* **36**, 1581–1597.

Hendrikson B.A., Conner D.A., Ladd D.J., Kendall D., Casanova J.E., Corthesy B., Max E.E., Neutra M.R., Seidman C.E., and Seidman J.G. (1995). Altered hepatic transport of immunoglobulin A in mice lacking the J chain. *J. Exp. Med.* **82**, 1905–1911.

Herzenberg L.A., Herzenberg L.A., Weir D.M., and Blackwell C. Eds. (1996). *Weir's Handbook of Experimental Immunology,* 5th ed. Vol.1. Blackwell, Oxford.

Hester R.B., Mole J.E., and Schrohenloher R.E. (1975). Evidence for the absence of noncovalent bonds in the Fc_{μ} region of IgM. *J. Immunol.* **114**, 486–491.

Hexam J.M., Carayannopoulos L., and Capra J.D. (1997). Structure and function in IgA. In: *Antibody Engineering.* (Ed. Capra J.D.). pp. 73–87. Karger, Basel.

Hill R.L., Delaney R., Fellows R.E., and Lebovitz H.E. (1966). The evolutionary origins of the immunoglobulins. *Proc. Natl. Acad. Sci. USA* **56**, 1762–1769.

Hilschmann N., and Craig L.C. (1965). Amino acid sequence studies with Bence-Jones proteins. *Proc. Natl. Acad. Sci. USA* **53**, 1403–1409.

Hirabayashi Y., Lecerf J.-M., Dong Z., and Stollar B.D. (1995). Kinetic analysis of the interactions of recombinant human Vpre-B and IgV domains. *J.Immunol.* **155**, 1218–1228.

Homans S.W., Dwek R.A., and Rademacher T.W. (1987). Solution conformations of N-linked oligosaccharides. *Biochemistry* **26**, 6571–6578.

Honjo T., and Matsuda F. (1995). Immunoglobulin heavy chain loci of mouse and human. In: *Immunoglobulin Genes,* 2nd ed. (Eds. Honjo T., Alt F.W.). pp. 145–171. Academic Press, San Diego.

Huber R., Deisenhofer J., Colman P.M., and Matsushima M. (1976). Crystallographic structure studies of an IgG molecule and an Fc fragment. *Nature* **264**, 415–420.

Hunkapiller T., and Hood L. (1989). Diversity of the immunoglobulin gene superfamily. *Adv. Immunol.* **44**, 1–63.

Huston J.S., Margolies M.N., and Haber E. (1996). Antibody binding sites. *Adv. Protein Chem.* **49**, 329–450.

Jefferis R., and Lund J. (1997). Glycosylation of antibody molecules: structural and functional significance. In: *Antibody Engineering* (Ed. Capra J.D.). pp. 111–128. Karger, Basel.

Johnstone A., and Thorpe R. (1982). *Immunochemistry in Practice.* Blackwell, Oxford.

Kabat E.A., and Mayer M.M. (1961) *Experimental Immunochemistry.* pp. 22–96. C.C Thomas, Springfield, IL.

Kabat E.A., Wu T.T., Perry H.M., Gottesman K.S., and Foeller C. (1991). *Sequences of Proteins of Immunological Interest,* 5th ed. Public Health Service, NIH, Washington, DC.

Kang A.S., Jones T.M., and Burton D.R. (1991). Antibody redesign by chain shuffling from random combinatorial immunoglobulin libraries. *Proc. Natl. Acad. Sci. USA.* **88**, 11120–11123.

Karasuyama H., Kudo A., and Melchers F. (1990). The proteins encoded by the V_{preB} and λ_5 pre-B cell specific genes can associate with each other and with μ heavy chain. *J. Exp. Med.* **172**, 969–972.

Karasuyama H., Rolink A., and Melchers F. (1993). A complex of glycoproteins is associated with V_{preB}/λ_5 surrogate light chain on the surface of m heavy chain-negative early precursor B cell lines. *J. Exp. Med.* **178**, 469–478.

Karasuyama H., Rolink A., and Melchers F. (1996). Surrogate light chain in B cell development. *Adv. Immunol.* **63**, 1–41.

Kemeny D.M. (1991). *A practical Guide to ELISA.* Pergamon Press, Oxford.

Kerr M.A. (1990). The structure and function of human IgA. *Biochem. J.* **271**, 285–296.

Kerr M.A., and Thorpe R. (Eds.) (1994). *Immunochemistry LabFax.* Bios Scientific Publishers, Oxford.

Kilian M., and Russell M.W. (1994). Function of mucosal immunoglobulins. In: *Handbook of Mucosal Immunology* (Ed. Ogra P.L. *et al.*). pp.127–137. Academic Press, San Diego.

Kim H.H., Matsunaga C., Yoshino A., Kato K., and Arata Y. (1994). Dynamical structure of the hinge region of immunoglobulin G as studied by ^{13}C nuclear magnetic resonance spectroscopy. *J. Mol. Biol.* **236**, 300–309.

Klafki H.W., Kratzin H.D., Pick A.I., Eckart K., Karas M., and Hilschmann N. (1992). Complete amino acid sequence determinations demonstrate identity of the urinary Bence Jones protein (BJP-DIA) and the amyloid fibril protein (AL-DIA) in a case of AL-amyloidosis. *Biochemistry* **31**, 3265–3272.

Knight A.M., Lucocq J.M., Prescott A.R., Ponnambalam S., and Watts C. (1997). Antigen endocytosis and presentation mediated by human membrane IgG1 in the absence of the Iga/Igb dimer. *EMBO J.* 16, 3842–3850.

Köhler G., and Milstein C. (1975). Continuous cultures of fused cells secreting antibody of predefined specificity. *Nature* **256**, 495–497.

Koshland M.E. (1985). The coming of age of the immunoglobulin J chain. *Annu. Rev. Immunol.* **3**, 425–453.

Krugmann S., Pleass R.J., Atkin J.D., and Woof J.M. (1997). Structural requirements for assembly of dimeric IgA probed by site-directed mutagenesis of J chain and a cysteine residues of the a-chain CH2 domain. *J. Immunol.* **159**, 244–249.

Lahav M., Arnon R., and Sela M. (1967). Biological activity of the cleavage product of human immunoglobulin G with cyanogen bromide. *J. Exp. Med.* **125**, 787–805.

Lam K.-P., Kühn R., and Rajewsky K. (1997). In vivo ablation of surface immunoglobulin on mature B cells by inducible gene targeting results in rapid cell death. *Cell* **90**, 1073–1083.

Lascombe M.-B., Alzari P.M., Poljak R.J., and Nisonoff A. (1992). Three-dimensional structure of two crystal forms of FabR19.9 from a monoclonal anti-arsonate antibody. *Proc. Natl. Acad. Sci. USA* **89**, 9429–9433.

Lassoued K., Illges H., Benlagha K., and Cooper M.D. (1996). Fate of surrogate light chains in B lineage cells. *J. Exp. Med.* **183**, 421–429.

Leatherbarrow R.J., Rademacher T.W., Dwek R.A., Woof J.M., Clark A., Burton D.R., Richardson N., and Feinstein A. (1985) Effector functions of a monoclonal aglycosylated mouse IgG2a: binding and activation of component C1 and interaction with human monocyte Fc receptor. *Mol. Immunol.* **22**, 407–415.

Li Y., Spellerberg M.B., Stevenson F.K., Capra J.D., and Potter K.N. (1996). The I binding specificity of human V_H4–34 (V_H4–21) coded antibodies is determined by both V_H framework region 1 and complementarity determining region 3. *J. Mol. Biol.* **256**, 577–589.

Lim K., Jameson D.M., Gentry C.A., and Herron J.N. (1995). Molecular dynamics of the anti-fluorescein 4–4–20 antigen-binding fragment. 2. Time-resolved fluorescence spectroscopy. *Biochemistry* **34**, 6975–6984.

Lis G., and Sharon N. (1993). Protein glycosylation. Structural and functional aspects. *Eur. J. Biochem.* **218**, 1–27.

Lüllau E., Heyse S., Vogel H., Marison I., von Stockar U., Kraehenbuhl J.-P., and Corthesy B. (1996). Antigen binding properties of purified immunoglobulin A and reconstituted secretory immunoglobulin A antibodies. *J. Immunol.* **271**, 16300–16309.

Lund J., Takahashi N., Pound J.D., Goodall M., Nakagawa H., and Jefferis R. (1995). Oligosaccharide-protein interactions in IgG can modulate recognition by Fcg receptors. *FASEB J.* **9**, 115–119,

Lund J., Tanaka T., Takahashi N., Sarmay G., Arata Y., and Jefferis R. (1990). A protein structural change in aglycosylated IgG3 correlates with loss of huFcγR1 and huFcγRIII biding and/or activation. *Mol. Immunol.* **27**, 1145–1153.

Manson M.M., Ed. (1992). *Immunochemical Protocols.* Humana Press. Totowa, NJ.

Margni R.A., and Binaghi R.A. (1988). Nonprecipitating asymmetric antibodies. *Annu. Rev. Immunol.* **6**, 535–554.

Marquart M., Deisenhofer J., Huber R., and Palm W. (1980). Crystallographic refinement and atomic models of the intact immunoglobulin molecule Kol and its antigen-binding fragment at 3.0 Å and 1.9 Å resolution. *J. Mol. Biol.* **141**, 369–391.

Mathov I., Plotkin L.I., Sqiquera L., Fossati C., Margni R.A., and Leoni J. (1995). *N*-glycanase treatment of $F(ab')_2$ derived from asymmetric murine IgG3 mAb determines the acquisition of precipitating activity. *Mol. Immunol.* **32**, 1123–1130.

Matsuda F, and Honjo T. (1996). Organization of the human immunoglobulin heavy-chain locus. *Adv. Immunol.* **62**, 1–29.

Melchers F. (1995). The role of B cell and pre-B-cell receptors in development and growth control of the B-lymphocyte cell lineage. In: *Immunoglobulin Genes,* 2nd ed (Eds. Honjo T., and Alt F.W.). pp. 33–56. Academic Press, San Diego.

Melchers F., Haasner D., Grawunder U., Kalberer C, Karasuyama H., Winkler T., and Rolink A.G. (1994). Roles of IgH and L chains and of surrogate H and L chains in the development of cells of the B lymphocyte lineage. *Ann. Rev. Immunol.* **12**, 209–225.

Melchers F., Karasuyama H., Haasner D., Bauer S., Kudo A., Sakaguchi N., Jameson B., and Rolink A. (1993). The surrogate light chains in B-cell development. *Immunol. Today* **14**, 60–68.

Mestecky J., and McGhee J.R. (1987). Immunoglobulin A (IgA): molecular and cellular reactions involved in IgA biosynthesis and immune response. *Adv. Immunol.* **40**, 153–245.

Metzger H. (1970). Structure and function of γM macroglobulin. *Adv. Immunol.* **12**, 57–116.

Michaelsen T.E. (1990). Fragmntation and conformational changes of IgG subclases. In: *The Human IgG Subclasses: Molecular Analysis of Structure, Function and Regulation* (Ed. Shakib F.). pp. 31–41. Pergamon Press, Oxford.

Mizuochi T., Hamakop J., Nose M., and Titani K. (1990). Structural changes in the oligosaccharide chains of IgG in autoimmune MRL/Mp-*lpr/lpr* mice. *J. Immunol.* **145**, 1794–1798.

Mol C.D., Muir A.K.S., Lee J.S., and Anderson W.F. (1994). Structure of an immunoglobulin Fab fragment specific for poly(dG).poly(dC). *J. Biol. Chem.* **269**, 3605–3614.

Mostov K.E., Friedlander M., and Blobel G. (1984). The receptor for transepithelial transport of IgA and IgM contains multiple immunoglobulin-like domains. *Nature* **308**, 37–43.

Mouthon L., Lacroix-Desmazes S., Nobrega A., Barreau C., Coutinho A., and Kazatchkine M.D. (1996). The self-reactive antibody repertoire of normal human serum IgM is acquired in early childhood and remains conserved throughout life. *Scand. J. Immunol.* **44**, 243–251.

Muraoka S., and Shulman M.J. (1989). Structural requirements for IgM assembly and cytolytic activity. Effects of mutations in the oligosaccharide acceptror site at Asn402. *J. Immunol.* **142**, 695–701.

Neuberger M.S. (1997). Antigen receptor signaling gives lympocytes a long life. *Cell* **90**, 971–973.

Nezlin R. (1977). *Structure and Biosynthesis of Antibodies.* Plenum Press, New York.

Nezlin R. (1990). Internal movements in immunoglobulin molecules. *Adv. Immunol.* **48**, 1–40.

Nezlin R. (1994). Immunoglobulin structure and function. In: *Immunochemistry* (Eds. van Oss C.J., and van Regenmortel M.H.V.). pp. 3–45. Marcel Dekker, New York.

Nezlin R. (1997). Immunoadsorbents prepared on small particles of dialdehyde cellulose (Gurvich's cellulose immunoadsorbents). In: *Immunology Methods Manual* (Ed. Lefkovits I.). pp. 517–520. Academic Press, San Diego.

Nezlin R., Zagyansky Y.A., Käiväräinen A.I., and Stefani S.B. (1973). Properties of myeloma immunoglobulin E (Yu). Chemical, fluorescent and spin-label studies. *Immunochemistry* **10**, 681–688.

Nezlin R., Rokhlin O.V., Vengerova T.I., and Machulla H.K.G. (1975). Allotype markers of kappa L chain of rat Ig localized in the constant region of the chain. *Immunochemistry* **11**, 517–518.

Niles M.J., Matsuuchi L., and Koshland M.E. (1995). Polymer IgM assembly and secretion in lymphoid and nonlymphoid cell lines: evidence that J chain is required for pentamer IgM synthesis. *Proc. Natl. Acad. Sci. USA* **92**, 2884–2888.

Nisonoff A., Hopper J.E., and Spring S.B. (1975).*The Antibody Molecule.* Academic Press, San Diego.

Novotny J., and Haber E. (1985). Structural invariants of antigen binding: comparison of immunoglobulin $V_{L-}V_H$ and V_L-V_L domain dimers. *Proc. Natl. Acad. Sci. USA* **82**, 4592–4596.

Padlan E.A. (1990). X-ray diffraction studies of antibody constant regions. In: *Fc Receptors and the Action of Antibodies* (Ed. Metzger H.). pp. 12–30. American Society of Microbiology, Washington, D.C.

Padlan E.A. (1994). Anatomy of the antibody molecule. *Mol. Immunol.* **31**, 169–217.

Padlan E.A. (1996). X-ray crystallography of antibodies. *Adv. Protein Chem.* **49**, 57–133.

Padlan E.A., Cohen G.H., and Davies D.R. (1987). Studies of the tertiary and quaternary structure of antibody constant domain. In: *Biological Organization: Macromolecular Interactions at High Resolution* (eds. Burnett R.M., and Vogel H.J.), pp. 193–214. Academic Press.

Panka D.J. (1997). Glycosylation is influential in murine IgG3 self-association. *Mol. Immunol.* **34**, 593–598.

Parekh R., Isenberg D., Rook G., Roitt I., Dwek R.A., and Rademacher T. (1989). A comparative analysis of disease-associated changes in the glycosylation of serum IgG. *J. Autoimmunity* **2**, 2101–2104.

Parekh R., Roitt I., Isenberg D., Dwek R.A. and Rademacher T. (1988). Age-related galactosylation of the *N*-linked oligosaccharides of human serum IgG. *J. Exp. Med.* **167**, 1731–1736.

Parham P. (1983). On the fragmentation of monoclonal IgG1, IgG2a, and IgG2b from BALB/c mice. *J. Immunol.* **131**, 2895–2902.

Perkins S.J., Nealis A.S., Sutton B.J., and Feinstein A. (1991). Solution strucuture of human and mouse immunoglobulin M by synchotron X-ray scattering and molecular graphics modeling. A possible mechanism for complement activation. *J. Mol. Biol.* **221**, 1345–1366.

Pillai S., and Baltimore D. (1987). Formation of disulphide-linked $\mu_2\ \omega_2$ tetramers in pre-B cells by the 18K ω-immunoglobulin chain. *Nature* **329**, 172–174.

Poljak R. (1991). Structure of antibodies and their complexes with antigens. *Mol. Immunol.* **28**, 1341–1345.

Poljak R., Amzel L.M., Avey H.P., Chen B.L., Phizackerley R.P., and Saul F. (1973). The three-dimensional structure of the Fab' fragment of a human immunoglobulin at 2.8 Å resolution. *Proc. Natl. Acad Sci. USA* **70**, 3305–3310.

Porter R.R. (1959). The hydrolysis of rabbit γ-globulin and antibodies with crystalline papain. *Biochem. J.* **73**, 119–126.

Potter M. (1972). Immunoglobulin-producing tumors and myeloma proteins in mice. *Physiol. Rev.* **52**, 631–719.

Potter M., and Smith-Gill S.J. (1990). Physiology of Immunoglobulins. In: *Immunophysiology.* (Eds. Oppenheim J.J., and Shevach E.M). pp. 129–151. Oxford University Press, Oxford.

Putnam F.W., Titani K., Wikler M., and Shinoda T. (1967). Structure and evolution of kappa and lambda light chains. *Cold Spring Harbor Symp.* **32**, 9–27.

Rademacher T.W., Parekh R.B., and Dwek R.A. (1988). Glycobiology. *Annu. Rev. Biochem.* **57**, 785–838.

Radic M.Z., Mascelli M.A., Erikson J., Shan H., and Weigert M. (1991). Ig H and L chain contributions to autoimmune specificities. *J. Immunol.* **146**, 176–182.

Randall T.D., Brewer J.W., and Corley R.B. (1992). Direct evidence that J chain regulates the polymeric structure of IgM in antibody-secreting B cells. *J. Biol. Chem.* **267**, 18002–18007.

Randall T.D., King L.B., and Corley R.B. (1990). The biological effects of IgM hexamer formation. *Eur. J. Immunol.* **20**, 1971–1979.

Reth M. (1989). Antigen receptor tail clue. *Nature* **338**, 383–384.

Reth M. (1995). Antigen receptors on B lymphocytes. In: *Immunoglobulin Genes,* 2nd ed. (Eds Honjo T., and Alt F.W.). pp. 125–142. Academic Press, San Diego.

Rodwell J.D., and Karush F. (1978). A general method for the isolation of the V_H domain from IgM and other immunoglobulins. *J. Immunol.* **121**, 1528–1531.

Roes J., and Raewsky K. (1993). Immunogloobulin D (IgD)-deficient mice reveal an auxiliary receptor function for IgD in antigen-mediated recruitment of B cells. *J. Exp. Med.* **177**, 45–55.

Rose D.R., Strong R.K., Margolies M.N., Gefter M.L., and Petsko G.A. (1990). Crystal structure of the antigen-binding fragment of the murine anti-arsonate monoclonal antibody 36–71 at 2.9-Å resolution. *Proc. Natl. Acad. Sci. USA.* **87**, 338–342.

Roux K.H., and Metzger D.W. (1982). Immunoelectron microscopic localization of idiotypes and allotypes on immunoglobulin molecules. *J. Immunol.* 129, 2548–2553.

Roux K.H. (1984). Direct demonstration of multiple V_H allotypes on rabbit Ig molecules : allotype charactersistics and Fab arms rotational flexibility revealed by immunoelectron microscope. *Eur. J. Immunol.* **14**, 459–464.

Roux K.H., Shuford W.W., Finley J.W., Esselstyn J., Pankey S., Raff H.V., and Harris L.J. (1994). Characterization of biosynthetic IgG oligomers resulting from light chain variable domain duplication. *Mol. Immunol.* **31**, 933–942.

Rowe D.S., and Fahey J.L. (1965). A new class of human immunoglobulins. I. A unique myeloma protein. *J. Exp. Med.* **121**, 171–17 .

Rudd P.M., Leatherbarrow R.J., Rademacher T.W., and Dwek R.A. (1991). Diversification of the IgG molecule by oligosaccharides. *Mol. Immunol.* **28**, 1369–1378.

Ryazantsev S., Tishchenko V., Vasiliev V., Zav'yalov V., and Abramov V. (1990). Structure of human myeloma IgG3 Kuc. *Eur. J. Biochem.* **190**, 393–399.

Sanz E., and de la Hera A. (1996). A novel anti-$V_{pre\text{-}B}$ antibody identifies immunoglobulin-surrogate receptors on the surface of human pro-B cells. *J. Exp. Med.* **183**, 2693–2698.

Satow Y., Cohen G.H., Padlan E.A., and Davies D.R. (1986). Phosphocholine binding immunoglobulin Fab McPC603. An X-ray diffraction study at 2.7 Å. *J. Mol. Biol.* **190**, 593–604.

Savvidou G., Klein M., Horne C., Hofmann T., and Dorrington K.J. (1981). A monoclonal immunoglobulin G1 in which some molecules possess glycosylated light chains. I. Site of glycosylation. *Mol. Immunol.* **18**, 793–805.

Sensel M.G., Coloma M.J., Harvill E.T., Shin S.-U., Smith R.I.F., and Morrison S.L. (1997). Engineering novel antibody molecules. In: *Antibody Engineering* (Ed. Capra J.D.). pp. 129–158. Karger, Basel.

Shakib F. (Ed.) (1990). *The Human IgG Subclasses: Molecular Analysis of Structure, Function and Regulation.* Pergamon Press, Oxford.

Sheriff S., Jeffrey P.D., and Bajorath J. (1996). Comparison of C_H1 domains in different classes of murine antibodies. *J. Mol. Biol.* **263**, 385–389.

Shinjo F., Hardy R.R., and Jongstra J. (1994). Monoconal anti-l5 antibody FS1 identifies a 130 kDa protein associated with $\lambda 5$ and $V_{pre\text{-}B}$ on the surface of early pre-B cell lines. *Int. Immunol.* **6**, 393–399.

Sharon N., and Lis G. (1997). Glycoproteins: structure and function. In: *Glycosciences* (Eds. Gabius H.-J., and Gabius S.). pp. 133–162. Chapman & Hall, London.

Silverton E.W., Navia M.A., and Davies D. (1977). Three-dimensional structure of an intact human immunoglobulin. *Proc. Natl. Acad. Sci. USA* **74**, 5140–5144.

Sitia R., Neuberger M., Alberini C., Bet P., Fra A., Valetti C., Williams G., and Milstein C. (1990). Developmental regulation of IgM secretion: the role of the carboxyl-terminal cysteine. *Cell* **60**, 781–790.

Smith R.I.F., Coloma M.J., and Morrison S.L. (1995). Addition of a m-tailpiece to IgG results in polymeic antibodies with enhanced effector functions including complement-mediated cytolysis by IgG4. *J. Immunol.* **154**, 2226–2236.

Solheim B.G., and Harboe M. (1972). Reversible dissociation of reduced and alkylated IgM subunits to half subunits. *Immunochemistry* **9**, 623–634.

Steiner L., and Lopes A.D. (1979). The crystallizable human myeloma protein Dob has a hinge-region deletion. *Biochemistry* **18**, 4054–4067.

Stevens F.J., Westholm F.A., Solomon A., and Schiffer M. (1980). Self-association of human immunoglobulin κI light chains: role of the third hypervariable region. *Proc. Natl. Acad. Sci. USA* **77**, 1144–1148.

Stevens F.J., Solomon A., and Schiffer M. (1991) Bence-Jones proteins: a powerful tool for the fundamental study of protein chemistry and pathophysiology. *Biochemistry* **30**, 6803–6805.

Stevens F.J., Myatt E.A., Chang C.-H., Westholm F.A., Eulitz M., Weiss D.T., Murphy C., Solomon A., and Schiffer M. (1995). A molecular model for self-assembly of amyloid fibrils: immunoglobulin light chains. *Biochemistry* **34**, 10697–10702.

Strong R.K., Campbell R., Rose D.R., Petsko G.A., Sharon J., and Margolies M.N. (1991). Three-dimensional structure of murine anti-*p*-azophenylarsonate Fab 36–71. 1. X-ray crystallography, site-directed mutagenesis, and modeling of the complex with hapten. *Biochemistry* 30, 3739–3748.

Sutton B.J., and Phillips D.C. (1983). The three-dimensional structure of the carbohydrate within the Fc fragment of IgG. *Biochem. Soc. Transact.* **11**, 130–132.

Sykulev Y.K., and Nezlin R. (1982). Spin labeling of immunoglobulin M and E carbohydrates. *Immunol. Lett.* **5**, 121–126.

Sykulev Y.K., and Nezlin R. (1990). The dynamics of glycan-protein interactions in immunoglobulins. Results of spin label studies. *Glycoconjugate J.* **7**, 163–182.

Takahashi N., Tetaert D., Debuire B., Lin L.-C., and Putman F.W. (1982). Complete amino acid sequence of the δ heavy chain of human IgD. *Proc. Natl. Acad. Sci. USA* **79**, 2850–2854.

Takahashi T., Iwase T., Takenouchi N., Saito M., Kobayashi K., Moldoveanu Z., Mestecki J., and Moro I. (1996). The joining (J) chain is present in invertebrates that do not express immunoglobulins. *Proc. Natl. Ac. Sci. USA* **93**, 1886–1891.

Taniguchi T., Mizuochi T., Baele M., Dwek R.A., Rademacher T.W., and Kobata A. (1985). Structures of the sugar chains of rabbit immunoglobulin G: occurence of asparagine-linked sugar chains in Fab fragment. *Biochemistry* **24**, 5551–5557.

Tao M.-H., and Morrison S.L. (1989). Studies of aglycosylated chimeric mouse-human IgG. Role of the carbohydrate in the structure and effector functions mediated by the human IgG constant region. *J. Immunol.* **143**, 2595–2601.

Terashima M., Kim K.-M., Adachi T., Nielsen P.J., Reth M., Köhler G., and Lamers M.C. (1994). The IgM antigen receptor of B lymphocytes is associated with prohibitin and a prohibitin-related protein. *EMBO J.* **13**, 3782–3792.

Timofeev V.P., Dudich I.P., Sykulev Y.K., and Nezlin R.S. (1978). Rotational correlation times of IgG and its fragments spin-labeled at carbohydrate or protein moieties. Spatially fixed position of the Fc carbohydrates. *FEBS Lett.* **89**, 191–195.

Toraño A., and Putnam F.W. (1978). Complete amino acid sequence of the $\alpha 2$ heavy chain of a human IgA2 immunoglobulin of the A2m(2) allotype. *Proc. Natl. Ac. Sci. USA* **75**, 966–969.

Toraño A., Tsuzukida Y., Li Y.V., and Putnam F.W. (1977). Location and structural significance of the oligosaccharides in human IgA1 and IgA2 immunoglobulins. *Proc. Natl. Ac. Sci. USA* **74**, 2301–2305.

Torres R.M., Flaswinkel H., Reth M., and Rajewsky K. (1996). Aberrant B cell development and immune response in mice with a compromised BCR complex. *Science* **272**, 1804–1808.

Tramontano A., and Lesk A.M. (1992). Common features of the conformations of antigen-binding loops in immunoglobulins and application to modeling loop conformations. *Proteins Struct. Funct. Genet.* **13**, 231–245.

Tormo J., Stadler E., Skern T., Auer H., Kanzler O., Betzel C., Blaas D., and Fita I. (1992). Three-dimensional structure of the Fab fragment of a neutralizing antibody to human rhinovirus serotype 2. *Protein Sci.* **1**, 1154–1161.

Underdown B.J., and Schiff J.M. (1986). Immunoglobulin A: strategic defence initiative at the mucosal surface. *Annu. Rev. Immunol.* **4**, 389–417.

Underdown B.J., and Mestecky J. (1994). Mucosal immunoglobulins. In: *Handbook of Mucosal Immunology,* (Ed. Ogra P.L. *et al),* pp. 79–97. Academic Press, San Diego.

van Oss C.J. (1994). Nature of specific ligand-receptor bonds, in particular the antigen-antibody bond. In: *Immunochemistry* (Eds. van Oss C.J., and van Regenmortel M.H.V.) pp. 581–614. Marcel Dekker, New York.

Vassalli P., Tartakoff A., Pink J.R.L., and Jaton J.-C. (1980). Biosynthesis of two forms of IgM heavy chains by normal mouse B lymphocytes. Membrane and secretory IgM. *J. Biol. Chem.* **255**, 11822–11827.

Venkitaraman A.R., Williams G.T., Dariavach P., and Neuberger M.S. (1991). The B-cell antigen receptor of the five immunoglobulin classes. *Nature* **352**, 777–780.

Wels J.A., Word C.J., Rimm D., Der-Balan C.P., Martinez H.M., Tucker P.W., and Blattner F.R. (1984). Structural analysis of the murine IgG3 constant region gene. *EMBO J.* **3**, 2041–2046.

Wienands J., and Reth M. (1992). Glycosyl-phosphatidylinositol linkage as a mechanism for cell-surface expression of immunoglobulin D. *Nature* **356**, 246–248.

Wiersma E.J., Chen F., Bazin R., Collins C., Painter R.H., Lemiex R., and Shulman M.J. (1997). Analysis of IgM structures involved in J chain incorporation. *J. Immunol.* **158**, 1719–1726.

Wiersma E.J., and Shulman M.J. (1995). Assembly of IgM. Role of disulfide bonding and noncovalent interactions. *J. Immunol.* **159**, 5265–5272.

Wieser P., Müller R., Braun U., and Reth M. (1997). Endosomal targeting by the cytoplasmic tail of membrane immunoglobulin. *Science* 276, 407–409.

Williams A.F., and Barclay A.N. (1988). The immunoglobulin superfamily—domains for cell surface recognition. *Annu. Rev. Immunol.* **6**, 381–405.

Williams A.F., and Barclay A.N. (1989). The immunoglobulin superfamily. In: *Immunoglobulin Genes* (Eds. Honjo T., and Alt F.W.). pp. 361–387. Academic Press, San Diego.

Wines B.D., and Easterbrook-Smith S.B. (1991). The Fab/c fragment of IgG produced by cleavage at cyanocysteine residues. *Mol. Immunol.* **28**, 855–863.

Winter G., and Milstein C. (1991). Man-made antibodies. *Nature* **349**, 293–299.

Wormald M.R., Rudd P.M., Harvey D.J., Chang S.-C., Scragg I.G., and Dwek R.A. (1997). Variations in oligosaccharide-protein interactions in immunoglobulin G determine the site-specific glycosylation profiles and modulate the dynamic motion of the Fc oligosaccharides. *Biochemistry* 36, 1370–1380.

Wormald M.R., Wooten E.W., Bazzo R., Edge C.J., Feinstein A., Rademacher T.W., and Dwek R.A. (1991). The conformational effects of *N*-glycosylation on the tailpiece from serum IgM. *Eur. J. Biochem.* **198**, 131–139.

Wrigley N.G., Brown E.B., and Skehel J.J. (1983). Electron microscopy evidence for the axial rotation and interdomain flexibility of the Fab regions of IgG. *J. Mol. Biol.* **169**, 771–774.

Wright A., Tao M.-H., Kabat E.A., and Morrison S.L. (1991). Antibody variable region glycosylation: position effects on antigen binding and carbohydrate structure. *EMBO J.* **10**, 2717–2723.

Wu P., Rice K.G., Brand L., and Lee Y.C. (1991). Differential flexibilities in three branches of an N-linked triantennary glycopeptide. *Proc. Natl. Acad. Sci. USA* **88**, 9355–9359.

Wu T.T., and Kabat E.A. (1970). An analysis of the sequences of the variable regions of Bence Jones proteins and myeloma light chains and their implication for antibody complementarity. *J. Exp. Med.* **132**, 211–250.

Zheng Y., Shopes B., Holowka D., and Baird B. (1992). Dynamic conformations compared for IgE and IgG1 in solution and bound to receptors. *Biochemistry* **31**, 7446–7456.

CHAPTER 2

Animal and Human Immunoglobulins

Immunoglobulin molecules are present in the circulation of all the existing vertebrates, with the exception of the most primitive ones, jawless lampreys and hagfishes (Turner, 1994). From sharks and skates to mammals, immunoglobulins have the same main principles of organization. These molecules are usually heterodimers of heavy (H) and light (L) polypeptide chains. Both types of chains are composed of variable (V) and constant (C) parts with characteristic immunoglobulin domain features. The hypervariable regions constructing the antigen-combining sites have a very similar main chain structure (Du Pasquier, 1993; Hsu and Steiner, 1992; Barre *et al.*, 1994). Most of the recent knowledge on the immunoglobulin structure and function was obtained from studies of human and mouse immunoglobulin molecules. During the past decades, new information has also been accumulated on the properties of immunoglobulins of various other species.

I. LOW VERTEBRATES

A. Fishes

Sharks and skates, representatives of the elasmobranches, are the most primitive vertebrates able to respond specifically to different antigenic stimuli by antibody biosynthesis. Their ancestors diverged from ancestors of other vertebrates about 400 million years ago.

1. Cartilaginous Fishes

In **sharks** and other cartilaginous fishes, the predominant type of immunoglobulins is pentameric IgM but a significant amount of IgM monomers are also present in the circulation. The total level of IgM in shark serum is very high, accounting for 30–50% of all serum proteins. The μ chains are composed of one variable and four constant domains, and their variable genes are 60% identical with the corresponding mammalian genes. IgM molecules are present in soluble or transmembrane forms. Three types of light chains are found in the cartilaginous fishes (Rast *et al.*, 1994). One of them resembles mammalian κ light chains with about 60% identity (Greenberg *et al.*, 1993) and another is more closely related to the λ chains (Hohman *et al.*, 1993). Most probably, the light chain divergence into the κ and λ types occurred before the appearance of the elasmobranchs.

Sharks can synthesize antibodies to a variety of antigens. However, there is no increase in the level of antibodies or their affinity to antigens as a result of immunization (Marchalonis *et al.*, 1993). Nearly all V_H regions belong to one family, with small structural differences between its members (90% identity).

The organization of immunoglobulin genes of cartilaginous fishes is different from that of mammalian immunoglobulin genes (Warr, 1995). In their genome, there are many closely linked clusters of the heavy μ chain (V_H–D_H1–D_H2–J_H–C_H) and the light chain (V_L–J_L–C_L) genes. There are about 200 λ gene clusters in the sandbar shark genome (Hohman *et al.*, 1993). In the horned shark genome, the μ chain clusters are placed at multiple loci and on different chromosomes. Another unique feature of the variable region genes in sharks is the presence of their joining in the germ line: about half of the μ variable region genes are partially (*VD–J*) or fully (*VDJ*) joined in the horned shark genome. The *V* and *J* genes of the light chains are also mainly fused in the germ line (Fig. 22) (Rast *et al.*, 1994,1995).

The origin of the diversity of shark IgM antibodies is different from that of mammals, where rearrangement of variable gene segments plays a very important role in the generation of antibody diversity (combinatorial diversity). The rearrangement of shark variable gene fragments takes place only within the

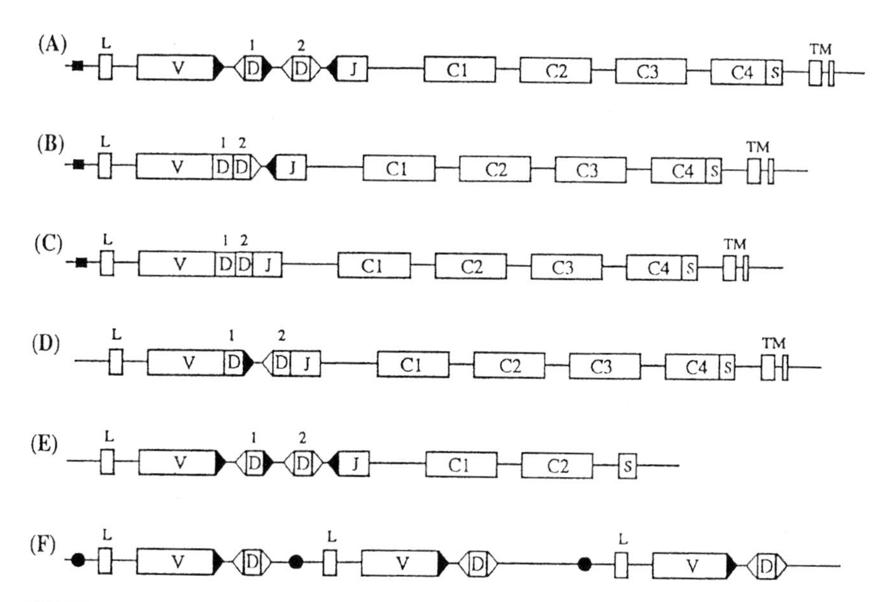

FIGURE 22 Immunoglobulin heavy chain gene organization in nonmammalian species. (A) Unarranged clusters in *Hydrolagus, Heterodontus;* and *Raja;* (B) partially rearranged cluster in *Heterodontus* and *Raja.*; (C) fully rearranged segments in *Hydrolagus* and *Heterodontus;* (D) partially repeting V–D organization in *Raja;* (E) IgX cluster in *Raja;* (F) tandem repeating V–D organization in *Latimeria.* L, leader sequence; V, variable; D, diversity; J, joining segments; C1–4, constant region exons; S, secretory; TM, transmembrane exons; ■, decamer-nonamer promoter element; •, immunoglobulin octamer promoter element; RSS spacers, ◁▷; 12 bp and ▶◀, 23 bp. (Rast *et al.*, 1995. Reprinted with permission.)

clusters and not between them. Conventional evolutionary processes, somatic mutations and junctional events are responsible for the diversification of variable genes in the shark genome.

Several other immunoglobulin molecules were discovered in sharks. An immunoglobulin (NAR) was found in the nurse shark, peptide chains of which are composed of one variable and five constant regions (Greenberg *et al.*, 1995). The NAR (nurse antigen receptor) protein is a secreted protein and exists as a dimer in serum. Extensive sequence diversity was found in its variable regions, which probably results from rearrangements and somatic hypermutations. Another antibody molecule (IgNARC) of the nurse shark is composed of light and heavy chains. The heavy chains comprise a variable domain and six constant domains (Fig. 23) (Greenberg *et al.*, 1996). Two amino–terminal constant domains of the IgNARC heavy chains are very similar to the constant domains of the skate IgX heavy chain, but the four carboxy-terminal constant heavy chain domains are homologous to the last four constant domains of the NAR protein.

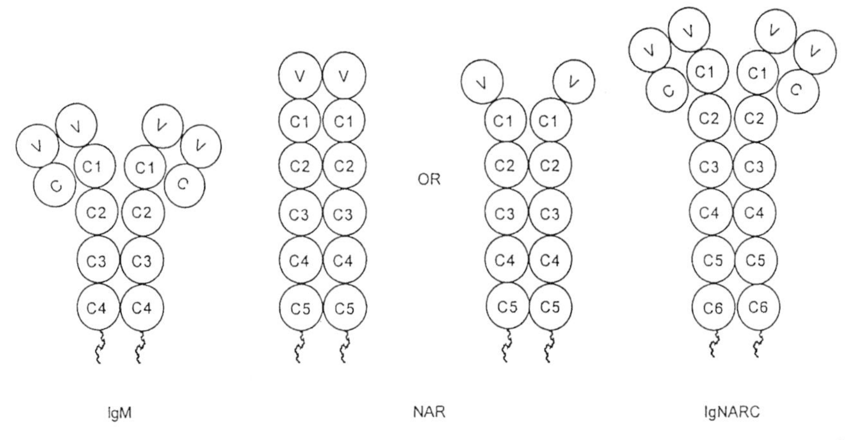

FIGURE 23 IgM, NAR, and IgNARC nurse shark immunoglobulin molecules. (Greenberg *et al.*, 1996. Reprinted with permission.)

An IgW molecule was detected in carcharhine sharks (Bernstein *et al.*, 1996). Its heavy chain is composed of a variable domain and six constant domains (782 amino acid residues total) and resembles mu chains of the skate and the NAR protein. The IgW variable regions are more similar to their constant regions than was found in other immunoglobulin molecules. Such high homology between variable and constant domains could be expected for a primordial immunoglobulin. The IgW chain has amino acid residues, which are responsible for contacts of the heavy chains with the light chains, and the IgW molecules probably exist in circulation as complexes of the heavy and light chains.

Skates, also one of the cartilaginous fishes, possess two classes of immunoglobulins. One is similar to IgM and has a typical pentameric structure. The second one (IgX) is a lower molecular weight immunoglobulin with a molecular weight of about 320,000. This molecule is a noncovalent dimer composed of two monomers with a molecular weight of about 150,000 each. The heavy chain of this molecule is smaller than that of the μ chain (45,000–50,000) and has its own specific antigenic determinants (Kobayashi *et al.*, 1984). IgX has no relationship to any of the immunoglobulin isotypes of the higher vertebrates. The variable region gene segments are organized in clusters (V_x, D_x1, D_x2, J_x, C_x1 and C_x2), which are dispersed throughout the genome. The gene clusters of IgM and IgX are not contiguous and there is no class switching between the isotypes (Rast *et al.*, 1995).

2. Bony Fishes

In the bony fishes, the predominant immunoglobulin is a IgM type tetramer with noncovalently associated subunits. In the channel catfish, two antigenic variants of the light chains (F and G) were found. They differ by molecular weight and have approximately equal similarity to the κ and λ chains of higher vertebrates (Lobb *et al.*, 1984). The amino acid similarity between the constant parts of both light chains is less than 35%. Three different heavy chain variants have been described (Lobb and Olson, 1988). The membrane form of IgM has a unique structure in all bony fishes: its transmembrane segment is directly linked to the $C_\mu3$ domain and not to $C_\mu4$ as in other species (Wilson *et al.*, 1990). Therefore, the membrane-linked heavy chains of these animals are shorter than the secreted heavy chains. A chimeric heavy chain was found in the channel catfish genome that has similarities to IgD (Wilson *et al.*, 1997). The constant region of this chain is composed from the C_μ domain, seven constant domains that are homologous to the δ chain, and secretory or membrane tails. This finding suggests that IgD appeared early in vertebrate evolution.

The organization of the heavy chain genes in the bony fishes is in general similar to that of mammals. These animals are probably the first to have evolved single C_H genes with many V_H genes and about nine functional J_H segments. Catfishes have at least eight different V_H-gene families (total about 120 V_H segments). However, the light chain gene segments of the channel catfish are organized in clusters, each of them composed of V_L, J_L, and C_L segments. In the catfish genome, there are at least 50 different F light chain clusters and 15 G light chain clustetrs. Each cluster of the catfish G light chains contains two V_L segments and two J_L segments (one J_L is probably a pseudo-gene) together with one C_L segment. By contrast, the F clusters are composed of single V_L, J_L, and C_L gene segments. The V_L segments are located in opposite transcriptional polarity relative to the J_L–C_L segments. Therefore, the V_L segments rearrange to J_LC_L segments by inversion (Ghaffari and Lobb, 1993, 1997).

B. Amphibians

Amphibians posses several immunoglobulin isotypes of the heavy and light chains. In the clawed-toad frog (*Xenopus laevis*), the most studied amphibian representative, three heavy chain variants, IgM, IgY, and IgX, were found (Du Pasquier *et al.*, 1989). The first one is homologous to the mammalian IgM (31–47% identity between the corresponding μ domains). The *Xenopus* C_μ domains are as similar to the mouse $C_\gamma2a$ domains as to other C_μ domains. These findings point out that the divergence of the μ and γ chains occurred not long before the separation of the amphibian and the mammal lineages (Schwager *et al.*, 1988). The IgX molecule is probably an analog of mammalian IgA (Mußmann *et al.*, 1996). The IgY heavy chain (upsilon or υ chain) has a molecular weight of about 80 kDa and can be found in the membrane form similar to the μ chain. The heavy chains of all isotypes are composed of four domains. The IgX with a heavy chain of molecular weight 69 kDa is a polymeric molecule but antigenically distinct from IgM. The C_H1 domain of *Xenopus*, axolotl, and chicken IgY, as well as the C_H1 domain of *Xenopus* IgX, human IgE, and rabbit IgG, has an additional intradomain disulfide bridge (Fig. 6) (Fellah *et al.*, 1993). Two terminal amino acids of amphibian and chicken $C_\upsilon4$ domain are –Gly–Lys. The same terminal dipeptide is characteristic for the mammalian domains $C_\gamma3$ and human $C_\varepsilon4$ of the secreted immunoglobulins. These data are consistent with the hypothesis that the υ chain is the ancestor of the mammalian γ and ε heavy chain isotypes (Parvari *et al.*, 1988)

Three varieties of light chains with different molecular weights have been distinguished in *Xenopus* by mobility in electrophoresis (Du Pasquier *et al.*, 1989). Two of them are antigenically distinct (designated as rho [ρ] and sigma [σ]) and resemble the mammalian $\varkappa$ light chains slightly more than the λ light chains (Zezza *et al.*, 1991). The ρ light chains associate preferentially with the μ chains and the ϱ light chains associate with the υ heavy chains of IgY. A third light chain isotype related to the mammalian λ chains has been characterized (Haire *et al.*, 1996). The light chains of the bullfrog immunoglobulins are not covalently bonded to the heavy chains because a cysteine residue, usually forming the disulfide heavy–light bridge, participates in an extra intrachain disulfide bridge in the constant region of the light chains (Mikoryak and Steiner, 1988).

The organization of the heavy and light chain genes are similar in *Xenopus laevis* and mammals. These frogs possess 11 diversified V_H families with a total of 80–100 V_H genes and about 17 D_H and 10 J_H gene elements. The *V* and *C* genes coding the ϱ and σ light chains are present at two separate loci. Both *C* genes have the 29% residue identity and 29–33% homoloy with C_L genes of shark, chicken, or mammals (Rast *et al.*, 1995). The constant region of the σ light chains is coded by two different C_σ genes that differ in several amino acid positions. There are six families of V_L genes, two J_L, and two C_L genes for the third *Xenopus* light chain (Haire *et al.*, 1996). The genomic complexity of the *Xenopus* immune system and the number of elements needed for combinatorial diversity is at least not less than that found in mammals. However, the antibodies formed by this animal are restricted in heterogeneity (Hsu *et al.*, 1991) and their affinity maturation is limited. The point mutation rate of complementary-determing regions (CDR) of the V_H genes is only slightly below that found for mice. There is also no shortage of junctional diversity. One plausible explanation for this discrepancy may be the absence of an effective mechanism for selecting mutants due to the lack of germinal centers in these animals (Wilson *et al.*, 1992).

C. Reptiles

Immunoglobulins of reptiles are less studied than those of amphibians. In turtles three types of immunoglobulin molecules with different molecular weights have been described. The first is a pentameric IgM-like molecule composed of five subunits. Two others are antigenically related monomeric molecules (IgY) with υ heavy chains of a molecular weight of about 67.5 and 35 kDa (Leslie and Clem, 1972). The heavy chain of the smaller molecule is lacking its C-terminal part and contains less carbohydrate than the other one. Some turtles have only the truncated form of IgY. Both isoforms of IgY were also found in ducks and geese.

Multiple V_H genes and one C_μ gene were found in the genome of red-eared turtles (Turchin and Hsu, 1996). Turtle V_H sequences show 63–79% identity with V_H sequences from human and mouse. The sequenced parts of the turtle μ chain are most similar to duck (42% identity), chicken (41%), and mouse (39%) μ chains. No affinity maturation was found in the course of a turtle immune response.

II. BIRDS

A. Chickens

Among avian immunoglobulins the most studied are those of turkey and chicken. Like all other vertebrates, both these fowls possess the high molecular weight pentameric IgM with mu chains that have a molecular weight about 71 kDa. In contrast to mammalian IgM, chicken IgM has significantly less carbohydrate (hexose content is about 3%). In chickens there is also a monomeric immunoglobulin with a heavy chain of 67 kDa built from one variable and four constant domains. This molecule is a homologue of amphibian and reptile IgY. To precipitate with antigen, IgY antibodies require the presence of a high salt concentration (1.5 M NaCl). Chickens also possess a minor immunoglobulin isotype, the C_H region of which has a significant homology to the constant region of mammalian IgA (Mansikka, 1992). This immunoglobulin molecule has four C_H domains. The chicken $C_\alpha 2$ domain is absent in the mammalian IgA molecules. In serum, the chicken IgA is mainly monomeric. IgA was found in larger quantities in various chicken secretions and has a tetrameric structure but in bile it is a dimer molecule. The secreted form of IgA is associated with a secretory component (Vaerman, 1994). IgD and IgE molecules were not found in chickens.

Nearly all light chains of chicken immunoglobulins resemble λ chains of mammals and their constant region has 61% homology with that of human and mouse λ chains. However, chicken λ chains have an unblocked NH_2-terminal and a cysteine residue at the COOH-terminal, features associated with the $\varkappa$ light chains (Reynaud *et al.*, 1983).

The organization of the chicken immunoglobulin genes is different from that of mammals (Fig. 24) (Bezzubova and Buerstedde, 1994; Weill and Reynaud,

1995). The light chain locus contains only one V_λ and one J_λ functional segment as well as one C_λ gene. Upstream of the functional V_λ segment, 25 highly homologous pseudo-V_λ gene segments (ψV_λ) have been found. These closely linked ψV_λ gene segments are homologous to the functional V_λ segment but lack the recombination signal sequences and the regulatory elements and cannot be rearranged. Similarly, in the heavy chain locus, there is only one functional V_H and one J_H segment, as well as 16 functional D segments accompanied by 80–100 pseudo-V_H segments that are fused with D segments (ψVD_H). Very early during ontogeny, the functional V(D)J gene segments are rearranged, creating limited diversity at the junctions of the gene segments. The diversification of the rearranged variable genes is generated mainly by gene conversion in the bursa of Fabricius during proliferation of B cells (McCormack and Thompson, 1990). At this stage, DNA segments from the variable pseudo-genes acting as donors are transferring into unique functional variable genes, which serve as acceptor sites. A single functional variable gene can receive DNA segments from up to six ψV genes. Four IgM chain and 12 IgY allotypes were detected among chicken strains (Ratcliffe, 1996).

B. Ducks

Ducks possess four immunoglobulin isotypes, IgM, a secretory immunoglobulin that resembles IgM, IgY, and IgY(ΔFc), a truncated variant of IgY (Higgins and

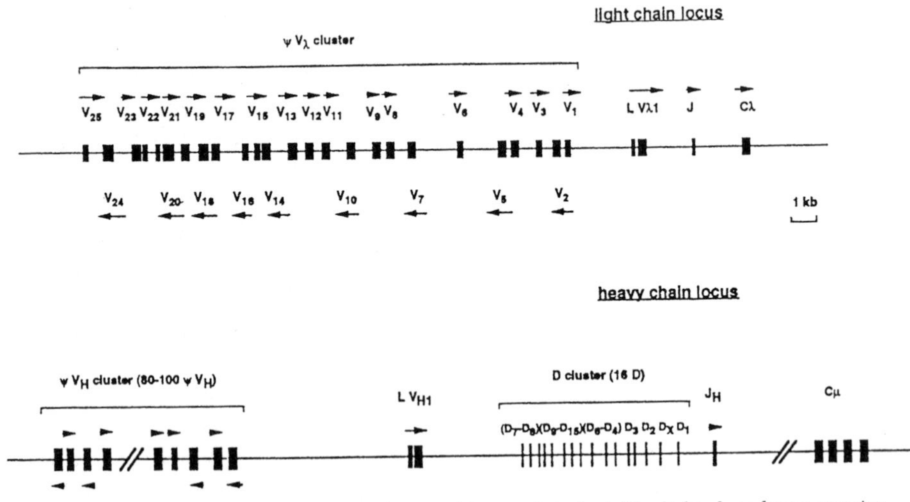

FIGURE 24 Organization of chicken light and heavy chain loci. The light chain locus contains single V_λ, J, and C_λ gene segments downstream of 25 ψV pseudogenes. The heavy chain locus contains single C_μ, J_H, and functional V_H segments, a cluster of 16 D segments, and a group of 80–100 ψV_H peseudo-genes. Arrows indicate transcriptional polarities. (Weill and Reynaud, 1995. Reprinted with permission.)

Warr, 1993). Duck antibodies usually have weak precipitation or agglutinating activity, even in the presence of a high salt concentration. The duck serum IgM is a minor immunoglobulin class built from four or five monomers. The secretory IgM found in bile in a high concentration is antigenically very similar to the serum IgM. Two other isotypes (IgY and ΔIgY) are related to each other. Although the heavy chain of IgY (υ chain) is composed of one variable and four constant domains, the heavy chain of ΔIgY ($\Delta\upsilon$ chain) has only two constant domains that are identical to C_H1 and C_H2 of IgY (Warr *et al.*, 1995). The lack of domains corresponding to the Fc fragment explains why the ΔIgY molecule is deficient in effector functions, such as complement fixation, opsonization, and tissue-sensitizing activity, that are characteristic to the IgY molecules.

As was already mentioned, IgY molecules are similar to both mammalian IgG and IgE according to some structural and functional criteria. The analysis of amino acid sequences of the immunoglobulin heavy chains shows that the relationship of the υ, γ, and ε chains to each other is much closer than that of any of these chains to μ or α chains. Furthermore, the terminal secretory segment has the same length for υ, γ, and ε chains (only two amino acids, with an exception of the rodent ε chains) and the transmembrane segments of these chains are similar in length and sequence. All these data speak in favor of the hypothesis that IgY is the immediate progenitor of IgE and IgG (Warr *et al.*, 1995).

The heavy chains of duck IgY and ΔIgY are products of one gene, which has a peculiar structure (Fig. 25) (Magor *et al.*, 1994). The υ gene contains seven exons and through alternate pre-mRNA processing pathways encodes three distinct heavy chains, the full υ chain, the $\Delta\upsilon$ chain, and a membrane form of the υ chain. A unique terminal exon in the intron between $C_\upsilon2$ and $C_\upsilon3$ domains, encoding only two C-terminal amino acids, is used for production of the truncated $\Delta\upsilon$ chain.

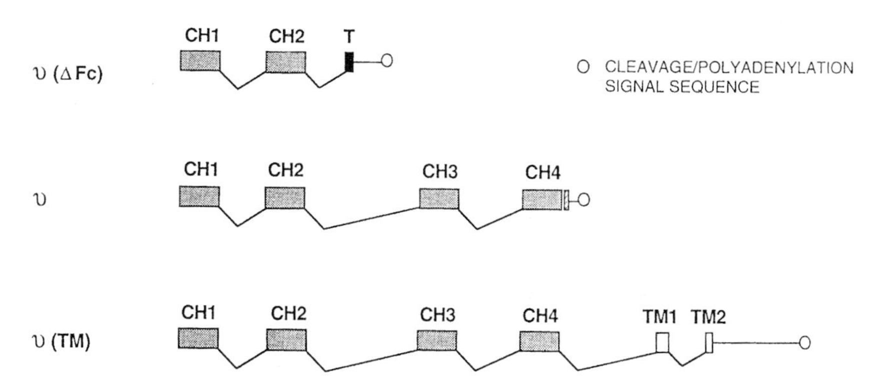

FIGURE 25 The duck υ constant region gene encodes three heavy chain constant regions by alternative pathways of RNA processing. T, terminal exon encodes two C-terminal residues that occur only in ΔFc truncated chain, TM1 and TM2, transmembrane exons. (Magor *et al.*, 1994. Reprinted with permission from The American Association of Immunologists.)

III. MAMMALS

Most mammalian species possess five major immunoglobulin isotypes, IgG, IgM, IgA, IgD, and IgE. However, some species lack one of these immunoglobulins. For example, rabbits probably have no IgD. By contrast, an unusual immunoglobulin molecule devoid of light chain was recently discovered in the circulation of camels and llamas. Significant variations were found in mammals for mechanisms for the generation of antibody diversity.

A. Laboratory Animals

1. Rabbits

In the not so far past, rabbits were one of the most popular models for immunological studies, particularly for the elucidation of the immunoglobulin structure and the genetic mechanisms of antibody biosynthesis. Studies of rabbit immunoglobulins in 1950–1960s have contributed significantly to our knowledge of the structure and genetics of immunoglobulin molecules. The technique of partial proteolysis of immunoglobulin molecules to obtain Fab and Fc fragments was developed first in the experiments with rabbit IgG. The development of methods of mild reduction of rabbit immunoglobulin molecules and isolation of their soluble heavy and light chains permitted performance of detailed studies of the chain properties and structure. The first data on the primary structure of rabbit immunoglobulin chains and the discovery of homologous regions led to the development of the important hypothesis that during evolution heavy and light chains evolve by duplication of a primordial gene coding-peptide 110 amino acid residues long (Hill *et al.*, 1966). The studies on allotypes of rabbit immunoglobulin peptide chains, beginning with the classical works of Oudin (1956, 1960), led to the very important findings that significantly influence our conceptions concerning the genetic mechanisms involved in control of antibody biosynthesis.

Rabbits possess four immunoglobulin classes—IgG, IgM, IgA, and IgE. Unlike other mammalian species, which usually have several IgG subclasses, there is only one rabbit IgG variant with one interheavy disulfide bridge. In contrast to other mammals, rabbits (as well all other lagomorphs) have several isotypes of IgA that are expressed differently in various lymphoid tissues (Knight and Tunyaplin, 1995). Most of the light chains of the rabbit immunoglobulins are of the $\varkappa$ type and only a minority of them are related to the λ type (about 10–20%). Most of $\varkappa$ chains have an unusual interdomain disulfide bridge between the constant and variable domains that links positions 80 and 171. The general scheme of the rabbit IgG molecule that demonstrates the arrangement of disulfide bridges is presented in Figure 26.

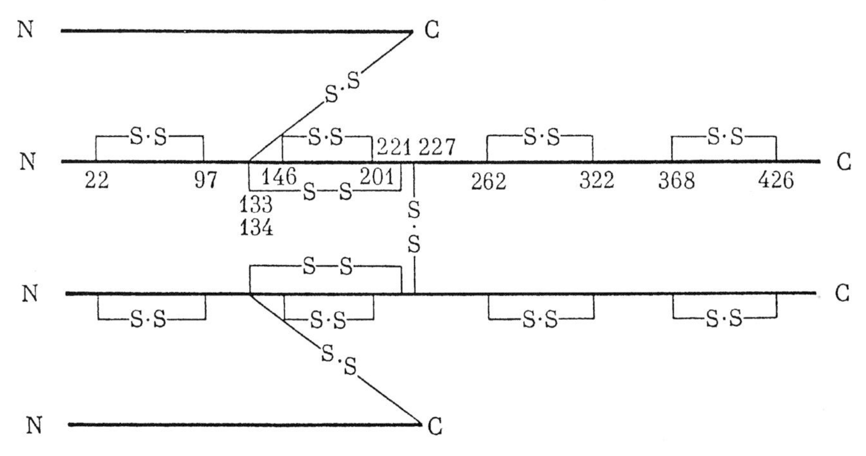

FIGURE 26 Rabbit IgG molecule. The arrangement of disulfide linkages of heavy chains are shown. (O'Donnell *et al.*, 1970. Reprinted with permission.)

The organization of the rabbit immunoglobulin genes was studied in the past decades in detail (Knight and Crane, 1994). One of the important factors facilitating these studies was the presence of several sets of antigenic allotypic markers on the heavy and light chains (Roux and Mage, 1996). Some of these determinants are located on the variable regions of the heavy chains (**a**, **x**, **y**, and **w** specificities), whereas the others are found on the constant regions of the heavy chains (the specificities **d** on C_γ, **n** on C_μ, and **f** and **g** on distinct subclasses of C_α). The allotypic antigenic specificity of the **d**-type correlates with the methionine–threonine variation in position of 225 of the γ chain hinge and the **e**-type specificity correlates with the threonine–alanine variation in position 301 ($C_\gamma 2$). The **b** determinants characterize allelic forms of the constant $\varkappa$ regions and the **c** determinants belong to the constant λ regions. Simultaneous presence of markers on the variable and constant regions of the rabbit γ chains allowed demonstration of the linkage of the variable and constant genes of the heavy chains and the discovery of a high frequency of germline recombination between them.

The region on chromosom 16, where the rabbit heavy chain locus locates, contains 100 or even more V_H gene segments, which are very similar to each other and are related to human V_H3 gene segments, 11 conserved *D* segments and 6 *J* gene segments (Fig. 27). However, only four variable genes participate in *V(D)J* rearrangements, and one of them, V_H1 gene, is involved in 80% of all rearrangements. About 80% of all rabbit immunoglobulin molecules belong to the allotype **a** group.

In the heavy chain locus, there are three single constant region genes, C_γ, C_μ, C_ε, and 13 functional nonallelic C_α genes. The light chain chromosomal region

HEAVY CHAIN LOCUS

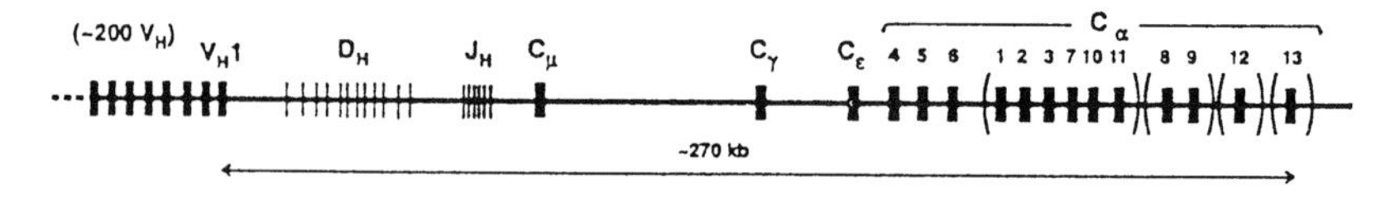

κ LIGHT CHAIN LOCUS

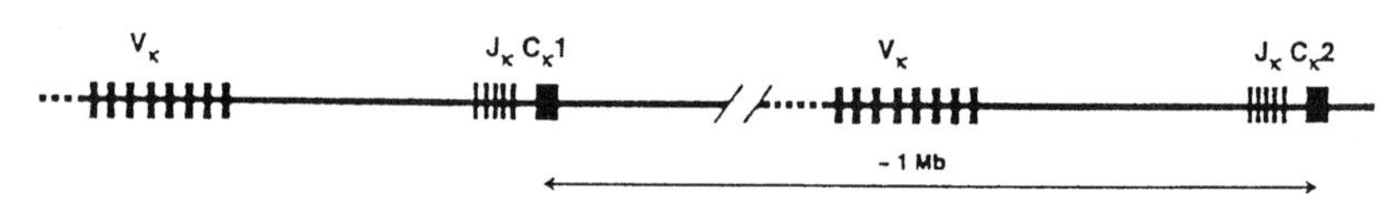

λ LIGHT CHAIN LOCUS

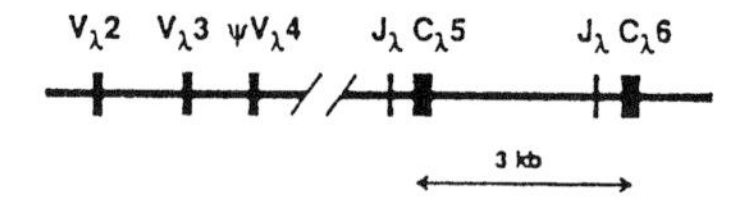

FIGURE 27 Organization of the rabbit heavy chain, $\varkappa$ and λ light chain loci. (Knight and Crane, 1994. Reprinted with permission.)

has two kappa constant genes ($C_\varkappa 1$ and $C_\varkappa 2$). Each of them has its own set of J segments. Five J segments link with the $C_\varkappa 1$ gene and three with the $C_\varkappa 2$ gene. Both constant genes probably also have their own assortment of variable genes (total about 30). In serum of normal rabbits, all kappa chains belong to the $\varkappa_1$ subtype. Rabbit genome contains more than five V_λ segments and up to eight λ constant genes (Duvoisin *et al.*, 1988; Hayzer and Jaton, 1989). Two constant λ genes are functional and each are associated with a *J* segment, but others are pseudo-genes.

As only the V_H1 gene and to a lesser extent three other variable genes participate in *V(D)J* rearrangement, combinatorial joining of variable gene segments has a minor role in the generation of antibody diversity in rabbits. After formation of *V(D)J* genes, they are diversified through a somatic gene conversion-like event using as donors the upstream variable genes (Becker and Knight, 1990; Knight and Crane, 1994). The latter are homologous in the framework regions but much more variable in their CDRs. The process of gene conversion probably occurs in the B cells located in gut-associated lymphoid tissue (appendix and sacculus rotundus), which is a bursa equivalent. The *D* segments as a part of the *VDJ* gene are probably diversified by a somatic hypermutation mechanism rather than by gene conversion. The origin of the V_L variability is less studied, but probably gene conversion as well as other mechanisms, such as variations in the *V–J* junction, contribute to the formation of rabbit light chain repertoire (Hayzer and Jaton, 1989). The role of somatic hypermutation in diversification of the first 120 base pairs downstream of *VDJ* genes was recently established (Lanning and Knight, 1997).

2. Guinea Pigs

Guinea pigs are another widely used laboratory animal. They possess IgM, serum and secretory IgA, IgE, and two subclasses of IgG, IgG1 and IgG2, which differ in their structure and effector functions (Nussenzweig and Benacerraf, 1967). The IgG1 molecules have one interheavy disulfide bridge but IgG2 molecules possess three such linkages. The predominant isotype, IgG2, has more effective "opsonic" activity and IgG1 molecules are responsible for local and systematic anaphylactic reactions, which result in releasing histamine from mast cells.

IgG1 fixes complement only through the alternate pathway, whereas IgG2 uses the conventional mechanism for complement fixation. These functional differences are due to the different structure of their Fc portion. The IgG1 antibody molecules are more flexible and can form dimers with antigen, whereas the IgG2 molecules have constraints in the movements of their Fab arms. The IgG1 complexes with antigen are larger. The difference in flexibility is linked with the dissimilar structure of the $\gamma 1$ and $\gamma 2$ chain hinge regions (Cebra *et al.*, 1977). Most light chains of guinea pig immunoglobulins are of the κ type (70%). The lambda chains occur in three isotypic forms that differ in the sequence of their constant regions.

3. Mice

Most modern immunological knowledge is due to the studies performed on mice. The availability of well studied inbred mouse strains, as well as homogeneous myeloma proteins and monoclonal antibodies, greatly facilitated structural and functional studies of murine immunoglobulins. The vast majority of three-dimensional data on immunoglobulin molecules was obtained using murine monoclonal immunoglobulins and their fragments.

Mice possess all five major immunoglobulin classes, IgM, IgG, IgA, IgD, and IgE. Four IgG subclasses are described, IgG1 and IgG2a (major subclasses), and IgG2b and IgG3 (minor ones), that differ in the structure of the constant region of their heavy chains and also in some functional properties. The heavy chains of IgG3 are linked by two interchain disulfide bridges, whereas all other molecules of IgG have three such bridges.

The molecules of both murine IgG2 subclasses fix complement through the classical pathway but IgG1 and IgG3 complement fixing activity is weak or absent completely. IgG1 molecules mediate passive cutaneous anaphylactic reactions (PCA) in mice, but the IgG2 molecules are capable of mediating PCA in guinea pigs. The IgG2a and IgG2b isotypes are efficient in antibody-dependent cytotoxity. The IgG3 antibodies are associated with anticarbohydrate activity, and IgG1 is a predominant isotype against parasites and viruses and effectively stimulates phagocytosis. The murine IgA molecules in serum are mostly dimers. All IgG molecules are able to pass the placenta.

Allelic variants (allotypes) of the murine heavy chains were extensively studied in the 1960s (Potter and Lieberman, 1967; Herzenberg *et al.*, 1968). Allotypes are known for all mouse heavy chains with the exception of IgG3, alleles of which are not found in inbred strains (Stall, 1996). Many allotypic determinants are dependent from the structure of the Fc fragment (for example, several determinants of IgG2a), whereas some others such as some determinants of IgD or IgM molecules are linked with Fab fragments. It was shown that the tertiary structure of immunoglobulin molecules plays an important role in antigenic activity of murine allotypic determinants. Using the allotypic markers, it was shown that the genes coding the constant regions of all murine heavy chains are tightly linked.

During the 1970–1980s, detailed information was obtained on the organization of all three families of mouse immunoglobulin genes (Max, 1993). The heavy chain structural genes have been mapped on chromosome 12 (12F1), the $\varkappa$ light chain genes on chromosome 6 (6C2), and the λ light chain genes on chromosome 16. The general organization of these genes is similar but not fully identical (Lai *et al.*, 1989). Each gene family is composed of DNA segments coding variable region and constant region genes. All these genes are distributed in a tandem fashion, being separated by DNA stretches of different length. The variable region segments localize on the 5′ end of the locus and the constant region genes are place closer to the 3′ end.

In the murine serum, only 5% of immunoglobulins are of the λ type. There are three nonallelic subtypes of the λ chains with a distinct constant region sequence (Blomberg and Solomon, 1996). The most frequent of them is $\lambda 1$ (80–90%), which differs markedly from the other two in the structure of the constant region. The λ chains of inbred mice have only very limited sequence variability in their variable regions and this correlates well with the small number of the V_λ genes in the mouse genom (Eisen and Reilly, 1985). The variable regions of the λ chains are encoded by two separate gene segments, V and J, which rearrange during the development of B lymphocytes to code the entire V_λ region. The λ J and C structural genes are arranged in clusters (Fig. 28). There are two J–C clusters, each contains two J–C pairs; J_4–C_4 appears to be a pseudo-gene. Two ($V_{\lambda 2}$ and $V_{\lambda x}$), from three variable λ genes, are located up-

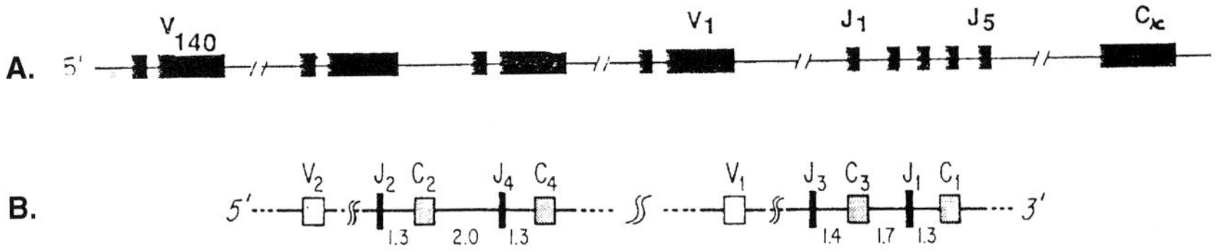

A. 5' V_{140} V_1 J_1 J_5 $C_\varkappa$

B. 5' V_2 J_2 C_2 J_4 C_4 V_1 J_3 C_3 J_1 C_1 3'
1.3 2.0 1.3 1.4 1.7 1.3

FIGURE 28 Organization of murine light chain loci $\varkappa$ light chain locus (A). λ light chain segments of BALB/c mouse (B). Distances are in kilobases. (Eisen and Reilly, 1985. Reprinted with permission from Annual Reviews, Inc.)

stream from the *J–C* clusters and the third one ($V_{\lambda 1}$), which differs significantly from $V_{\lambda 2}$ and $V_{\lambda x}$, is located between two pairs of *J–C* clusters. The *J–C* clusters are similar to each other and probably arose by a duplication of an ancestral cluster. Some wild mice have larger numbers of λ genes.

The organization of the mouse $\varkappa$ light chain locus is different (Fig. 28) (Lai *et al.*, 1989). There is only a single gene for the constant $\varkappa$ region ($C_\varkappa$). Upstream from the constant $\varkappa$ gene, five $J_\varkappa$ gene segments are located; one of them, $J_{\varkappa 3}$, is a pseudogene. Numerous variable $\varkappa$ genes are located further upstream. They are organized in several clusters and the members of each cluster are closely related by their sequence. The largest $V_\varkappa$ family, $V_\varkappa 4$, includes approximately 50 gene segments. Part of the variable $\varkappa$ gene segments are pseudo-genes. The total number of variable $\varkappa$ gene segments and pseudo-genes was estimated at 140 (Kirschbaum *et al.*, 1996).

The organization of the mouse heavy chain locus is even more complex. First, there are eight tightly linked genes that code the constant region of heavy chains. They are located in a tandem array and organized as follows: 5'J_H–(6.5 kb)–C_μ–(4.5 kb)–C_δ–(55 kb)–$C_\gamma 3$–(34 kb)–$C_\gamma 1$–(21 kb)–$C_\gamma 2b$–(15 kb)–$C_\gamma 2a$–(14 kb)–C_ε–(12 kb)–C_α–3'. Second, the variable regions of the heavy chains are coded by three gene segments (V_H, D_H, and J_H), which recombine during B-cell development in a complete variable heavy chain gene. The first step of recombination is the joining of *D* and J_H segments, followed by the joining of DJ_H with V_H segments. The complete variable heavy chain genes in the course of differentiation assemble with one of the constant heavy chain genes, forming the $V_H DJ_H C_H$ gene, which codes the entire heavy chain. There are five J_H segments; one of them is a pseudo-gene. The 12 identified D_H gene segments of different sizes are located upstream of the J_H segments. The V_H gene segments are arranged in several discrete clusters (Brodeur *et al.*, 1988)., Members of each cluster have a significant nucleotide sequence similarity. The number of V_H segments of the studied clusters is shown in Table 7. The total number of V_H segments is about 200 (Brodeur, 1996). A high percentage of pseudo-genes was found among V_H genes (Givol *et al.*, 1981). The organization of the V_H clusters, particularly their sizes, is different among various inbred mouse

TABLE 7 Mouse V_H Gene Families[a]

V_H gene families	Number of V_H gene segments per family in the germline of BALB/c mice
Subgroup I	
V_H2 Q52	15
V_H3 6–60	5–8
V_H8 3609	7–10
V_H12 CH27	1
Subgroup II	
V_H1 J558	100
V_H9 GAM 3.8	5–7
V_H14 SM7	3–4
Subgroup III	
V_H4 X-24	2
V_H5 7183	12
V_H6 J606	10–12
V_H7 S107	2–4
V_H10 MRL-DNA4	2–5
V_H11 CP3	1–6
V_H13 3609N	1

[a]Modified from Kofler *et al.*, 1992; (subgroups according to Tutter and Riblet, 1989).

strains. This difference reflects the processes of duplication, deletion, and sequence divergence during the evolution of the mouse V_H locus (Tutter and Riblet, 1989).

4. Rats

The immunoglobulin system of rats, a widespread laboratory animal, is very similar to the mouse one. All five major immunoglobulin classes (IgM, IgG IgA, IgD, and IgE) were described and four variants of IgG identified in rats, IgG1, IgG2a, IgG2b, and IgG2c (Bazin *et al.*, 1988; Gutman 1996). Serum IgA molecules are mainly dimeric. A major IgG subclass, IgG2a, possesses homocytrotropic properties and all IgG2 molecules are able to activate complement. The IgG subclasses have different susceptibility to proteolytic digestion (Nezlin *et al.*, 1973; Rousseaux *et al.*, 1980).

Allotypic variants were found for γ 2b (Igh-2 locus, *a* and *b* variants), γ 2c (*Igh-3* locus, *a* and *b* variants), and α (*Igh-1* locus, *a* and *b* variants) heavy chains (Bazin *et al.*, 1974; Beckers and Bazin, 1975). About 95% of rat light chains are of κ type. Two allotypic variants are known for the rat κ chains (*Ig-κ* locus, *a* and *b* variants) (Nezlin and Rokhlin, 1976), which have multiple amino acid differences in their constant regions (Vengerova *et al.*, 1972). Eleven substitutions were found among 107 residues of the $C_\varkappa$ domains of both variants. (Gutman *et al.*, 1975). $\varkappa$ chains of only one allotype, Ig-$\varkappa$ a or Ig-$\varkappa$ b, are present in all studied wild and laboratory rats (Rokhlin and Nezlin, 1974).

The loci of the rat immunoglobulin genes have a similar organization in rat and mouse (Brüggemann *et al.*, 1986; Brüggemann, 1988). The rat heavy chain locus is localized on chromosome 6, the $\varkappa$ chain locus on chromosome 4 and the λ chain locus on chromosome 11. The rat γ2b gene is equivalent to the mouse γ2a/γ2b genes and the rat γ2c gene is homologous to the mouse γ3 gene. Mouse and rat heavy chain genes evolved probably from three common $C\gamma$ genes. More recently, there were two duplication events in the rat and mouse heavy chain locuses: one yielded rat γ2a and γ1 genes and another yielded mouse γ2a and γ2b genes (Brüggemann, 1988).

In all studied rat species and subspecies only one $C_\varkappa$ gene was found (Frank *et al.*, 1987). There are about 10–15 V_λ gene segments (Aguilar and Gutman, 1992) and two C_λ genes in the rat genome. One of the λ constant genes (C_λ1) is probably a pseudo-gene (Steen *et al.*, 1987).

5. Hamsters

Three immunoglobulin classes in the hamster have been described, IgG, IgA, and IgM. Two well studied IgG subclasses, IgG1 and IgG2, differ in their functional capacity to effect passive cutaneous anaphylaxis and complement fixation (Coe, 1978). A minor third IgG subclass (IgG3) with high affinity to protein A has been described. From five studied inbred Syrian hamster strains, animals of three strains have no detectable amounts of IgG3 at all. It was shown that the IgG3 deficiency is a result of a single gene defect that developed in three strains during inbreeding (Coe *et al.*, 1995).

B. Farm Animals

Immunoglobulins of domestic farm animals are less well characterized in general than that of laboratory animals. However, during the past years intensive structural and genetic studies were performed and much new information is now known, which allows comparison of the immunoglobulin system of farm animals with that of the mouse and human (Butler, 1997).

1. Cattle

Cattle possess IgM, IgG, IgA, and IgE molecules. The light chains of most bovine immunoglobulin molecules are of the λ type. The bovine IgG class contains three well studied subclasses, IgG1, IgG2a, which constitutes half of total

IgG, and IgG2b. Probably a fourth IgG subclass also exists (Knight *et al.*, 1988). The IgG1 and IgG2 molecules fix bovine complement, while only IgG1 fixes guinea pig complement. Both these IgG subclasses cause passive cutaneous anaphylaxis in bovine skin. There is an additional disulfide intrachain bridge in the C_H1 domain of both bovine IgG subclasses (the same as in the γ chain of rabbit and goat and in the sheep $\gamma 2$ chain). The bovine $\gamma 1$ and $\gamma 2$ chains have different sequences of their hinge regions (Symons *et al.*, 1989). Whereas the IgG1 molecule possesses three interheavy chain disulfide bonds, the IgG2 molecule has only one such bond (as in rabbit and goat IgG and sheep IgG2).

Bovine IgG2a occurs in two allotypic forms, A1 and A2. The A1 and A2 heavy chains have multiple amino acid differences, particularly in the hinge region (only 71% homology), as well as different susceptibility to digestion with pepsin (Kacskovics and Butler, 1996). According to sequence studies, the variable regions of the bovine heavy and light chains have the same canonical repertoire of loop conformations as the variable regions of the murine and human immunoglobulins (Armour *et al.*, 1994).

Two main variable λ chain gene segment families ($V_\lambda 1$ and $V_\lambda 2$) were identified in cattle. Most of the studied variable segments are related to the $V_\lambda 1$ family (Sinclair *et al.*, 1995). The rearrangement in the V_λ gene locus is limited and this process does not contribute significantly to λ chain diversity. Only a single *J* gene segment participates in the rearrangement at the λ locus. The major mechanism of the diversification of the cattle λ chains is gene conversion. The bovine germ line contains many V_λ pseudo-genes: from 20 sequenced germ line genes, 14 are pseudo-genes. In the rearranged variable λ genes, a significant number of donor sequences were found from one or more V_λ pseudo-genes (Parng *et al.*, 1996). Cattle, similar to swine and rodents, have only one C_α gene in their genome, which has a high homology to the swine α constant gene, with about 75% sequence similarity (Brown *et al.*, 1997).

The bovine immunoglobulin repertoire is generated from a single diversified V_H gene family, which appears to be a homologue of the human $V_H II$ gene family. Somatic hypermutation plays a main role in diversifying the bovine heavy chain variable genes (Aitkin *et al.*, 1997; Saini *et al.*, 1997).

2. Sheep

Sheep immunoglobulins have much in common with those of other ruminant species. Four classes of ovine immunoglobulins are characterized, IgM, IgG1, IgG2, and IgE. Most sheep immunoglobulin molecules are of the lambda type. The major differences between the $\gamma 1$ and $\gamma 2$ heavy chains occur in their hinge regions: the hinge of the $\gamma 2$ chain is shorter, and it has only one interchain disulfide bridge. The $\gamma 2$ chain also lacks the high affinity FcγRI receptor motif at the beginning of the $C_\gamma 2$ domain sequence (Clarkson *et al.*, 1993). There are differences between the sheep $C_\mu1$–$C_\mu2$ joining segment and the same segments of the μ chains of other mammals. This segment, considered an analogue of the hinge region, is much shorter in the sheep μ chain, which could cause diminished flexibility of the whole sheep IgM molecule (Patri and Nau, 1992).

The sheep V_H gene repertoire derives from a small germ-line gene family, which only has about 10 members (Dufour *et al.*, 1996). The sheep V_λ genomic pool is larger, up to 90–100 V_λ genes (Reynaud *et al.*, 1995). The recombined variable genes are diversified by somatic hypermutation occuring in ileal Peyer's patches. This process is important for development of the primary antibody repertoire. Foreign antigens do not influence somatic hypermutations.

3. Pigs

Four major classes of swine immunoglobulins, IgM, IgG, IgA, and IgE were recognized serologically. Half of the immunoglobulin molecules belong to the $\varkappa$ type and another half to the λ type (Butler and Brown, 1994). Porcine serum IgA molecules are mainly dimers.

Single copies of C_α and C_ε genes and several C_γ genes present in the pig genom. Five putative IgG subclasses were identified, which can be grouped in two clusters. One contains IgG1 and IgG3 and the other IgG2a, IgG2b, and IgG4 subclasses (Kacskovics *et al.*, 1994). Between the heavy chains of IgG subclasses, there is a high sequence similarity, which is also valid for the hinge regions. A significant homology was found between the pig and human γ and λ chains but the homology between pig and human $\varkappa$ chains is less evident (Lammers *et al.*, 1991).

Swine V_H genes appear to belong to a single family of not more than 20 members. They are homologous to the human V_HIII gene family rabbit and chicken V_H genes (Sun *et al.*, 1994). There is only a single J_H gene in the swine genome. The constant gene for δ chains (C_δ) was not found in swine nor as in rabbit and cattle genomes (Butler *et al.*, 1996). The gene conversion is thought to be a mechanism of diversification of the pig V_H gene repertoire.

4. Horses

Four major equine immunoglobulin isotypes are known, IgM, IgG, IgA, and IgE. Horse IgG comprises several subclasses, IgG2a, IgG2b, IgG2c, IgG1 ("aggregated immunoglobulin"), and IgG(T). There is two to three times as much carbohydrate in IgG(T) molecules than in other IgG molecules and about two-thirds of the carbohydrate chains localize in the Fd portion of the IgG(T) heavy chain. By papain digestion, horse IgG releases monovalent Fab fragments, whereas IgG(T) gives bivalent $F(ab')_2$ (Weir and Porter, 1966). The sequence of the horse ε chain is most similar to that of the human ε chain, followed by the sheep ε chain. The greatest similarity was found for the last two constant domains. The variable region sequence of the studied IgE gene is most homologous to bovine V_H followed by sheep V_H (Navarro *et al.*, 1995). Over 90% of horse light chains are of the λ type. The number of the variable λ germ line gene segments was estimated to be from 20 to 30. They may rearrange with any one of three J_λ–C_λ genes. The fourth J_λ–C_λ gene is probably a pseudo-gene (Home *et al.*, 1992).

5. Camels and Llamas

Three IgG subclasses were identified in camel and llama sera (Hamers-Casterman *et al.*, 1993). One of them, IgG1, is similar to typical IgG of other animals and its molecules are composed of heavy and light chains. The molecules of two other subclasses, IgG2 and IgG3, representing up to 75% of all serum molecules reacting with protein A, are composed only of one pair of the heavy chains. The γ2 and γ3 chains completely lack the C_H1 domain. Their hinge regions are of different length: the γ2 chain hinge region is equal to that of the human γ2 or γ4 chains and the hinge of the camel γ 3 chain is much longer. All camelid IgG classes can react with a large number of antigens and hence the heavy chain antibody can form an extensive repertoire of antigen-combining sites.

In the camelid V_H, there are several substitutions in positions that are usually involved in V_H–V_L association. These substitutions make the potential V_H contact surface with V_L more hydrophilic. One of the more important is a substitution of the conserved Leu-45 residue by arginine or cysteine. The change of Gly-44 and Trp-47 into Glu-44 and Gly-47 probably also contributes to changes in the V_H surface responsible for contacts with V_L. In camel V_H, the Leu-11 residue, which usually is involved in the interactions with the C_H1 domain, is changed to serine. The CDR3 loop of camel heavy chains is unusually large, up to 24 residues. (Muyldermans *et al.*, 1994). All these substitutions contribute to the increased solubility of isolated camel V_H fragments. The camelid V_H does not form dimers after isolation but retains the ability to combine with antigens and behave like the smallest antigen-binding units (Spinelli *et al.*, 1996; Desmyter *et al.*, 1996). In the recent attempts to "camelize" human V_H to obtain soluble and active minimal antigen-binding fragments, the peculiar structural properties of camelid V_H were used (Davies and Riechmann, 1996; Riechmann, 1996).

C. Pets

Cat serum contains IgG, IgM, IgA, and IgE isotypes and several feline IgG subclasses have been demonstrated (IgG1, IgG2, IgG3, and IgG4) (Baldwin and

Denham, 1994). The cat IgG subclasses react differently with protein A. Protein G has an unusual low affinity to feline IgG (Grant, 1995).

Dogs possess all major classes of immunoglobulins, IgG, IgM, IgA, and IgE. Recently, IgD-like molecules were also identified in dog serum (Yang *et al.*, 1995). The λ light chains prediminate in the feline and dog serum immunoglobulins ($\lambda/\kappa \approx 3{:}1$).

IV. HUMAN IMMUNOGLOBULINS

A. General Considerations

Immunoglobulins are among the best-studied proteins of the human body. Several factors favored the intensive studies of human immunoglobulin structure and function. The most important one is the strong demand of clinical medicine to understand the role of antibodies in the origin of various human diseases and first of all in autoimmune pathology.

The structural and functional studies of human immunoglobulins are greatly facilitated by the fact that myeloma proteins, which are products of malignant plasma cells and present in myelomatosis patients, are monoclonal immunoglobulins or their peptide chains. Most studies of the primary structure of immunoglobulin chains in 1960s–1970s were performed on myeloma proteins or Bence Jones proteins, which are homogeneous light chains. These data are summarized by Kabat *et al.*, (1991). Since then, the amount of new sequences has at least doubled and several types of electronic mail archive servers have been prepared, which help to find any sequence information and takes about a couple of minutes (Johnson *et al.*, 1996). Based on sequence studies, variable regions of immunoglobulin chains are classified into subgroups or families. Usually the members of a family are 75% identical, whereas members of different families have much less similarity.

In discussions of immunoglobulin structure and function, human immunoglobulins are often used as a reference prototype. A good example is the human IgG1κ(Eu) (Fig. 29), the first immunoglobulin whose chemical structure was completely studied (Edelman, 1970). The numeration of amino acid residues of the IgG1κEu peptide chains is used for comparison of the primary structure of peptide chains of other immunoglobulins. Another numbering system (according to Kabat *et al.*, 1991) delineates the CDR framework regions and defines the domain boundaries. The structural and functional properties of human immunoglobulins are discussed in nearly all other chapters of this book, often as reference standard.

As in other mammals, the human immunoglobulin family comprises five classes three major, IgG, IgM, and IgA, and two minor, IgD and IgE. The

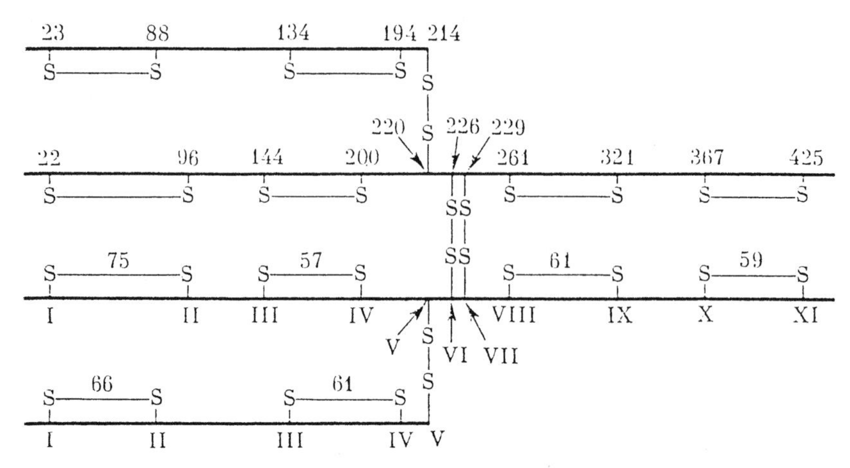

FIGURE 29 Scheme of the human IgG1*ϰ* Eu molecule. Arabic numbers denote position of half-cystines in peptide chains, roman numbers show numerical order of the same residues. (Gall and Edelman, 1970. Reprinted with permission from American Chemical Society.)

immunoglobulin molecules belong either to the *ϰ* or *λ* type with the *ϰ*—*λ* chain ratio approximately equal to 65:35. There are four isotypes of the heavy *γ* chains (Table 2) and two isotypes of *α* heavy chains that differ by the structure of their constant parts and by their functional and antigenic properties. Protein antigens induce as a rule the IgG1 response and to a smaller extent IgG3 and IgG4, whereas the response to polysaccharide antigens is mainly IgG2.

Several serological markers characterize four isotypes of the human lambda chains (Sittisombut, 1996). These markers are associated with the presence of distinctive amino acids in constant regions of the *λ* chains. Oz and Kern markers are associated with differences at positions 152 and 190, respectively, whereas the Mcg marker reflects differences at positions 112, 114, and 163 (Kabat's numbering). Four *λ* chain isotypes, $Mcg^+Kern^+Oz^-$, $Mcg^-Kern^-Oz^-$, $Mcg^-Kern^-Oz^+$, and $Mcg^-Kern^+Oz^-$ encoded by $C_\lambda1$, $C_\lambda2$, $C_\lambda3$, and $C_\lambda7$ constant region genes, are identified according to these markers. A new *λ* chain isotype, Mcp, was identified more recently by sequencing a Bence Jones *λ* protein, which serologically was Mcg^-Oz^- (Niewold *et al.*, 1996).

B. Human Allotypes

After the first description of human *γ* chain allotypic variants about 40 years ago (Grubb, 1956), allotypes of human *α*, *ε*, and *ϰ* chains were also found (Table 8) (Sittisombut, 1996). For the detection of human allotypes, a serological method, the hemagglutination inhibition assay, is usually used. Amino acid substitutions correlating with one or another variants were identified for some

TABLE 8 Notation of Human Immunoglobulin Allotypes (World Health Organization)

Location	Symbols		Domain
IgG1	G1m (a)	Glm (1)	C_H3
	(x)	(2)	C_H3
	(f)	(3)	C_H1
	(z)	(17)	C_H1
IgG2	G2m (n)	G2m (23)	C_H2
IgG3	G3m (b0)	G3m (11)	C_H3
	(b1)	(5)	C_H2
	(b3)	(13)	C_H3
	(b4)	(14)	C_H2
	(b5)	(10)	C_H3
	(c3)	(6)	C_H3
	(c5)	(24)	C_H3
	(g)	(21)	C_H2
	(s)	(15)	C_H3
	(t)	(16)	C_H3
	(u)	(26)	C_H2
	(v)	(27)	C_H3
IgA2	A2m (1)		C_H2
	A2m (2)		C_H3
IgE	Em (1)		
$\varkappa$ chain	Km (1)		$C_\varkappa$
	Km (2)		$C_\varkappa$
	Km (3)		$C_\varkappa$

of the γ chain allotypes and for $\varkappa$ chain allotypes (Table 9). The Gm allotypic epitopes are located on the constant part of the heavy chains of IgG and they serve as the true markers for the constant domains of different IgG subclasses. Allotypic epitopes are mainly conformational ones and dependent on the three-dimensional structure of the corresponding domains. For example, allotypic epitopes localized in the Fd fragment of γ chains are found only if the whole IgG molecule is studied, not just the isolated Fd.

Some allotypic Gm epitopes can be found only on a part of molecules of a IgG subclass, but at the same time they are present on all molecules of one or two other subclasses. Such allotypic variants are designated as isoallotypes. For example, allotype "non-z" [nG1m(z)] is present only on $C_\gamma 1$ of IgG1 molecules without the G1m(17) allotype but is also present on all IgG3 molecules. The "non-g" epitope [nG3m(g)] exists in all IgG2 molecules but only in IgG3 molecules that have no G3m(21) epitope.

Particular combinations of human allotypes are inherited as fixed complexes or haplotypes (Lefranc and Lefranc, 1990). It was shown that haplotype

TABLE 9 Structural Correlation with Gim and Km Allotypes

Allotype	Amino acid residues		Codon	
G1m	356	358		
a	Asp	Leu	GAT	CTG
a−	Glu	Met	GAG	ATG
G1m	214			
f	Lys		AAA	
z	Arg		AGA	
G1m	431			
x	Gly		GGA	
x−	Ala		GCA	
Km	153	191		
1,2	Ala	Leu	CGG	GAG
1	Val	Leu	CAG	GAG
3	Ala	Val	CGG	CAG

Grubb, 1994. Reprinted with permission from Marcel Dekker, NY.

frequencies vary from population to population. For example, frequency of the G1mza;G3mg haplotype is 0.007 in Nigeria, 0.407 in Japan, and 0.187 in The Netherlands (Grubb, 1994). There are haplotypes that are found only in Caucasian populations or that are characterstic mainly for Mongoloid or Negroid populations. Allotypic epitopes, therefore, are very useful markers in population studies and in forensic medicine.

C. Human Immunoglobulin Genes

The organization of variable and constant genes encoded the human immunoglobulin chains is now known in detail (Cook *et al.*, 1994; Zachau, 1996; Matsuda and Honjo, 1996). In the human genome, as in the genomes of other mammals, there are three gene complexes localized on different chromosomes that code heavy chains and κ and λ light chains. Each of the complexes includes locuses of variable and constant genes. The first of them contains groups of two (V_L–J_L) or three (V_H–D–J_H) variable gene segments, whereas the second one contains constant genes.

1. Heavy Chain Locus

The heavy chain locus is mapped to chromosome 14(q32). The V_H and C_H gene locuses are tightly linked and located in the order of $5'$–V_H–D–J_H–C_H–$3'$ (Fig.

30). The V_H locus is composed of 81 V_H segments in the smallest haplotype with an additional 10 V_H polymorphic segments (Matsuda and Honjo, 1996). Twenty-four translocated V_H segments ("orphons") with several *D* segments were identified on chromosomes 15 and 16 and the total number of V_H segments in the human genome is about 115. Approximately 50 of them are functional genes (Cook *et al.*, 1994). The size of the V_H locus is 1.0–1.1 Mb depending on the haplotypes. The locus is polymorphic, what is partly linked with deletions or insertions of V_H segments and the presence of V_H alleles. The significant part (about 44%) of all sequenced V_H segments are pseudo-genes (Matsuda and Honjo, 1996).

According to the nucleotide sequences obtained in earlier studies, all V_H segments were classified into three families, V_H1, V_H2 and V_H3. They correspond to the three V subgroups, V_HI, V_HII, and V_HIII, based on the amino acid sequences (Pascual and Capra, 1991). When sequences obtained later were compared, four other families were revealed (V_H4–V_H7). Using this classification, V_H segments with 80% or greater similarity belong to the same family, whereas V_H segments that have less than 70% similarity form different V_H families. The number of members in each family is varied, the largest are V_H3 and V_H1 and the smallest is V_H6 (a single segment). The V_H4 sequences are the most conserved and probably this V_H family evolved more recently. The conserved residues specific to the V_H family located in all three framework regions (Kabat *et al.*, 1991). The regions with higher conservation within the same family are codons 6–24 in the first framework region and codons 66–85 in the third framework region (Matsuda and Honjo, 1996). The V_H segments of one family are not located in one cluster but are dispersed across the entire V_H complex. This is in contrast to the situation in the mouse genome.

In human genome, about 30 functional *D* segments have been found. They can be grouped into six families that are arranged in four tandem clusters. Near the C_μ gene, nine J_H segments are located. From them three are pseudo-genes. The C_H gene family is organized as follows: 5′–J_H–(8 kb)–C_μ–(5 kb)–C_δ–(60 kb)–$C_\gamma3$–(26 kb)–$C_\gamma1$–(19 kb)–ψC_ε–(13 kb)–$C\alpha1$–(35 kb)–$\psi C\gamma$–(40 kb)–$C_\gamma2$–(18 kb)–$C\gamma4$–(23 kb)–C_ε–(10 kb)–$C\alpha2$–3′. The C_μ and C_ε genes are composed of four exons and than other constant genes of three exons.

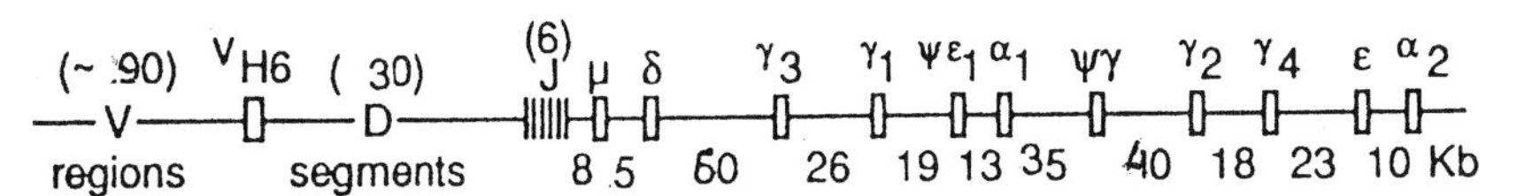

FIGURE 30 Organization of human immunoglobulin heavy chain genes. Distances separating gene segments are shown in kilobases. Pseudo-genes are designated by ψ. (Adapted from Pascual and Capra, 1991.)

2. $\varkappa$ Chain Locus

The human $\varkappa$ light chain locus is 1.8 Mb long and located on chromosome 2(p11–12). It is composed of one $C_\varkappa$ gene, five $J_\varkappa$ segments and two contigs of $V_\varkappa$ gene segments, which are arranged on opposite 5′, 3′-polarity (Zachau, 1995, 1996). Probably nearly all functional $V_\varkappa$ genes are now identified. From known $V_\varkappa$ genes, 32 are functional, 16 posses minor defects, and 25 are pseudogenes. Three genes have functional as well as slightly defective alleles. Some $V_\varkappa$ genes ("orphons") are located on chromosomes 1, 22, and others. The sequence of 24 orphons is known. Some individuals have lost one of the $V_\varkappa$ gene contigs (36 $V_\varkappa$ genes) but despite this, their immune response is apparently normal.

3. λ Chain Locus

The human λ light chain locus on chromosome 22q11.2 covers 0.8 Mb of DNA and is composed of a V_λ segment region and seven J_λ–C_λ clusters (Selsing *et al.*, 1989; Blomberg and Solomon, 1996). About 10 V_λ families (total 70 V_λ segments, of which about two-thirds are functional) are localized 14 kb upstream from the J_λ–C_λ gene region and composed of 7 J_λ and 7 C_λ genes (Combriato and Klobeck, 1991). Genes of different V_λ families are arranged in three distinct clusters in contrast to the dispersed organization of the human V_H and $V_\varkappa$ genes (Chuchana *et al.*, 1990; Williams *et al.*, 1996). The most frequently used V_λ genes are located proximal to the J_λ–C_λ genes. The V_λ gene segments from cluster A, which is closest to J_λ–C_λ, account for 62% of the expressed repertoire. From 30 functional V_λ segments, three encode half of the expressed V_λ repertoire (Ignatovich *et al.*, 1997). Two orphon V_λ genes are found on chromosome 8(q11.2) (Frippiat *et al.*, 1997).

Four J_λ–C_λ genes (J_λ–$C_\lambda 1$, J_λ–$C_\lambda 2$, J_λ–$C_\lambda 3$, and J_λ–$C_\lambda 7$) are functional and code four human J_λ–C_λ isotypes, which can be differentiated serologically (Niewold *et al.*, 1996). Three others represent pseudo-genes ($\psi C\lambda 4$, $\psi C\lambda 5$, and $\psi C\lambda 6$) and one of them ($\psi C\lambda 6$) encodes a λ chain with a truncated constant region (Stiernholm *et al.*, 1995). The λ gene locus also include two genes coding surrogate λ chains ($V_{pre\text{-}B}$ and $V_{\lambda 5}$).

V. ORIGIN OF ANTIBODY DIVERSITY IN MICE AND HUMANS

In the 1960s immunologists came across a paradox that was very hard to explain from the point of view of classical biological ideas, particularly from the viewpoint of the formula "one gene—one peptide chain." It was found that in rabbits different allotypic specificities of the N-terminal part of the heavy chains can combine with a single C-terminal part ("Todd's phenomenon"). Similarly, it was noted that the different sets of nonallelic sequences in the N-terminal parts of the murine $\varkappa$ chains can unite with a single C-terminal sequence (C. Milstein's findings).

To explain this paradox, a theory has been proposed that special mechanisms are responsible for the formation of immunoglobulin chains. It was based first of all on the very important finding that immunoglobulin chains are composed from two different parts—the N-terminal variable and the C-terminal constant (Hilschmann and Craig, 1965). According to this theory, "the variable portion of these molecules results from genetic material, which is present in the germ line and which is combined with the common gene during the differentiation of the immunologycally competent cells" (Dreyer and Bennett, 1965). The similar concept graphically illustrated on Figure 31 was formulated in our review of that time (Gurvich and Nezlin, 1965). The studies performed in the next years substantiated the multigenic theory of the immunoglobulin chain biosynthesis.

In the past two decades, the origin of the antibody diversity in higher vertebrates, including mice and humans, was mainly elucidated. First, primary information already exists in the germ line genome in the form of different numbers of gene segments, which are components of variable genes (V–D–J segments). Second, these segments can recombine in different ways (combinatorial diversity). Third, this recombination is imprecise, with deletions or insertions at joint regions, and the products of the association are more or less different

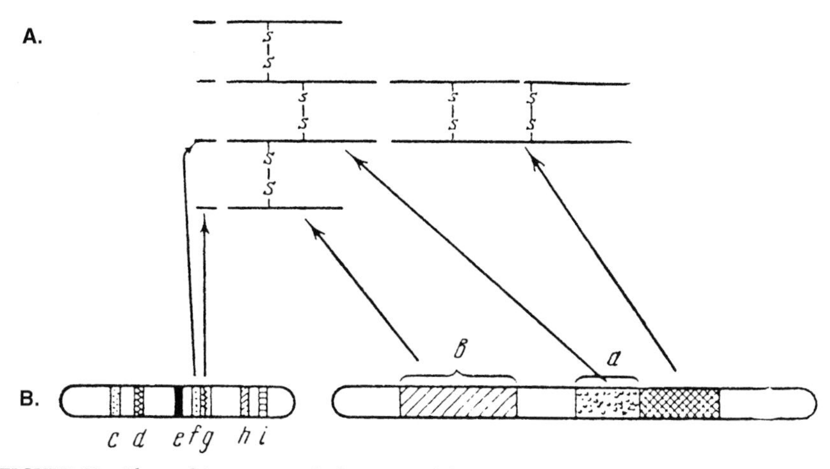

FIGURE 31 The multigene control of immunoglobulin peptide chains as it was seen in 1965. (Gurvich and Nezlin, 1965.) Different parts of the IgG peptide chains are coded by sets of variable (c–i) and constant (a,b) genes.

from each other (junctional diversity). Finally, the joined variable genes are hypermutated during the maturation of the antibody response and in the latest steps of this process the mutated antibody molecules are selected by antigen.

A. V(D)J Recombination

The somatic process of the germ line V segment association with the formation of a complete gene occurs in the early steps of lymphocyte development and it generates the major part of the antibody diversity of the naive repertoire in mice and humans (Tonegawa, 1983). About 10^{10} different antibody molecules are produced at this step of the diversification process (Milstein and Neuberger, 1996). The main part of combining site diversity is caused by the variation in shapes and sizes of the CDR3 loops.

The recombination is an ordered process. Its first step is the D to J_H association, which takes place in pro-B lymphocytes. The second one occurs in pre-B lymphocytes and consists of joining V_H with DJ_H. After the joining of V_HDJ_H with the C_μ region, the complete μ chain is expressed on the surface of the B lymphocytes as a complex with the surrogate light chains. Only after the formation of this early B-cell receptor protein, does the rearrangement of light chain V gene segments (V_L and J_L) begin (Gellert, 1994; Okada and Alt, 1995; Sleckman *et al.*, 1996; Lansford *et al.*, 1996).

All germ line V-region gene segments, which participate in recombination, are flanked by recombination signal sequences (RSSs). A RSS consists of a heptamer (CACAGTG) and a nonamer (ACAAAAACC) separated by 12 ± 1 or 23 ± 1 basepart (bp) nucleotide spacer sequences. D segments are flanked by 12 ± 1 spacers and V and J segments are flanked by the larger spacer sequence (23 ± 1) from both sides (Early *et al.*, 1980). A gene segment with the 12-bp spacer can only join with a segment flanked by the 23-bp spacer (12–23 bp rule). After signal sequences are recognized, both strands of DNA are cleaved between RSSs and coding sequences. Both ends of coding sequences form closed structures (hairpins). Intervening sequences are excised precisely and form circular pieces of DNA by head-to-head ligation (deletional rearrangement). Less frequently, during D–J joining the intervening sequence is inverted but not deleted and the 5′ RSS of D is used for joining instead of the 3′ RSS (inversional rearrangement).

The joining of the cleaved ends of two coding sequences occurs with losses of the terminal nucleotides and template-independent addition of nucleotides (N-regions). Another way the modification of the joining ends occurs is a template-dependent nucleotide addition (P-nucleotides). Nontemplated N-sections, which are G/C-rich and could be up to 15 nucleotides long, are built by terminal deoxynucleotidyl transferase (TdT) expressed predominantly in early lymphocytes. The templated P-insertions are shorter, usually 1–2 nucleotides only. They probably arise when hairpins are opened asymmetrically and coding ends have different ends. The addition of bases to the shorter end results in the appearance of complementary P-nucleotides (Gellert, 1994).

A complex of various factors are involved in V(D)J recombination. Some of them are lymphocyte-specific and some are not. RAG-1 and RAG-2 phosphorylated nuclear proteins are products of *RAG-1* and *RAG-2* genes, which are co-expressed only at the early steps of lymphocyte development (Oettinger *et al.*, 1990). They participate in recognizing RSS and in introduction of DNA double-strand breaks (McBlane *et al.*, 1995). Both proteins are also needed for efficient joining of coding ends (Ramsden *et al.*, 1997).

The product of the *scid* gene, which is a generally expressed factor, is also involved in V(D)J recombination. Mice with the mutation of the *scid* gene located on chromosome 16 suffer from severe combined immune deficiency (scid) and their lymphocyte cannot complete V(D)J recombination normally (Bosma and Carroll, 1991). The molecular properties of the scid factor are unknown. Other generally expressed factors that participate in V(D)J recombination are involved in repair of double-strand DNA break. Some of them are components of the DNA-dependent protein kinases (DNA-PKs), Ku80 and Ku70, and the DNA-PK catalytic subunit. Ku70–deficient cells are unble to support V(D)J recombination (Gu *et al.*, 1997). DNA ligase I, and XRCC4 factors that also involved in V(D)J recombination are now being studied intensively (Gellert, 1992; Sleckman *et al.*, 1996; Bogue and Roth, 1996).

The use of different variable region gene segments is not random and some of them are expressed much more frequently than others (Milstein *et al.*, 1992). The reasons for that are not always completely clear. The variation in spacer length or localization of variable gene segments relative to *D* and *J* segments could be among the decisive factors. The use of the variable region segments is independent of antigenic selection (Tuaillon *et al.*, 1994).

The most diversified part of the variable region is the CDR3 loop. It is formed by the *D* gene segment and the V_H-D and D-J_H junction sections. According to some calculations, about 10^{14} peptides can be created by gene segments forming CDR3 (Sanz, 1991). The diversity of CDR3 is generated by junctional variations, including joining with loss of some bases and formation of new codons by intracoding recombination and random addition of N nucleotides. Other mechanisms include *D–D* segment fusion and addition of P nucleotides.

B. Somatic Hypermutation

Further diversification of recombined *V(D)J* genes is a result of somatic hypermutation, which occurs in germinal centers of the secondary lymphoid follicles located in the peripheral lymph nodes of mouse and human (Milstein and Rada, 1995; Neuberger and Milstein, 1995; Wagner and Neuberger, 1996). By this process, affinity of antibodies can be increased up to 100-fold. The target of hypermutation is a region around the rearranged variable region gene. The constant region genes and unrearranged variable region genes are usually unmutated. Usually up to 24 amino acid substitutions can be found in the coding part of the variable genes in their framework and hypervariable regions as well (Milstein and Neuberger, 1996). They are as a rule single nucleotide substitutions and transitions occur more often than transversions. Nucleotide insertions or deletions are rarely found. The hypermutation process is not random and some bases are mutated more often than others (Padlan, 1997). It was noted that the second base of the serine codons AGC/T is preferentially targeted. There is a clear predominance of mutations of A versus T and G versus C bases, which indicates that mutations modify only one DNA strand of the double helix.

The diversity that results from the rearrangement of germline variable genes is located predominantly in the center of the combining site. Somatic hypermutation spreads diversity to regions at the periphery of the combining site. Therefore, both processes of diversification are complementary and result in more even distribution of sequence diversity (Tomlinson *et al.*, 1996). Only a part of the mutations improves antigen binding, and about three-fourths of them are neutral.

The molecular mechanism of somatic hypermutation is still largely unknown. According to recent data, hypermutation of immunoglobulin genes is linked to the initiation of transcription and depends on the presence of immunoglobulin enhancer sequences (Betz *et al.*, 1994; Peters and Storb, 1996). Perhaps the proposed mutator factor is a part of a transcription initiation complex and is brought into the immunoglobulin gene in the course of transcription (Storb, 1996). For recruiting somatic hypermutation, the variable segment sequence is not necessary and bacterial genes or a globin gene can also be effectively hypermutated if they are in transgenic constructs under the control of the immunoglobulin $\varkappa$ chain promoter and enhancer (Betz *et al.*, 1994; Yélamos *et al.*, 1995).

C. Class Switching

Another type of recombination is heavy chain class switching. This process consists of replacing one constant region gene (usually C_μ) with another one with a different biological function. In this way, the immune response changes the functional properties of the synthesized antibody but retains its antigen specificity (Harriman *et al.*, 1993; Siebenkotten and Radbruch, 1995; Zhang *et al.*, 1995).

The signal for class switching appears to be cross linking the B-cell surface receptor CD40 induced by its ligand, which is expressed by the activated T lymphocytes (Banchereau *et al.*, 1994). The direction of class switching is regulated by cytokines that can stimulate or suppress the production of a particular subclass. Interleukin-4 induces switching to IgG1 and IgE and interleukin 13 to IgE and IgG4. Interferon-γ causes the selective expression of IgG2a, whereas transforming growth factor-β induces switching to IgG2b and IgA.

Class switch recombination occurs in switch (S) regions, which consist of conserved tandem repetitive sequences 2–10 kb in size and located 5′ of constant genes (with exception of C_δ and pseudo-genes). The most common repeats are pentameric sequences GAGCT and GGGGT but there is no evidence that they are essential for recombination. The mechanism of class switching consists in looping-out and deletion of the sequence between the *VDJ* variable gene and a new constant gene. The deleted sequence was isolated in a form of circular DNA (Schwedler *et al.*, 1990). The switch circles are between 65 and 200 kb long, depending on the location of the breakpoints (Fig. 32). The enzymatic factors that participate in the class switching machinery are unknown.

VI. EVOLUTIONARY ASPECTS

The specific acquired immunity can be found only in vertebrate species. Nearly all vertebrates possess the necessary genetic information for biosynthesis of immunoglobulin and T-cell receptor proteins, including structural genes and genetic elements for recombination of variable gene segments into functional variable genes. However, most primitive jawless vertebrates, lampreys, and hagfishes are unable to synthesize antibodies or T-cell receptors. Hence, the very important events leading to the appearance of the immunoglobulin superfamily happened in the relatively short period of about 50–60 million years, which separates jawless vertebrates from the next evolutionary step, sharks and skates.

The genetic information necessary for synthesis of immunoglobulin-type domains was present in the eukaryotic and probably in the the protokaryotic genome from a time long before the appearance of vertebrates. However, the machinery responsible for the formation of the the antigen-specific part of antigen receptors can be found only in vertebrates, beginning with the elasmobranches.

A. Immunoglobulin-Type Domains in Prokaryotes and Invertebrates

It is well established that immunoglobulin-like folds are widely distributed not only among vertebrates but also among insects and other primitive animals and

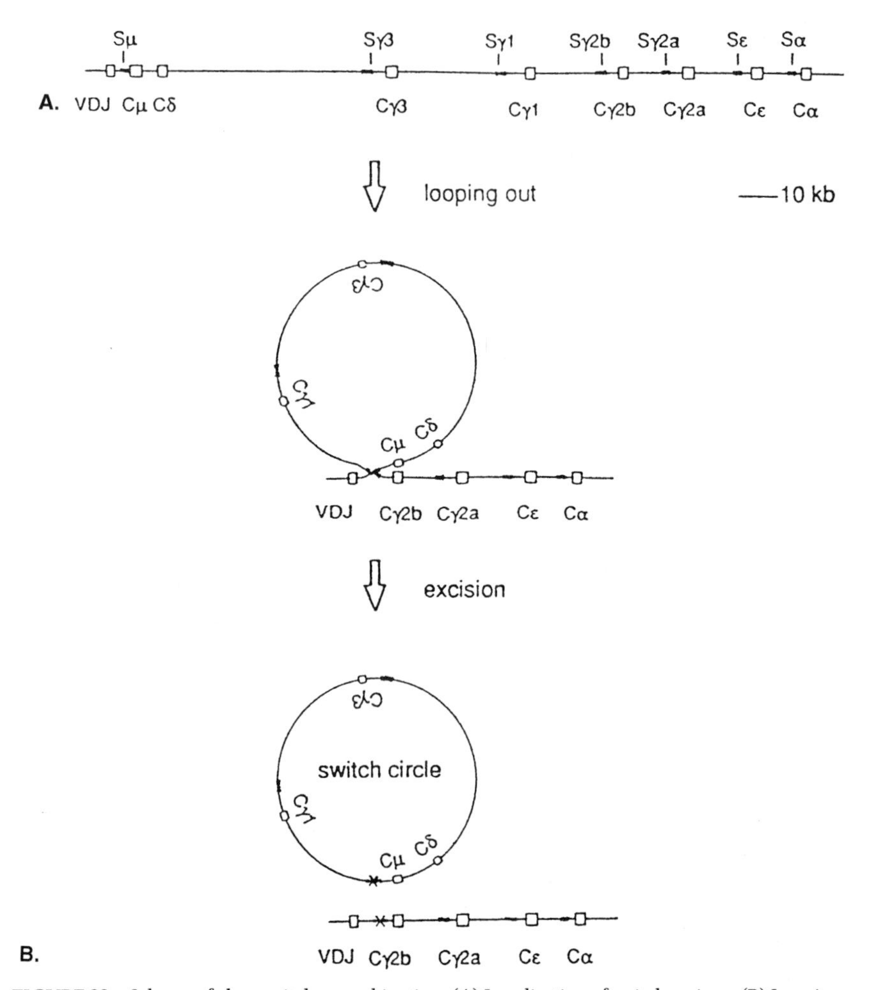

FIGURE 32 Scheme of class switch recombination. (A) Localization of switch regions. (B) Looping-out and deletion of the sequence between the variable and Cγ2b genes (Schwedler *et al.*, 1990. Reprinted with permission from Macmillan Magazines, Ltd.)

even among protocaryotes. Proteins that are built from immunoglobulin domains belong to a large group of proteins, which is known as the immunoglobulin superfamily (Williams, 1982; Williams and Barclay, 1988; Hunkapiller and Hood, 1989). Immunoglobulin domains are integral parts not only of proteins that participate in immunological recognition but they are also structural elements of many other types of protein molecules such as various receptors, cell adhesion, and muscle proteins.

There are two main criteria for a protein to be incorporated in the immunoglobulin superfamily. First, the protein should have a sequence(s) about 100 residues long with clear homology to immunoglobulin or immunoglobulin-related domains (with a characteristic intrachain disulfide bridge and some other conserved residues). Second, this sequence(s) should have the key structural characteristics of the immunoglobulin fold (the β-strand/loop organization).

1. Prokaryotes

Several bacterial proteins were found to be built from immunoglobulin-like domains. Among them are periplasmic chaperones in Gram-negative prokaryotes that assist in the assembly of adhesive pili. One of the members of this protein family, PapD of *Escherichia coli*, is composed of two domains, each consisting of two antiparallel β-sheets (Fig. 33). The overall topology of the domains is similar to that of immunoglobulin domains (Holmgren and Bränden, 1989; Holmgren *et al.*, 1992). However, PaPD has practically no sequence homology with immunoglobulin domains. An intrachain disulfide bridge is formed between residues of one strand and not between two strands, as was shown for most other members of the immunoglobulin superfamily. The PapD sequence can be aligned with the sequence of the CD5 lymphocyte antigen. After the alignment, about 25% of residues of both proteins are identical. It was proposed that PapD and CD5 domains belong to a separate domain group, members of which resemble immunoglobulin domains. Based on a detailed sequence analysis, a prediction was made that two other prokaryote proteins, IgA Fc receptor of a streptococcus and endoglucanase of *Cellumonas fimi*, contain immunoglobulin-like domains (Bateman *et al.*, 1996). It was suggested that these bacterial proteins diverged from an ancestor common to the eukarytic immunoglobulin superfamily domains.

However, it is hard to prove whether these bacterial proteins are products of divergent evolution from a common ancestor or rather of convergent evolution to the same three-dimensional structure.

2. Invertebrates

There is good evidence that invertebrates have proteins that are related to the immunoglobulin superfamily. The glycoprotein Thy-1, the first of such proteins, was isolated from squid as well as from mouse brain (Williams and Gagnon, 1982). It is present on the surface of thymocytes, neuronal cells, and some other cells and is probably one of the most abundant surface glycoproteins (10^6 molecules per thymocyte). The Thy-1 protein part has a molecular weight of 12.5 kDa and its length (111 residues) is equal to that of an immunoglobulin domain. The Thy-1 sequence is homologous to those of the variable immunoglobulin domains and at the same time it has a good homology

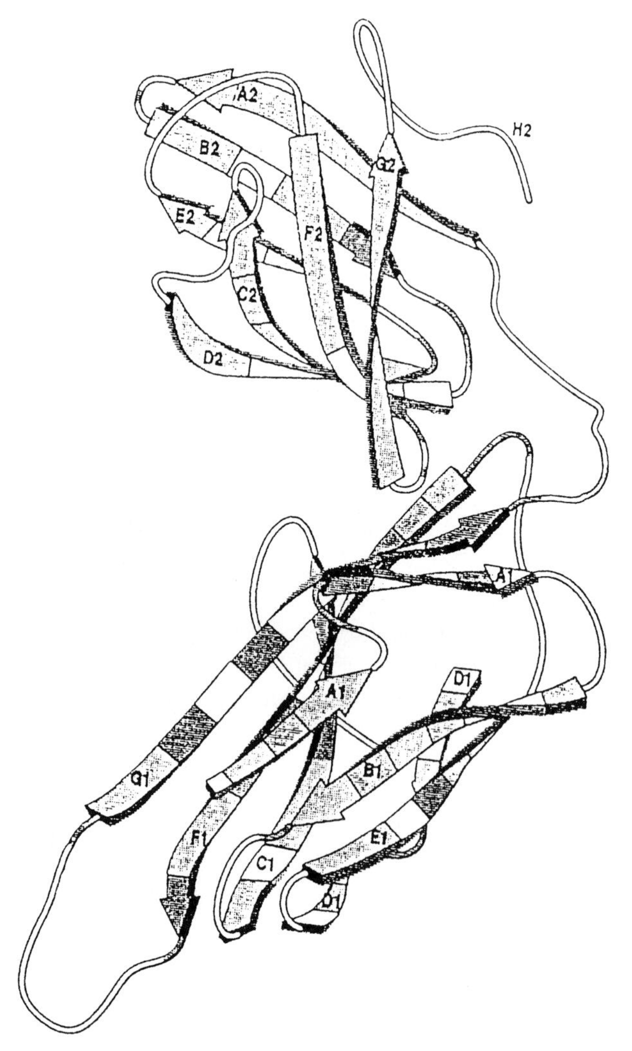

FIGURE 33 Model of the crystal structure of PapD, a member of periplasmic chaperon. Arrows indicate β-strands. (Holmgren *et al.*, 1992. Reprinted with permission.)

with constant domains. Thy-1 is attached to cell membranes via a glycophospholipid tail (Tse *et al.*, 1985). Its functions are still unknown.

Fasciclin II, a glycoprotein that participates in neuronal recognition, was identified in the grasshopper embryo. This protein contains two types of domains: five of them belong to the immunoglobulin superfamily (C2 type) and other two are similar to the fibronectin type III domains (Fig. 34) (Harrelson

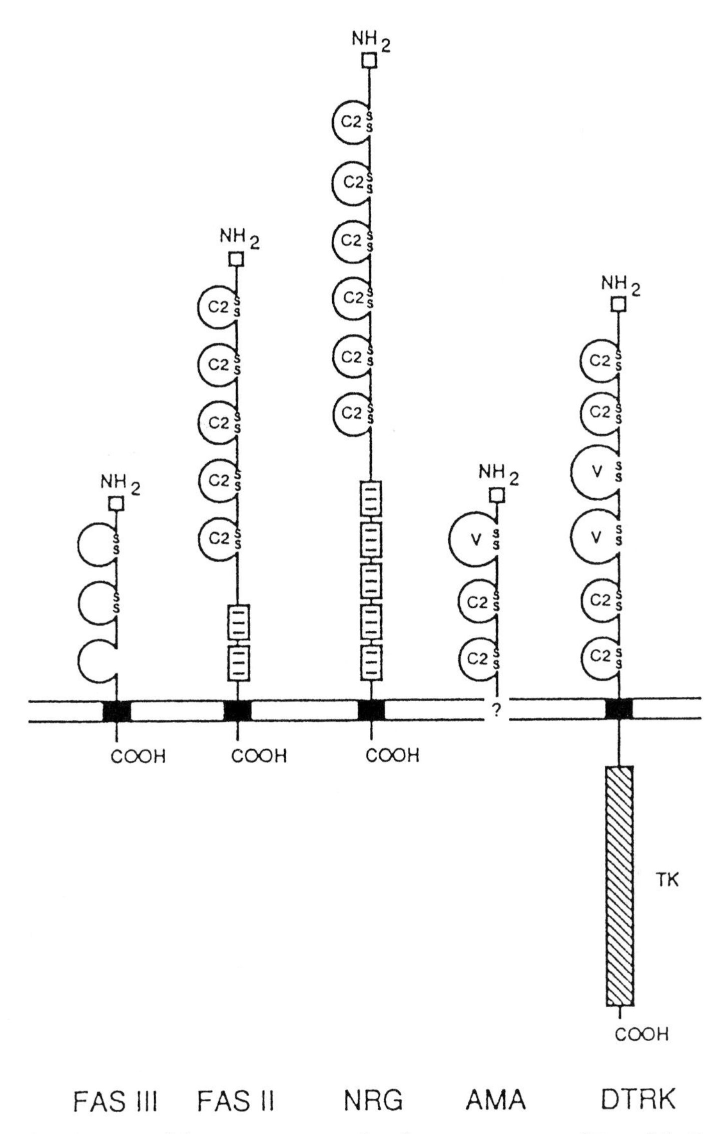

FIGURE 34 Structure of the proteins expressed in the nervous system of *Drosophila*. Boxes with three bars indicate fibronectin type III domains, and the shaded box in DTRK indicate the tyrosine kinase domain (TK). Open boxes indicate signal sequences. C2 and V, immunoglobulin constant type C2 and variable domains; FAS III, fasciclin III; FAS II, fasciclin II; NGR, neuroglian; AMA, amalgam; DTRK, gp160$^{D\ trk}$. (Pulido *et al.*, 1992. Reprinted with permission.)

and Goodman, 1988). Immunoglobulin-like domains have two cysteines separated by 50 residues and are surrounded by many other conserved amino acids. Fasciclin II is highly related to vertebrate neural adhesion molecules, which are also composed of C2-type domains. Its greatest overall similarity is with NCAM adhesion molecules.

A protein coded by the drosophila *amalgam (ama)* gene is another well-characterized invertebrate protein with immunoglobulin-like domains (Seeger *et al.*, 1988). It contains three domains with high homology to other members of the superfamily (Fig. 34). Each of them is approximately 100 residues long with two conserved cysteine residues. The first domain is a V-type and the other two are C2-type. The precise function of the ama-protein is not known but it probably participates in cell surface recognition. Another drosophila protein, Dtrk, with high homology to immunoglobulin domains is a tyrosine kinase receptor (Pulido *et al.*, 1992). Its extracellular part is composed of six immunoglobulin-like units especially homologous to neural cell adhesion molecules (NCAMS). Each of them contains 70–100 residues, including two characteristically spaced cysteines. The Drtk protein is composed of four C2-type and two V-type domains (Fig. 34).

Two invertebrate immune defense molecules were identified as belonging to the immunoglobulin superfamily. One of them, called hemolin, is a bacteria-inducible hemolymph protein of the giant silk moth (Sun *et al.*, 1990). It contains four internal repeats 90–110 amino acids in length with two cysteines and a number of other residues, which are conserved among the members of the immunoglobulin superfamily. These repeats form C2-type immunoglobulin folds that reveal a close relation with amalgam and fasciclin II as well as with vertebrate neural cell adhesion molecules NCAMs. The molluscan defense glycoprotein (MDM) has high structural homology with the moth hemolin and NCAMs of invertebrates and vertebrates (Hoek *et al.*, 1996). It contains five C2-like domains that are 90–110 residues long and has two cysteine residues. Twitchin, a large muscle protein from a nematode (Fong *et al.*, 1996), contains 30 immunoglobulin I-type domains , which are very similar to those of muscle vertebrate proteins telokin and titin.

B. Evolution of Immunoglobulin Domains

These data provide evidence that immunoglobulin domains appeared early in evolution and are the building parts of proteins participating in invertebrate immune defense, muscle contraction, and cell adhesion. The ancestral immunoglobulin domain gene was probably evolved from genes coding a protein of half-domain size. A homology found in V_H between segments of 32 residues surrounding both cysteine residues points to such possibility (Bourgois, 1975).

However, in invertebrates two important features of vertebrate immune system are absent—the V(D)J segmentation and the machinery that is responsible for the rearrangement of immunoglobulin gene segments. They both appeared in a relatively short time before the appearance of jawed fishes.

According to a plausible hypothesis, an important role in the evolution of the vertebrate immune system was played by mobile discrete DNA sequences or transposons that can transport themselves to various locations within the genome (Sakano *et al.*, 1979). The insertion of these transposable elements into genes coding ancestral domains could lead to the disruption of gene integrity and the appearance of gene segments that can recombine by the enzymatic machinery, also introduced by a transposon. The gene segmentation has a obvious evolutionary advantage because the recombination process creates junctional diversity.

The transposon elements end in inverted terminal repeats used for DNA cleavage and excision. They also provide the transposase activity necessary for transposon movements to another location within the genome. The intervening segments, which are located between *V(D)J* gene segments, are flanked by RSSs. There is a sequence similarity between the RSS heptamer and the termini of a repetitive element of the Tc1 family of invertebrate transposons (Dreyfus, 1992). This fact supports the previously mentioned hypothesis that the crucial evolutionary event in the history of the vertebrate immune system was the interruption by a transposon of an ancestral domain located on the cell surface (Thompson, 1995). Intron sequences have penetrated into genes coding immunoglobulin domains not only at early stages of vertebrate evolution, but probably also later. Domains of various proteins belonging to the immunoglobulin superfamily are encoded by two and even three exons. For example, the genes of V1 and C2 domains of the CTX protein, a glycoprotein of *Xenopus* cortical thymocytes (Du Pasqier and Chrétien, 1996), C2 domains of NCAM adhesion protein (Owens *et al.*, 1987), and the first V2-like domain of the CD4 protein are split into two exons. The gene of the second V2 domain of the melanoma-associated glycoprotein MUC18 consists of three exons divided by two introns of different length (Sers *et al.*, 1993).

All these data are summarized in Figure 35 in which a hypothetical model of the evolution of V and C domains is presented (Du Pasquier and Crétien, 1996). According to this model, the evolutionary first type of immunoglobulin domains was the C2-type domain followed by the V2-like domains. The V1 and C1 domains appeared significantly later, as components of proteins participating in the vertebrate immune system.

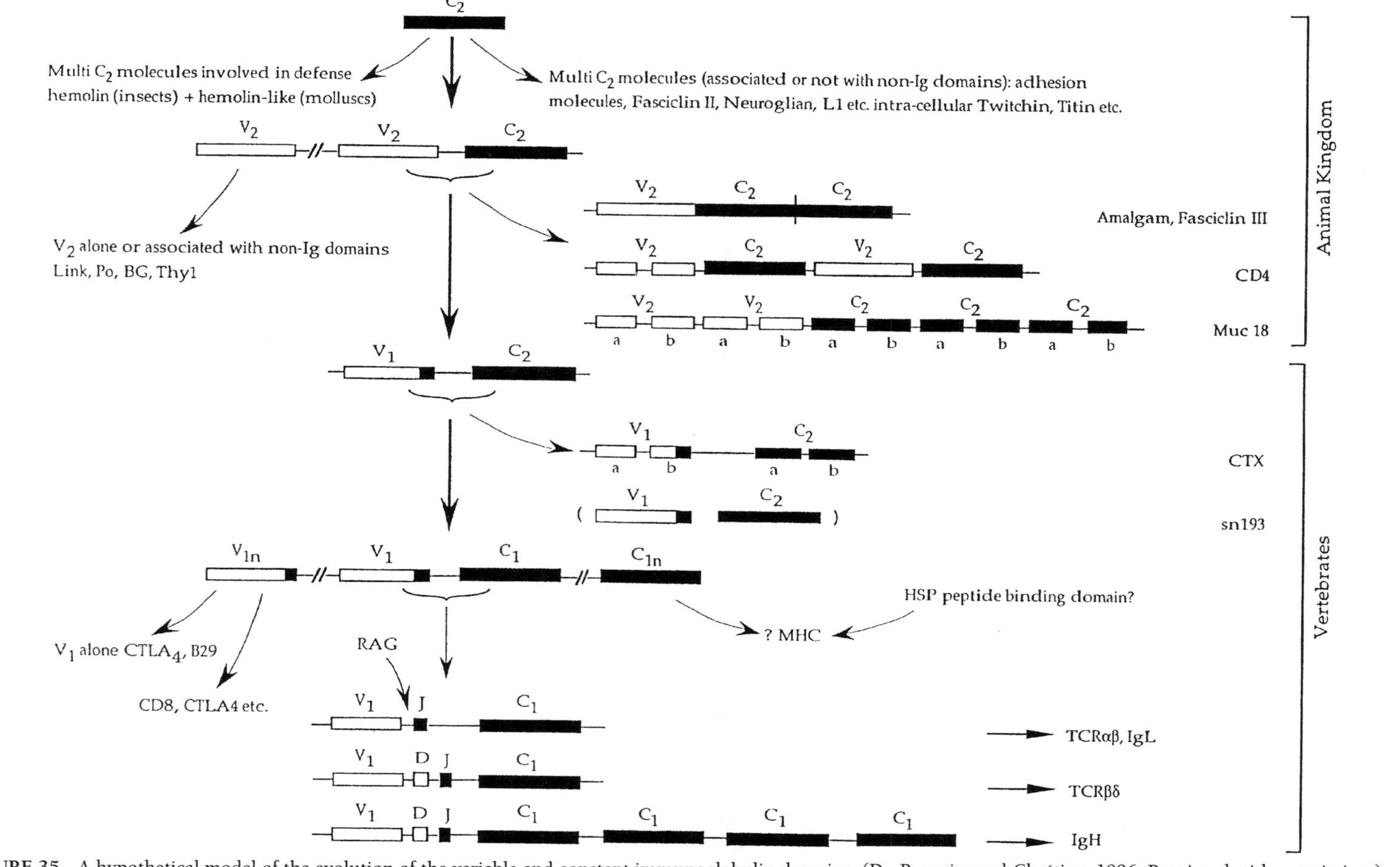

FIGURE 35 A hypothetical model of the evolution of the variable and constant immunoglobulin domains. (Du Pasquier and Chrétien, 1996. Reprinted with permission.)

REFERENCES

Aitken R., Gilchrist J., and Sinclair M.C. (1997). A single diversified V_H gene family dominates the bovine immunoglobulin repertoire. *Bioch. Soc. Transact.* **25**, 326S.

Aguilar B.A., and Gutman G.A. (1992). Transcription and diversity of immunoglobulin lambda chain variable genes in the rat. *Immunogenetics* **37**, 39–48.

Armour K.L., Tempest P.R., Fawcett P.H., Fernie M.L., King S.I., White P., Taylor G., and Harris W.J. (1994). Sequences of heavy and light chain variable regions from four bovine immunoglobulins. *Mol. Immunol.* **31**, 1369–1372.

Baldwin C.I., and Denham D.A. (1994). Isolation and characterization of three subpopulation of IgG in the common cat *(Felis catus). Immunology* **81**, 155–160.

Banchereau J., Bazan F., Blanchard D., Brière F., Galizzi J.P., van Kooten C., Liu Y.J., Rousset F., and Saeland S. (1994). The CD40 antigen and its ligand. *Annu. Rev. Immunol.* **12**, 881–922.

Barre S., Greenberg A.S., Flajinik M.F., and Chothia C. (1994). Structural conservation of hypervariable regions in immunoglobulin molecules. *Nature Struct. Biol.* **1**, 915–920.

Bateman A., Eddy S.R., and Chothia C. (1996). Members of the immunoglobulin superfamily in bacteria. *Protein Sci.* **5**, 1939–1941.

Bazin H., Beckers A., Vaerman J.-P., and Heremans J.F. (1974). Allotypes of rat immnoglobulins. I. An allotype at the α-chain locus. *J.Immunol.* **12**, 1035–1039.

Bazin H., Pear W.S., and Sumelgi J. (1988). Louvin rat immunocytomas. *Adv. Cancer Res.* **50**, 279–310.

Becker R.S., and Knight K.L. (1990). Somatic diversification of immunoglobulin heavy chain VDJ genes: evidence for somatic gene conversion in rabbits. *Cell* **63**, 987–997.

Beckers A., and Bazin H. (1975). Allotypes of rat immunogloblins—III: an allotype of the γ2b-chain locus. *Immunochemistry* **12**, 671–675.

Bernstein R.M., Schluter S.F., Shen S,. and Marchalonis J.J. (1996). A new high molecular weight immunoglobulin class from the carcharhine shark: implications for the properties of the primoridal immunolgobulin. *Proc. Natl. Ac. Sci. USA* **93**, 3289–3293.

Betz A.G., Milstein C., González-Fernández A., Pannell R., Larson T., and Neuberger M. (1994). Elements regulating somatic hypemutation of an immunoglobulin κ gene: critical role for the intron enhancer/matrix attachment region. *Cell* **77**, 239–248.

Bezzubova O.Y., and Buerstedde J.M. (1994). Gene conversion in the chicken immunoglobulin locus: a paradigm of homologous recombination in higher eukaryotes. *Experientia* **50**, 270–276.

Blomberg B.B., and Solomon A. (1996). The murine and human lambda light chain immunoglobulin loci: organization and expression. In: *Weir's Handbook of Experimental Immunology,* 5th ed. (Eds. Herzenberg L.A., Herzenberg L.A., Weir D.M., and Blackwell C.C.). Vol. 1, pp. 10.1–10.26. Blackwell Science, Cambridge, MA.

Bogue M., and Roth D.B. (1996). Mechanism of V(D)J recombination. *Curr. Opin. Immunol.* **8**, 175–180.

Bourgois A. (1975). Evidence for an ancestral immunoglobulin gene coding for half of a domain. *Immunochemistry* **12**, 873–876.

Bosma M.J., and Carroll A.M. (1991). The scid mouse mutant: definition, characterization, and potential uses. *Annu. Rev. Immunol.* **9**, 323–350.

Brodeur P.H. (1996). The Igh-V and Igk-V genes of the mouse. In: *Weir's Handbook of Experimental Immunology,* 5th ed. (Eds. Herzenberg L.A., Herzenberg L.A., Weir D.M., and Blackwell C.C.). Vol. 1, pp. 9.1–9.8. Blackwell Science, Cambridge, MA.

Brodeur P.H., Osman G.E., Mackle J.J., and Lalor T.M. (1988). The organization of the mouse *Igh-V* locus. Dispersion, interspersion, and the evolution of V_H gene family genes. *J. Exp. Med.* **168**, 2261–2278.

Brown W.R., Rabbani H., Butler J.E., and Hammarström L. (1997). Characterization of the bovine $C\alpha$ gene. *Immunology* **91**, 1–6.

Brüggemann M. (1988). Evolution of the rat immunoglobulin gamma heavy-chain gene family. *Gene* 74, 473–482.

Brüggemann M., Free J., Diamond A., Howard J., Cobbold S., and Waldmann H. (1986). Immunoglobulin heavy chain locus of the rat: striking homology to mouse antibody genes. *Proc. Natl. Acad. Sci. USA* **83**, 6075–6079.

Butler J.E. (1996). The swine Ig heavy chain locus has a single J_H and no identifiable IgD. *Intern. Immunol.* **8**, 1897–1904.

Butler J.E. (1997). Imunoglobulin geen organization and the mechanism of repertoire development. *Scand. J. Immunol.* **45**, 455–462.

Butler J.E., and Brown W.R. (1994). The immunoglobulins and immunoglobulin genes of swine. *Vet. Immunol. Immunopathol.* **43**, 5–12.

Butler J.E., Sun J-s. and Navarro P. (1996). The swine lg heavy cahin locus has a single J_H and no identifiable IgD. *Intern. Immunol.* **8**, 1897–1904.

Cebra J., Brunhouse R., Cordle C., Daiss J., Fechheimer M., Ricardo M., Thunberg A., and Wolfe P.B. (1977). Isotypes of guinea pig antibodies: restricted expression and bases for interactions with other molecules. *Progr. Immunol.* 3, 269–277.

Chuchana P., Blancher A., Brockly F., Allexandre D., Lefranc G., and Lefranc M.-P. (1990). Definition of the human immunoglobulin variable lambda (IGLV) gene subgrouops. *Eur. J. Immunol.* **20**,1317–1325.

Clarkson C.A., Beale D., Coadwell J.W., and Symons D.B.A. (1993). Sequence of ovine Ig $\gamma 2$ constant region heavy chain cDNA and molecular modelling of ruminant IgG isotypes. *Mol. Immunol.* **30**, 1195–1204.

Coe J.E. (1978). Humoral immunity and serum proteins in the Syrian hamster. *Fed. Proc.* **37**, 2030–2031.

Coe J.E., Schell R.F., and Ross M.J. (1995). Immune response in the hamster: definition of a novel IgG not expressed in all hamster strains. *Immunology* **86**, 141–148.

Combriato G., and Klobeck H.-G. (1991). V_λ and J_λ–C_λ gene segments of the human immunoglobulin λ light chain locus are separated by 14 kb and rearrange by a deletion mechanism. *Eur. J. Immunol.* **21**, 1513–1522.

Cook G.P., Tomlinson I.M., Walter G., Riethman H., Carter N.P., Buluwela L., Winter G., and Rabbits T.H. (1994). A map of the human immunoglobulin V_H locus completed by analysis of the telomeric region of chromosome 14q. *Nature Genet.* 7, 162–168.

Davies J., and Riechmann L. (1996). Single antibody domains as small recognition units: design and *in vitro* antigen selection of camelized, human VH domains with improved protein stability. *Protein Engin.* **9**, 531–537.

Desmyter A., Transue T.R., Ghahroudi M.A., Thi M.-H.D., Poortmans F., Hamers R., Muyldermans S., and Wyns L. (1996). Crystal structure of a camel single-domain V_H antibody fragment in complex with lysozyme. *Nature Struct. Biol.* **3**, 803–811.

Dreyer W.J., and Bennett J.C. (1965). The molecular basis of antibody fromation: a paradox. *Proc. Natl. Ac. Sci. USA* **54**, 864–869.

Dreyfus D.H. (1992). Evidence suggesting an evolutionary relationship between transposable elements and immune system recombination sequences. *Mol. Immunol.* **29**, 807–810.

Dufour V., Malinge S., and Nau F. (1996). The sheep Ig variable region repertoire consists of a single V_H family. *J. Immunol.* **156**, 2163–2170.

Du Pasqueir L. (1993). Evolution of the immune system. In: *Fundamental Immunology*, 3rd ed. (Ed. Paul W.E.). pp. 199–233. Lippinch-Raven, Philadelphia.

Du Pasqueir L., and Chrétien I. (1996). CTX, a new lymphocyte receptor in *Xenopus*, and the early evolution of Ig domains. *Res. Immunol.* **147**, 217–226.

Du Pasqueir L., Schwager J., and Flajnik M.F. (1989). The immune system of *Xenopus*. *Annu. Rev. Immunol.* 7, 251–274.

Duvoisin R.M., Hayzer D.J., Belin D., and Jaton J.-C. (1988). A rabbit Ig λ L chain C region gene encoding C21 allotypes. *J. Immunol.* **141**, 1596–1601.

Early P., Huang H., Davis M., Calame K., and Hood L. (1980). An immunoglobulin heavy chain variable region gene is generated from three segments of DNA: VH, D and JH. *Cell* **19**, 981–992.

Edelman G. M. (1970). The covalent structure of a human γG-immunoglobulin. IX. Functional implications. *Biochemistry* **9**, 3197–3204.

Eisen H.N., and Reilly E.B. (1985). Lambda chains and genes in inbred mice. *Annu. Rev. Immunol.* **3**, 337–365.

Fellah J.S., Kerfourn F., Wiles M.V., Schwager J., and Charlemagne J. (1993). Phylogeny of immunoglobulin heavy chain deduced from cDNA sequence. *Immunogenetics* **38**, 311–317.

Frank M.B., Best R.M., Baverstock, and Gutman G.A. (1987). The structure and evolution of immunoglobulin kappa chain constant region genes in the genus *Rattus*. *Mol. Immunol.* **24**, 953–961.

Frippiat J.-P., Dard P., Marsh S., Winter G., and Lefranc M.-P. (1997). Immunoglobulin lambda light chain orphons on human chromosome 8q11.2. *Eur. J. Immunol.* **27**, 1260–1265.

Fong S., Hamill S.J., Proctor M., Freund S.M.V., Benian G.M.,Chothia C., Bycroft M., and Clarke J. (1996). Structure and stability of an immunoglobulin domain from twitchin, a muscle protein of the nematode *Caenorhabditis elegans*. *J. Mol. Biol.* **264**, 624–639.

Gall W.E., and Edelman G.M. (1970). The covalent structure of a human γG-immunoglobulin. X. Intrachain disulfide bonds. *Biochemistry* **9**, 3188–3196.

Gellert M. (1992). Molecular analysis of V(D)J recombination. *Annu. Rev. Genet.* **22**, 425–446.

Gellert M. (1994). DNA double-strand breaks and hairpins in V(D)J recombination. *Semin. Immunol.* **6**,125–130.

Ghaffari S.H., and Lobb C.J. (1993). Structure and genomic organization of immunoglobulin light chain in the channel catfish. *J. Immunol.* **151**, 6900–6912.

Ghaffari S.H., and Lobb C.J. (1997). Structure and genomic organization of a second class of immunoglobulin light chain genes in the channel catfish. *J. Immunol.* **159**, 250–258.

Givol D., Zakut R., Effron K., Rechavi G., Ram D., and Cohen J.B. (1981). Diversity of germ-line immunoglobulin V_H genes. *Nature* **292**, 426–430.

Grant C.K. (1995). Purification and characterization of feline IgM and IgA isotypes and three subclasses of IgG. In: *Feline Immunology and Immunodeficiency* (Eds. Willet B.J., and Jarrett O.). pp. 95–107. Oxford University Press, Oxford.

Greenberg A.S., Steiner L., Kasahara M., and Flajnik M.F (1993). Isolation of a shark immunoglobulin light chain cDNA clone encoding a protein resembling mammalian κ chain light chains: implications for the evolution of light chains. *Proc. Natl. Ac. Sci. USA* **90**, 10603–10607.

Greenberg A.S., Avila D., Hughes M., Hughes A., McKinney E.C., and Flajnik M.F. (1995). A new antigen receptor gene family that undergoes rearrangement and extensive somatic diversification in sharks. *Nature* **374**, 168–173.

Greenberg A.S, Hughes A.L., Guo J., Avila D., McKinney E.C., and Flajnik M.F. (1996). A novel "chimeric" antibody class in cartilaginous fish: IgM may not be the primordial immunoglobulin. *Eur. J. Immunol.* **26**, 1123–1129.

Grubb R.E. (1956). Agglutination of eruthrocytes coated with "incomplete" anti-Rh by certain rheumatoid arthritis sera and some other sera. The existence of human sera groups. *Acta Pathol. Microbiol. Scand.* **39**, 195–197.

Grubb R. (1994). Human immunoglobulin allotypes and mendelian poymorphisms of the human immunoglobulin genes. In: *Immunochemistry* (Eds. van Oss C.J., and van Regenmortel M.H.V.). pp. 47–68. Marcel Dekker, New York, 1992.

Gu Y., Jin S., Gao Y., Weaver D.T., and Alt F.W. (1997). Ku70–deficient embryonic stem cells have increased ionizing radiosensitivity, defective DNA end-binding activity, and inability to support V(D)J recombination. *Proc. Natl. Acad. Sci. USA* **94**, 8076–8081.

Gurvich A.E., and Nezlin R.S. (1965). DNA and biosynthesis of antibodies and gamma-globulins (in Russian). *Progr. Biol. Chem.* (Moscow) 7, 150–175.

Gutman G.A. (1996). Rat immunoglobulins. In: *Weir's Handbook of Experimental Immunology,* 5th ed. (Eds. Herzenberg L.A., Herzenberg L.A., Weir D.M., and Blackwell C.C.). Vol. 1, pp. 23.1–23.11. Blackwell Science, Cambridge, MA.

Gutman G.A., Loh E., and Hood L. (1975). Structure and regulation of immunoglobulins: kappa allotypes in the rat have multiple amino acid differences in the constant region. *Proc. Natl. Acad. Sci. USA* **72**, 5046–5050.

Haire R.N., Ota T., Rast J.P., Litman R.T., Cahn F.Y., Zon L.I., and Litman G.W. (1996). A third light chain gene isotype in *Xenopus laevis* consists of six distinct V_L families and is related to mammalian λ genes. *J. Immunol.* **157**, 1544–1550.

Hamers-Casterman C., Atarhouch T., Muyldermans S., Robinson G., Hamers C., Bajyana Songa E., Bendahman N., and Hamers R. (1993). Naturally occuring antibodies devoid of light chains. *Nature* **363**, 446–448

Harrelson A.L., and Goodman C.S. (1988). Growth cone guidance in insects: fasciclin II is a member of the immunoglobulin superfamily. *Science* **242**, 700–708.

Harriman W., Völk H., Defranoux N., and Wabl M. (1993). Immunoglobulin class switch recombination. *Annu. Rev. Immunol.* **11**, 361–384.

Hayzer D.J., and Jaton J.-C. (1989). Inactivation of rabbit immunoglobulin chain variable region genes by the insertion of short interspersed elements of the C family. *Eur. J. Immunol.* **19**, 1643–1648.

Herzenberg L.A., McDevitt H.O., and Herzenberg L.A. (1968). Genetics of antibodies. *Annu. Rev. Genet.* **2**, 209–244.

Higgins D.A., and Warr G.W. (1993). Duck immunoglobulins: structure, functions and molecular genetics. *Avian Pathol.* **22**, 211–236.

Hill R.L., Delaney R., Fellows R.E., and Lebovitz H.E. (1966). The evolutionary origins of the immunoglobulins. *Proc. Natl. Acad. Sci. USA* **56**, 1762–1769.

Hilschmann N., and Craig L.C. (1965). Amino acid sequence studies with Bence-Jones proteins. *Proc. Natl. Acad. Sci. USA* **53**, 1403–1409.

Hoek R.M., Smit A.B., Frings H., Vink J.M., de Jong-Brink M., and Geraerts W.P.M. (1996). A new Ig-superfamily member, molluscan defence molecule (MDM) from *Lymnea stagnalis,* is down regulated during parasitosis. *Eur. J. Immunol.* **26**, 939–944.

Hohman V.S., Schuchman D.B., Schluter S.F., and Marchalonis J.J. (1993). Genomic clone for sandbar shark λ chain: generation of diversity in the absence of gene rearrangement. *Proc. Natl. Acad. Sci. USA* **90**, 9882–9886.

Holmgren A., and Bränden C.-I. (1989). Crystal strcuture of chaperone protein PapD reveals an immunoglobulin fold. *Nature* **342**, 248–251.

Holmgren A., Kuehn M.J., Bränden C.-I., and Hultgren S.J. (1992). Conserved immunoglobulin-like features in a family of periplasmic pilus chaperones in bacteria. *EMBO J.* **11**, 1617–1622.

Home W.A., Ford J.E., and Gibson D.M. (1992). L chain isotype regulation in horse. I. Characterization of Ig λ genes. *J. Immunol.* **149**, 3927–3936.

Hsu E., Lefkovits I., Flajnik M., and Du Pasquier L. (1991). Light chain heterogeneity in the amphibian *Xenopus. Mol. Immunol.* **28**, 985–994.

Hsu E., and Steiner L.A. (1992). Primary structure of immunoglobulins through evolution. *Curr. Opin. Struct. Biol.* **2**, 422–431.

Hunkapiller T., and Hood L. (1989). Diversity of the immunoglobulin gene superfamily. *Adv. Immunol.* **44**, 1–63.

Ignatovich O., Tomlinson I.M., Jones P.T., and Winter G. (1997). The creation of diversity in the human immunoglonbulin V_λ repertoire. *J. Mol. Biol.* **268**, 69–77.

Johnson G., Kabat E.A., and Wu T.T (1996). Kabat database of sequences of proteins of immunological interest. In: *Weir's Handbook of Experimental Immunology,* 5th ed. (Eds. Herzenberg L.A.,

Herzenberg L.A., Weir D.M., and Blackwell C.C.). Vol. 1, pp. 6.1–6.21. Blackwell Science, Cambridge, MA.

Kabat E.A., Wu T.T., Perry H.M., Gottesman K.S., and Foeller C. (1991). *Sequences of Proteins of Immunlogical Interest,* 5th ed. Public Health Service, NIH, Washington, DC.

Kacskovics I., and Butler J.E. (1996). Heterogeneity of bovine IgG2—VIII. The complete cDNA sequence of bovine IgG2a (A2) and and IgG1. *Mol. Immunol.* **33**, 189–195.

Kacskovics I., Sun J., and Butler J.E. (1994). Five putative subclasses of swine IgG identified from the cDNA sequences of a single animal. *J. Immunol.* **153**, 3565–3573.

Kirschbaum T., Jaenichen R., and Zachau H.G. (1996). The mouse immunoglobulin κ locus contains about 140 variable gene segments. *Eur. J. Immunol.* **26**, 1613–1620.

Knight K.L., and Crane M.A. (1994). Generating the antibody reportoire in rabbit. *Adv. Immunol.* **56**, 179–218.

Knight K.L., Suter M., and Becker R.S. (1988). Genetic engineering of bovine Ig. Construction and characterization of hapten-binding bovine/murine chimeric IgE, IgA, IgG1, IgG2 and IgG3 molecules. *J. Immunol.* **140**, 3654–3659.

Knight K.L., and Tunyaplin C. (1995) Immunoglobulin heavy chain genes of rabbit. In: *Immunoglobulin genes,* 2nd ed. (Ed. Honjo T., and Alt F.W.). pp. 289–314. Academic Press, San Diego.

Kobayashi K., Tomonaga S., and Kajii T. (1984). A second class of immunoglobulin other than IgM present in the serum of a cartilageneous fish, the skate, *Raja kenojei:* isolation and characterization. *Mol. Immunol.* **21**, 397–404.

Kofler R., Geley S., Kofler H., and Helmberg A. (1992). Mouse variable-region gene families: complexity, polymorphism and use in non-autoimmune responses. *Immunol. Rev.* **128**, 5–21.

Lai E., Wilson R.K., and Hood L.E. (1989). Physical maps of the mouse and human immunoglobulin-like loci. *Adv. Immunol.* **46**, 1–58.

Lammers B., Beaman K.D., and Kim Y.B. (1991). Sequence analysis of porcine immunoglobulin light chain cDNAs. *Mol. Immunol.* **28**, 877–880.

Lanning D.k., and Knight K.L. (1997). Somatic hypermutation. Mutations 3′ of rabbit VDJ H-chain genes. *J. Immunol.* **159**, 4403–4407.

Lansford R., Okada A., Chen J., Oltz E.M., Blackwell T.K., Alt F.W., and Rathbun G. (1996). Mechanism and control of immunoglobulin gene rearrangement. In: *Molecular Immunology,* 2nd ed. (Eds. Hames B.D. and Glover D.M.). pp. 1–100. IRL Press, Oxford.

Lefranc M.-P., and Lefranc G. (1990). Molecular genetics of immunoglobulin allotype expression. In: *The human IgG Subclasses: Molecular Analysis of Structure, Function and Regulation* (Ed. Shakib F.). pp. 43–78. Pergamon Press, New York.

Leslie G.A., and Clem L.W. (1972). Phylogeny of immunoglobulin structure and function. VI. 17S, 7.5S and 5.7S anti-DNP of the turtle, *Pseudamys scripta. J. Immunol.* **106**, 1656–1664.

Lobb C.J., Olson M.O.J., and Clem L.W. (1984). Immunoglobulin light chain classes in a teleost fish. *J. Immunol.* **132**, 1917–1923.

Lobb C.J., and Olson M.O.J. (1988). Immunoglobulin heavy chain isotypes in a teleost fish. *J. Immunol.* **141**, 1236–1245.

Magor K.E., Higgins D.A., Middleton D.L., and Warr G.W. (1994). One gene encodes the heavy chains for three differemt forms of IgY in the duck. *J. Immunol.* **153**, 5549–5555.

Mansikka A. (1992). Chicken IgA H chain. Implications concerning the evolution of H chain genes. *J.Immunol.* **149**, 855–861.

Marchalonis J.J., Hohman V.S., and Schluter S.F. (1993).Antibodies of sharks. Novel methods of generation of diversity. *Immunologist* **1**, 115–120.

Matsuda F., and Honjo T. (1996). Organization of the human immunoglobulin heavy-chain locus. *Adv. Immunol.* **62**, 1–29.

Max E.E. (1993). Immunoglobulins. Molecular genetics. In: *Fundamental Immunology,* 3rd ed. (Ed. Paul W.E.). pp. 315–382, Lippencott-Raven, Philadelphia.

McBlane J.F., Van Gent D.C., Ramsden D.A., Romeo C., Cuomo C.A., Gellert M., and Oettinger M.A. (1995). Cleavage at a V(D)J recombination signal requires only RAG-1 and RAG-2 proteins and occurs in two steps. *Cell* **83**, 387–395.

McCormack W.T., and Thompson C.B. (1990). Somatic diversification of the chicken immunoglobulin light-chain gene. *Adv. Immunol.* **48**, 41–67.

Mikoryak C., and Steiner L. (1988). Amino acid sequence of the constant region of immunoglobulin light chains from Rana catesbeiana. *Mol. Immunol.* **25**, 695–703.

Milstein C., Even J., Jarvis J.M., Gonzalez-Fernandez A., and Gherardi E. (1992). Non-random features of the repertoire expressed by the members of one $V_\varkappa$ gene family and of the V-J recombination. *Eur. J. Immunol.* **22**, 1627–1634.

Milstein C., and Neuberger M.S. (1996). Maturation of the immune response. *Adv. Prot. Chem.* **49**, 451–485.

Milstein C., and Rada C. (1995). The maturation of the antibody response. In: *Immunoglobulin Genes,* 2nd ed. (Eds. Honjo T., and Alt F.W.). pp. 57–81. Academic Press, San Diego.

Muβmann R., Du Pasquier L., and Hsu E. (1996). Is *Xenopus* IgX an analog of IgA? *Eur. J. Immunol.* **26**, 2823–2830.

Muyldermans S., Atarhouch T., Saldanha J., Barbosa J.A.R.G., and Hamers R. (1994). Sequence and structure of V_H domain from naturally occuring camel heavy chain immunoglobulins lacking light chains. *Protein Eng.* 7, 1129–1135.

Navarro P., Barbis D.P., Antczak D., and Butler J.E. (1995). The complete cDNA and deduced amino acid sequence of equine IgE. *Mol. Immunol.* **32**, 1–8.

Neuberger M.S., and Milstein C. (1995). Somatic hypermutation. *Curr. Opin. Immunol.* 7, 248–254.

Nezlin R., Krylov M.Y., and Rokhlin O.V. (1973). Different suceptibility of rat IgG2 subclasses to trypsin. *Immunochemistry* **10**, 651–652.

Nezlin R., and Rokhlin O.V. (1976). Allotypes of light chains of rat immunoglobulins and their application to the study of antibody biosynthesis. *Contemp. Topics Mol. Immunol.* **5**, 161–184.

Niewold T.A., Murphy C.L., Weiss D.T., and Solomon A. (1996). Characterization of a light chain product of the human JCλ7 gene complex. *J. Immunol.* **157**, 4474–4477.

Nussenzweig V., and Benacerraf B. (1967). Synthesis, structure and specificity of 7S guinea pig immunoglobulins. In: *Gamma Globulins. Structure and Control of Biosynthesis* (Ed. Killander J.). 3rd Nobel Symposium. pp. 233–250. Almquist & Wiksell, Stockholm.

O'Donnell J., Frangione B., and Porter R.R. (1970). The disulphide bonds of the heavy chains of rabbit immunoglobulin G. *Biochem J.,* **116**, 261–270.

Oettinger M.A., Schatz D.G., Gorka C., and Baltimore D. (1990). RAG-1 and RAG-2, adjacent genes that synergistically activate V(D)J recombination. *Science* **248**, 1517–1523.

Okada A., and Alt F.W. (1995). The variable region gene assembly mechanism. In: *Immunoglobulin Genes,* 2nd ed. (Eds. Honjo T., and Alt F.W.). pp. 205–234. Academic Press, San Diego.

Oudin J. (1956). L'allotypie de certains antigènes protéidiques du sérum. *C. R. Acad. Sci.* **242**,1–3.

Oudin J. (1960). Allotypy of rabbit serum proteins. I. Immunochemical analysis leading to the individualization of seven main allotypes. *J. Exp. Med.* **112**, 107–124.

Owens G.C., Edelman GM., and Cunningham B.A. (1987). Organization of the neural cell adhesion molecule (N-CAM) gene: alternative exon usage as the basis for different membrane-associated domains. *Proc. Natl. Acad. Sci. USA* **84**, 294–298.

Padlan E.A. (1997). Does base composition help predispose the complementarity-determining regions of antibodies to hypermutation? *Molec. Immunol.* 34, 765–770.

Parng C.-L., Hansal S., Goldsby R.A., and Osborne B.A. (1996). Gene conversion contributes to Ig light chain diversity in cattle. *J. Immunol.* **157**, 5478–5486.

Parvari R., Avivi A., Lentner F., Ziv E., Tel-Or S., Burstein Y., and Schecter I. (1988). Chicken immunoglobulin γ-heavy chains: limited VH gene repertoire, combinatorial diversification by D gene segmets and evolution of the heavy chain locus. *EMBO J.* 7, 739–744.

Pascual V., and Capra D.J. (1991). Human immunoglobulin heavy-chain variable region genes: organization, polymorphism, and expesiion. *Adv. Immunol.* **49**, 1–74.

Patri S., and Nau F. (1992). Isolation and sequence of a cDNA coding for the immunoglobulin μ chain of the sheep. *Mol. Immunol.* **29**, 829–836.

Peters A., and Storb U. (1996). Somatic hypermutation in immunoglobulin genes is linked to transcription initiation. *Immunity* **4**, 57–65.

Potter M., and Lieberman R. (1967). Genetics of immunoglobulins in the mouse. *Adv. Immunol.* **7**, 91–145.

Pulido D., Campuzano S., Koda T., Modolell J., and Barbacid (1992). *Dtrk*, a *Drosophila* gene related to the trk family of neurotropin receptors, encodes a novel class of neural cell adhesion molecule. *EMBO J.*, **11**, 391–404.

Ramsden D.A., Paull T.T., and Gellert M. (1997). Cell-free *V(D)J* recombination. *Nature* **388**, 488–491.

Rast J.P., Anderson M.K., Ota T., Litman R., Margittai M., Shamblott M.J., and Litman G.W. (1994). Immunoglobulin light chain multiplicity and alternative organizational forms in early vertebrate phylogeny. *Immunogenetics* **40**, 83–99.

Rast J.P., Anderson M.K., and Litman G.W. (1995). The structure and organization of immunoglobulin genes in lower vertebrates. In: *Immunoglobulin Genes,* 2nd ed. (Eds. Honjo T., and Alt F.W.). pp. 315–341. Academic Press, San Diego.

Ratcliffe M.J.H. (1996). Chicken immunoglobulin isotypes and allotypes. In: *Weir's Handbook of Experimental Immunology,* 5th ed. (Eds. Herzenberg L.A., Herzenberg L.A., Weir D.M., and Blackwell C.C.). Vol. 1, pp. 24.1–24.15. Blackwell Science, Cambridge, MA.

Reynaud C.-A., Dahan A., and Weill J.-C. (1983). Complete sequence of a chicken λ light chain immunoglobulin derived from the nucleotide sequence of its mRNA. *Proc. Natl. Acad. Sci. USA* **80**, 4099–4103.

Reynaud C.-A., Garcia C., Hein W.R., and Weill J.-C. (1995). Hypermutation generating the sheep immunogloblin repertoire is an antigen-independent process. *Cell* **80**, 115–125.

Riechmann L. (1996). Rearrangement of the former VL interface in the solution structure of a camelised, single antibody VH domain. *J. Mol. Biol.* **259**, 957–969.

Rokhlin O.V., and Nezlin R.S. (1974). RL allotypes of light chains of rat immunoglobulins: ratio in heterozygous rats and distribution among inbred and random-bred rats. *Scand. J. Immunol.* **3**, 209–214.

Rousseaux J., Biserte G., and Bazin H. (1980). The differential enzyme sensitivity of rat immunoglobulin G subclasses to papain and pepsin. *Mol. Immunol.* **17**, 469–482.

Roux K.H., and Mage R.G (1996). Rabbit immunoglobulin allotypes. In: *Weir's Handbook of Experimental Immunology,* 5th ed. (Eds. Herzenberg L.A., Herzenberg L.A., Weir D.M., and Blackwell C.C.). Vol. 1, pp. 26.1–26.17. Blackwell Science, Cambridge, MA.

Saini S.S., Hein W.R., and Kaushik A. (1997). A single predominantly expressed polymorphic immunoglobulin V_{H} gene family, related to mammalian group I, clan II, is identified in cattle. *Mol. Immunol.* **34**, 641–651.

Sakano H., Huppi P., Heinrich G., and Tonegawa S. (1979). Sequences at the somatic recombination sites of immunoglobulin light chain genes. *Nature* **280**, 288–294.

Sanz I. (1991). Multiple mechanisms participate in the generation of diversity of human H chain CDR3 regions. *J. Immunol.* **147**, 1720–1729.

Schwager J., Mikoryak C.A., and Steiner L.A. (1988). Amino acid sequence of heavy chain from *Xenopus laevis* IgM deduced from cDNA sequence: implications for evolution of immunoglobulin domains. *Proc. Natl. Acad. Sci. USA* **85**, 2245–2249.

Schwedler von U., Jäck H.-M., and Wabl M. (1990). Circular DNA is a product of the immunoglobulin class switch rearrangement. *Nature* **345**, 452–456.

Seeger M.A., Haffley L., and Kaufman T.C. (1988). Characterization of *amalgam:* a member of the immunoglobulin superfamly from Drosophila. *Cell* **55**, 589–600.

Selsing E., Durdik J., Moore M.W., and Persiani D.M. (1989). Immunoglobulin λ genes. In: *Immunoglobulin Genes* (Eds. Honjo T., and Alt F.W.). pp.111–122. Academic Press, San Diego.

Sers C., Kirsch K., Rothbächer U., Riethmüller G., and Johnson J.P. (1993). Genomic organization of the melanoma-associated glycoprotein MUC18: implications for the evolution of the immunoglobulin domains. *Proc. Natl. Acad. Sci. USA* **90**, 8514–8518.

Siebenkotten G., and Radbruch A. (1995). Towards a molecular understanding of immunoglobulin class switching. *Immunologist* **3**, 141–145.

Sinclair M.C., Gilchrist J., and Aitken R. (1995). Molecular characterization of bovine Vλ regions. *J. Immunol.* **155**, 3068–3078.

Sleckman B.P., Gorman J.R., and Alt F.W. (1996). Accessibility control of antigen-receptor variable-region gene assembly: role of *cis*-acting elements. *Annu. Rev. Immunol.* **14**, 459–481.

Sittisombut N. (1996). Human immunoglobulin isotypes and allotypes. In: *Weir's Handbook of Experimental Immunology*, 5th ed. (Eds. Herzenberg L.A., Herzenberg L.A., Weir D.M., and Blackwell C.C.). Vol. 1, pp. 25.1–25.9. Blackwell Science, Cambridge, MA.

Spinelli S., Frenken L., Bourgeois D., de Ron L., Bos W., Verrips T., Anguille C., Cambillau C., and Tegoni M. (1996). The crystal structure of a llama heavy chain variable domain. *Nature Struct. Biol.* **3**, 752–757.

Stall A.M. (1996). Mouse immunoglobulin allotypes. In: *Weir's Handbook of Experimental Immunology*, 5th ed. (Eds. Herzenberg, L.A., Herzenberg L.A., Weir D.M., and Blackwell C.C.). Vol. 1, pp. 27.1–27.16. Blackwell Science, Cambridge, MA.

Steen M.-L., Hellman L., and Pettersson U. (1987). The immunoglobulin lambda locus in rat consts of two Cλ genes and a single Vλ gene. *Gene* **55**, 75–84.

Stiernholm N.B., Verkoczy L.K., and Biernstein N.L. (1995). Rearrangement and expression of the human ψCλ6 gene segment results in a surface Ig receptor with a truncated light chain constant region. *J. Immunol.* **154**, 4583–4591.

Storb U. (1996). The molecular basis of somatic hypermutation of immunoglobulin genes. *Curr. Opin. Immunol.* **8**, 206–214.

Sun J., Kacskovics I., Brown W.R., and Butler J.E. (1994). Expressed swine V_H genes belong to a small V_H gene family homologous to human V_HIII. *J. Immunol.* **153**, 5618–5627.

Sun S.-C., Lindström I., Boman H.G., Faye I., and Schmidt O. (1990). Hemolin: an insect-immune protein belonging to the immunoglobulin superfamily. *Science* **250**, 1729–1732.

Symons D.B.A., Clarkson C.A., and Beale D. (1989). Structure of the bovine immunoglobulin constant region heavy chain gamma 1 and gamma 2 genes. *Mol. Immunol.* **26**, 841–850.

Thompson C.B. (1995). New insights into V(D)J recombination and its role in the evolution of the immune system. *Cell* **3**, 531–539.

Tomlinson I.M., Walter G., Jones G., Dear P.H., Sonnhammer E.L.L., and Winter G. (1996). The imprint of somatic hypermutation on the repertoire of human germline V genes. *J. Mol. Biol.* **256**, 813–817.

Tonegawa S. (1983). Somatic generation of antibody diversity. *Nature* **302**, 575–581.

Tse A.G.D., Barclay A.N., Watts A., and Williams A.F. (1985). A glycophospholipid tail at the carboxyl-terminus of the Thy-1 glycoprotein of neurons and thymocytes. *Science* **230**, 1003–1008.

Tuaillon N., Miller A.B., Longberg N., Tucker P.M., and Capra J.D. (1994). Biased utilization of D_{HQ52} and J_H4 gene segments in a human Ig transgenic minilocus is independent of antigenic selection. *J. Immunol.* **152**, 2912–2920.

Turchin A., and Hsu E. (1996). The generation of antibody diversity in the turtle. *J. Immunol.* **156**, 3797–3805.

Turner R.J., Ed. (1994). *Immunology: A Comparative Approach*. Wiley, Chichester, U.K.

Tutter A., and Riblet R. (1989). Conservation of an immunoglobulin variable-region gene family indicates a specific, noncoding function. *Proc. Natl. Acad. Sci. USA* **86**, 7460–7464.

Vaerman J-P. (1994). Phylogenetic aspects of mucosal immunoglobulins. In: *Handbook of Mucosal Immunology* (Ed. Ogra P.L. *et al.*). pp. 99–104. Academic Press, San Diego.

Vengerova T.I., Rokhlin O.V., and Nezlin R.S. (1972). Chemical differences between two allotypic variants of light chains of rat immunoglobulins. Peptide mapping and cyanogen bromide cleavage. *Immunochemistry* **9**, 1239–1245.

Wagner S.D., and Neuberger M.S. (1996). Somatic hypermutation of immunoglobulin genes. *Annu. Rev. Immunol.* **14**, 441–457.

Warr G.W. (1995). The immunoglobulin genes of fish. *Dev. Comp. Immunol.* **19**, 1–12.

Warr G.W., Magor K.E., and Higgins D.A. (1995). IgY: clues to the origins of modern antibodies. *Immunol. Today* **16**, 392–398.

Weill J.-C., and Reynaud C.-A. (1995). Generation of diversity by post-rearrangement diversification mechanisms: the chicken and the sheep antibody repertoires. In: *Immunoglobulin Genes*, 2nd ed. (Ed. Honjo T., and Alt. F.W.). pp. 267–288. Academic Press, San Diego.

Weir R.C., and Porter R.R. (1966). Comparison of the structure of the immunoglobulins from horse serum. *Biochem. J.* **100**, 63–68.

Williams A (1982). Surface molecules and cell interactions. *J. Theor. Biol.* **98**, 221–234.

Williams A , and Barclay A.N. (1988). The immunoglobulin superfamily—domains for cell surface recognition. *Annu. Rev. Immunol.* **6**, 381–405.

Williams A., and Gagnon J. (1982). Neuronal cell Thy-1 glycoprotein: homology with Ig. *Science* **216**, 696–703.

Williams S.C., Frippiat J.-P., Tomlinson I.M., Ignatovich O., Lefranc M.-P., and Winter G. (1996). Sequence and evolution of the human germline V_λ repertoire. *J. Mol. Biol.* **264**, 220–232.

Wilson M., Bengtén E., Miller N.W., Clem L.W., Du Pasquier L., and Warr G.W. (1997). A novel chimeric Ig heavy chain from a teleost fish shares similarities to IgD. *Proc. Natl. Acad. Sci. USA* **94**, 4593–4597.

Wilson M., Hsu E., Marcuz A., Courtet M., Du Pasquier L., and Steinberg C. (1992). What limits affinity maturation of antibodies in *Xenopus*—the rate of somatic mutation or the ability to select mutants? *EMBO J.* **11**, 4337–4347.

Wilson M.R., Marcuz A., van Ginkel F., Miller N.W., Clem L.W., Middleton D., and Warr G.W. (1990). The immunoglobulin M heavy chain constant region gene of the channel fish, *Ictalurus punctatus:* an unusual mRNA spice pattern produces the membrane form of the molecule. *Nucleic Acids Res.* **18**, 5227–5233.

Yang M., Becker A., Simons F.E.R., and Peng Z. (1995). Identification of a dog-IgD-like molecule by a monoclonal antibody. *Vet. Immunol. Immunopathol.* **47**, 215–224.

Yélamos J., Klix N., Goyenechea B., Lozano F., Chui Y.L., González-Fernández A., Pannell R., Neuberger M., and Milstein C. (1995). Targeting of non-Ig sequences in place of the V segment by somatic hypermutation. *Nature* **376**, 225–229.

Zachau H.G. (1995). The human immunoglobulin κ genes. In: *Immunoglobulin Genes*, 2nd ed. (Eds. Honjo T., and Alt F.). pp. 173–191. Academic Press. San Diego.

Zachau H.G. (1996). The human immunoglobulin κ genes. *Immunolgist* **4**, 49–54.

Zezza D.J., Mikoryak C., Schwager J., and Steiner L. (1991). Sequence of C region of L chains from *Xenopus laevis* Ig. *J. Immunol.* **146**, 4041–4047.

Zhang J., Alt F.W., and Honjo T. (1995). Regulation of class switch recombination of the immunoglobulin heavy chain genes. In: *Immunoglobulin Genes*, 2nd ed. (Eds. Honjo T., and Alt F.). pp. 235–264. Academic Press, San Diego.

CHAPTER 3

Engineering Antibody Molecules

I. PROBLEMS OF SERUM IMMUNOTHERAPY

Emil Behring and Shibasaburo Kitasato discovered in 1890 that the serum of animals immunized by tetanus or diphtheria toxins possesses antitoxic activity (Behring and Kitasato, 1890). Already in these first experiments, it was proved that the immune serum can be used to cure infected animals or to prevent infection. In the years that followed Behring successfully developed principles of specific serum therapy for patients using antitoxin prepared from immunized animals. As a result of these studies the first Nobel Prize for medicine was awarded in 1901 to E. Behring "for his work on serum therapy, especially its application against diphtheria" (*op. cit.* Kantha, 1991).

From the very beginning of the development of serum therapy, researchers realized that its side effect, so-called serum sickness, is due to the heterologous nature of animal immune sera. The first attempts to avoid this problem were to decrease the nonactive protein ingredients; for example, by acid proteolysis or by purification of the active component of the immune sera (i.e., antibodies). During World War II, a relatively simple method of human gammaglobulin purification by cold ethanol fractionation was developed (Cohn *et al.*, 1946). More recently, the production of immunoglobulin preparations for intravenous application was elaborated (Gronski *et al.*, 1991). Such preparations from pooled serum with the lowest possible anticomplementary activity and minimum aggregated proteins are widely used for treatment of primary or secondary immunodeficiencies, autoimmune diseases, and recurrent infections (Chapel, 1993; Rosen, 1993).

Although specific polyclonal sera were used for the treatment of particular infectious diseases and antitoxic therapy for a long time, new perspectives for specific immunotherapy appeared after the development by G. Köhler and C. Milstein of monoclonal antibody production using the hybridoma technique (Cambrosio and Keating, 1995). The rodent monoclonal antibodies are widely used against many specific targets, particularly in clinical oncology for target delivery, tumor imaging, and diagnosis and radiolabelled or toxin-conjugated tumor-specific monoclonal antibodies to cancer localization and therapy (Mach, 1995). However, the patient immune response to the injection of rodent monoclonal antibodies is the usual obstacle to their clinical usage.

The production of human monoclonal antibodies in hybridomas was not achieved until now. Therefore, various genetic engineering approaches were developed with the aim of diminishing immunogenic activity of rodent immunoglobulin molecules as well as creating molecules with new properties (Table 10). They are based on the expression of immunoglobulin or fused genes in different expression systems using mammalian or bacterial, yeast, insect and plant cells and special vectors. Each expression system has its advantages and limitations. Fully functional immunoglobulin molecules are formed only in mammalian cells, which have mechanisms for correct glycosylation and assembly of multichain molecules (Traunecker, 1997). Bacterial cells produce large amounts of functional Fab and single chain Fv (scFv) antibody fragments

TABLE 10 Methods of Antibody Engineering

Chimeric proteins
Humanization (CDR grafting)
Fv and single-chain Fv fragments
Phage-display libraries
Transgenic animals
Fusion proteins
Immunotoxins
Polymeric IgG molecules
Bispecific antibodies

(Skerra and Plückthun, 1988; Better *et al.*, 1988). However, bacterial cells are unable at all to add oligosaccharides to peptide chains. High level production of immunoglobulins was also achieved in insect cells with baculovirus expression vectors (Haseman and Capra, 1990; Putlitz *et al.*, 1990). The immunoglobulin molecules synthesized by insect and yeast cells are *N*-glycosylated but their glycosylation patterns are different from that of the antibodies synthesized by mouse cells. Antibodies were successfully produced in transformed tobacco leaves in high concentrations (Hiatt *et al.*, 1989). The yield of synthesized antibody ("plantibodies") may be as much as 1% of soluble plant proteins. However, the plant machinery for glycosylation is quite different from that of mammals and the plantibodies may have unusual properties.

II. CHIMERIC ANTIBODY MOLECULES AND HUMANIZATION

Several approaches are used to minimize immunogenicity of rodent monoclonal antibodies. One of them is the construction of **chimeric** molecules that have rodent variable regions or Fabs joined to constant regions of human immunoglobulins (Fig. 36). In the first experiments chimeric molecules were obtained that have mouse antihapten antibody activity with human effector functions like those of human IgM (Boulianne *et al.*, 1984) and IgE (Neuberger *et al.*, 1985). A number of chimeric antibodies, which are able to recognize cancer-specific cell antigens, were developed and clinical trials were performed (Mayforth, 1993; Kelley, 1996). As a rule, the chimeric antibodies are less immunogenic than their rodent counterparts and retain the effector functions. A good example is a study of Fab from the 7E3 monoclonal antibody against the platelet receptor that inhibits *in vivo* platelet thrombus formation (Knight *et al.*, 1995). A significant portion of patients exhibited an immune response after injections of the murine 7E3 Fab fragment. Even though the immune response was mainly directed against the 7E3 variable regions, the anti-7E3 immune reactions were dramatically reduced when the constant parts of Fab were replaced with human ones.

Immune response to administered antibodies could be further reduced by the so-called **humanization** procedure (complementary-determining region [CDR]-grafting,). According to this procedure, CDR parts of the murine or rat variable region taken from antibody molecules with known three-dimensional structure are transplanted on to the human heavy and light chain framework residues (Fig. 36). In the first study, CDRs from a rat antibody against a human lymphocyte antigen were introduced on to the human framework (Riechmann *et al.*, 1988). It already was clear in these experiments that the affinity of the reshaped antibodies depends also on residues that do not belong to CDR loops.

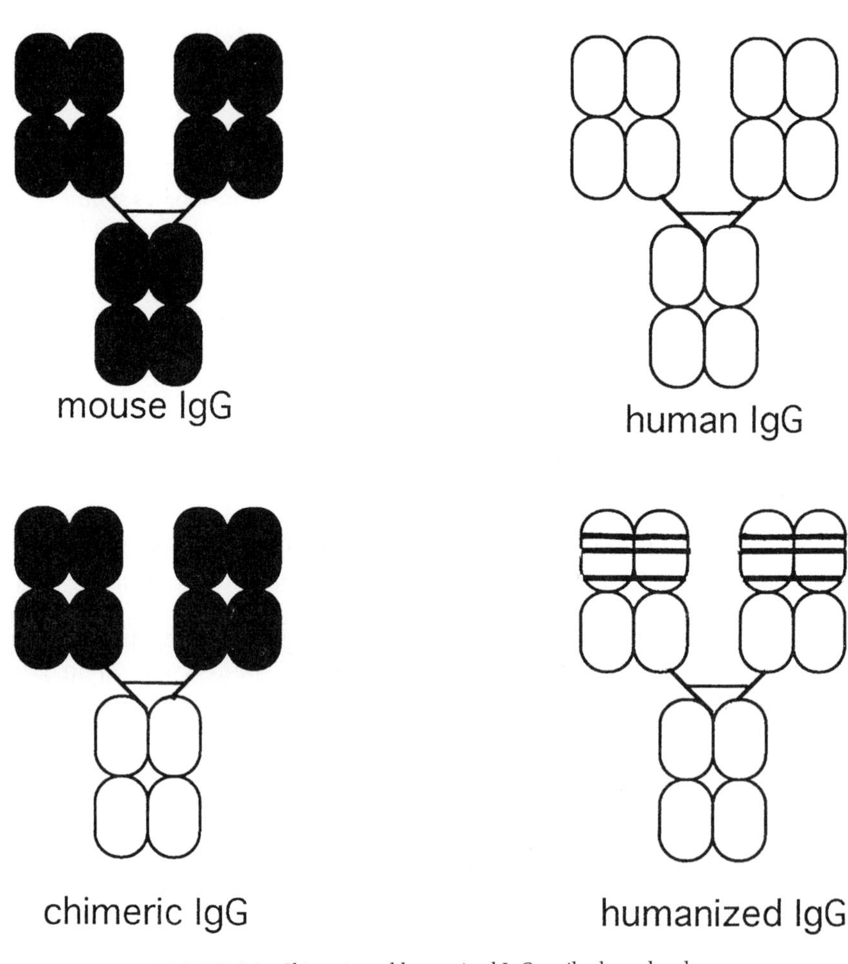

FIGURE 36 Chimeric and humanized IgG antibody molecules

A single change in the framework (Ser-27 → Phe) restores the ability of a reshaped human antibody to react with antigen more than 10-fold. By the later systematic studies, it was confirmed that substitutions in the β-sheet framework underlying the CDRs and not in contact with antigen directly can affect the conformation of hypervariable loops and the antibody affinity (Kettleborough *et al.*, 1991; Foote and Winter, 1992; Hawkins *et al.*, 1993; Xiang *et al.*, 1995). In most cases, it was found necessary to alter some human framework residues to murine in order to obtain original binding activity of the murine counterpart. Functional humanized Fab' fragments can be produced in bacterial cells in large amounts, up to 1–2 g/liter and used for preparation of bispecific $F(ab')_2$ (Carter *et al.*, 1992).

For reducing the immunogenicity of allogeneic variable domains, a procedure was suggested, according to which only the exposed residues in the framework regions are replaced (Padlan, 1991). By comparing murine and human Fv structures, it was found that only a few amino acid changes need to be made to convert the murine Fv surface to the human Fv surface because the surfaces of the variable regions are well conserved (Pedersen *et al.*, 1994). The "resurfaced" antibodies (Fig. 37) have identical antigen affinities to those of their parent murine antibodies (Roguska *et al.*, 1994). The experimental procedure for humanization of monoclonal antibodies (Güsow and Seemann, 1991; Emery and Harris, 1995) is still a long procedure and often yields antibody fragments with reduced affinity.

III. MINIMAL ANTIBODY FRAGMENT (Fv)

Another approach is the construction of the minimal Fv fragment with antigen-binding activity. There are only rare cases of the Fv fragment being isolated by proteolysis of the whole antibody molecule (Hochman *et al.*, 1973). If V_H and V_L fragments obtained as a result of the expression of antibody variable bacterial cells are recombined, the product, the Fv fragment, is unstable because the V_L and V_H units can easily dissociate. More useful are single-chain Fvs (scFvs), recombinant V_L and V_H fragments covalently tethered together by a polypeptide link and forming one polypeptide chain (Fig. 38) (Huston *et al.*, 1988). Polypeptides with the average length of 5–18 amino acids are usually used as links. They are rich in serine and glycine residues, to which introduce flexibility, and in charged glutamic acid and lysine residues, which improve solubility (Sharma and Rose, 1997). Such peptides lack an ordered three-dimensional structure and have probably reduced immunogenicity. For expression of Fv genes several system were effectively used including myeloma cells, insect, yeast and *Escherichia coli* cells (Huston *et al.*, 1995).

Due to small size, rapid clearance *in vivo,* stability, and easy engineering, scFvs have various potential applications in the diagnosis and treatment of diseases, particularly of cancer. In the first studies on the construction of scFvs, it was shown that scFvs have the same affinity and specificity for antigen as monoclonal antibodies, V genes of which were used for the engineering of Fvs (Huston *et al.*, 1988; Bird *et al.*, 1988). Since then dozens of scFvs with different specificities have been constructed (Raag and Whitlow, 1995). They are potentially useful in the imaging of tumors after radiolabeling and for genetic fusion to potent toxins (immunotoxins). Anti-idotypic single chain Fv fragments with the internal images of toxins can display a therapeutic activity against fungal pathogens (Magliani *et al.*, 1997). The monovalency of scFv is a disadvantage and attempts were made to create constructs with di- or multivalency with increasing combining efficiency (Cumber *et al.*, 1992; Kipriyanov *et al.*, 1996).

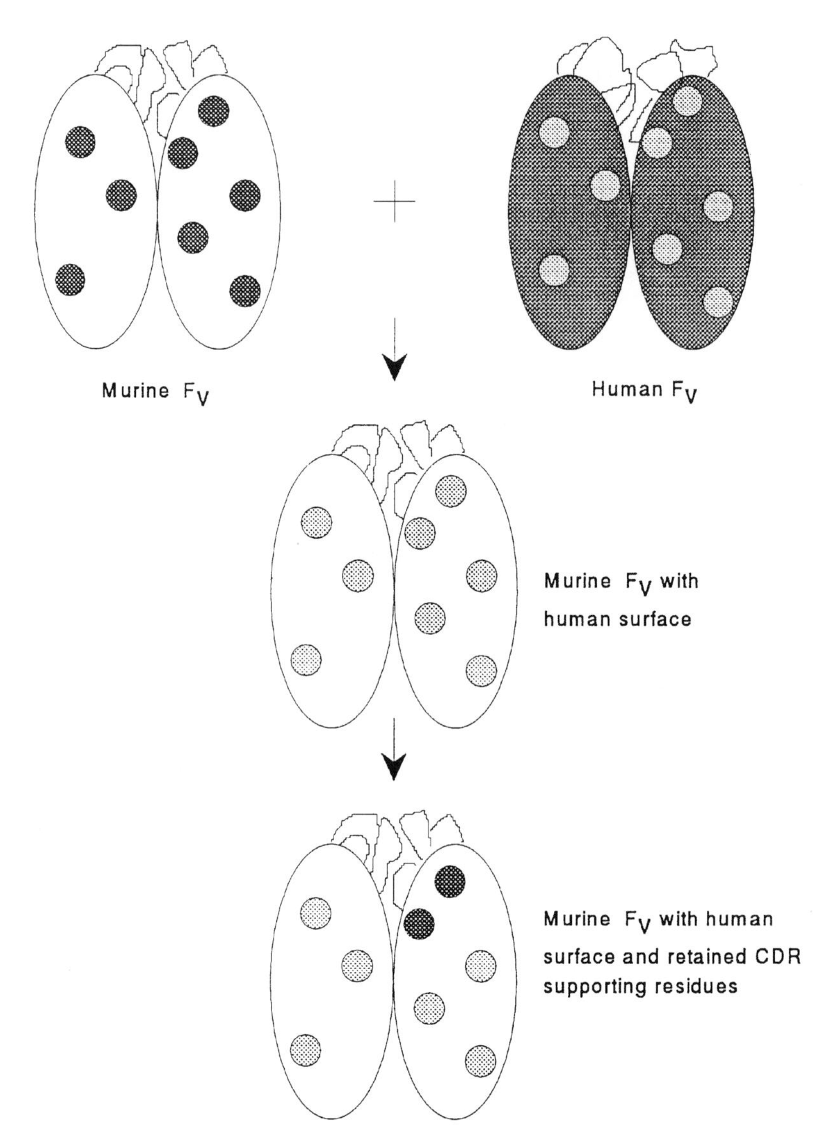

FIGURE 37 Humanization by resurfacing of murine Fv fragment. At first, the surface residues are changed from mouse (dark gray circles) to the closest human patterns (light gray circles). Later, surface residues, which perturb the CDRs according to modeling, are replaced with the original mouse residues to secure retention of antigen binding (Pedersen *et al.*, 1994. Reprinted with permission.)

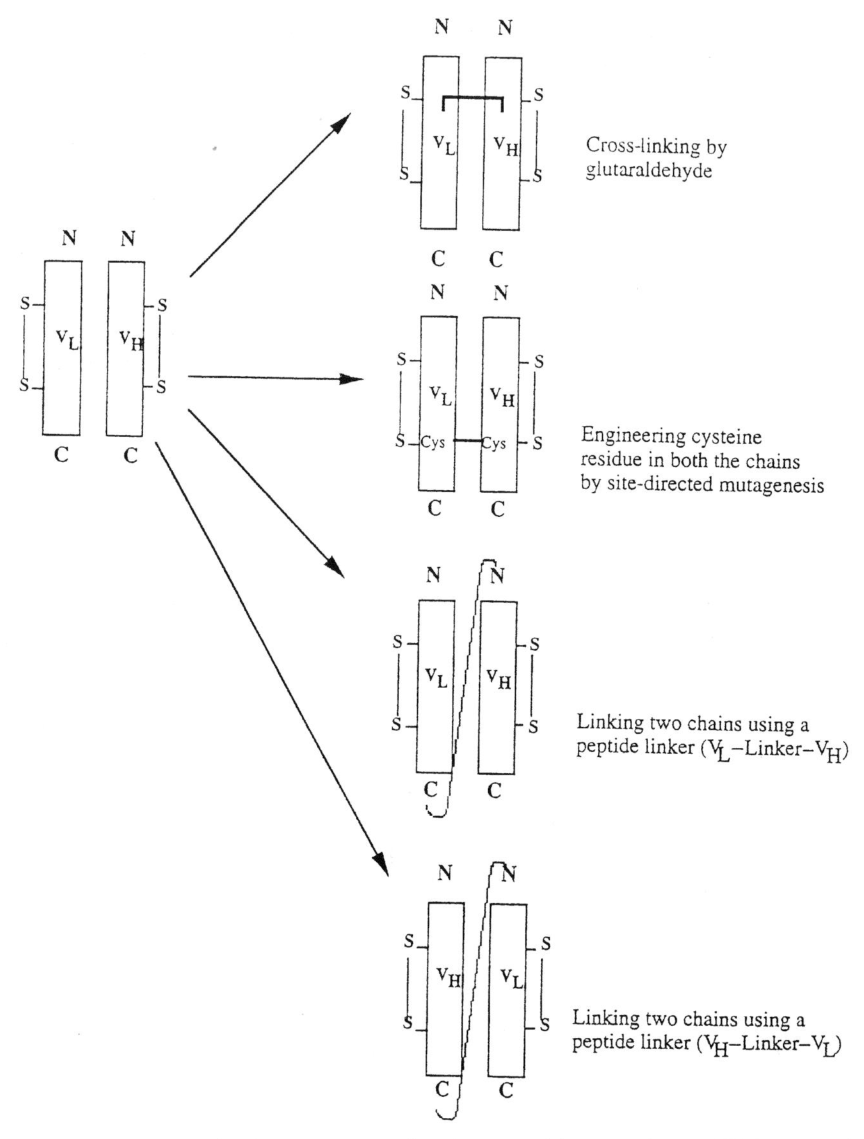

FIGURE 38 Variants for the construction of Fv fragments (Sharma and Rose, 1997. Reprinted with permission.)

IV. PRODUCTION OF HUMAN MONOCLONAL ANTIBODIES BY PHAGE-DISPLAY AND TRANSGENE TECHNOLOGIES

Each hybridoma cell produces only one pair of immunoglobulin chains. Methods were developed that in a relatively short time, allow expression of many variants of immunoglobulin chains in different combinations in bacterial cells. After it was shown that functional, properly folded and assembled Fab and scFv can be synthesized in *E. coli* (Skerra and Plückthun, 1988; Better *et al.*, 1988), expression of antibody chain libraries in bacterial cells was achieved (Sastry *et al.*, 1989; Ward *et al.*, 1989). To get antibody genes into bacterial cells, bacteriophages are chosen as cloning vectors. Antibody genes contain conserved sequences in the 5′ and 3′ portions of variable and constant region sequences. Polymerase chain reaction (PCR) allows specific amplification of antibody genes using primers directed to these conserved sequences and the construction of libraries of antibody heavy and light chain fragments (Ruiz *et al.*, 1997).

A. Phage-Display Technology

New perspectives for constructing human monoclonal antibodies and their derivatives were seen after phage-display technology was elaborated (McCafferty *et al.*, 1990; Kang *et al.*, 1991; Clackson *et al.*, 1991). Phage display is a selection technique, according to which an antibody fragment (scFv or Fab) is expressed on the surface of the filamentous phage fd. For this, the coding sequence of the antibody variable genes is fused with the gene that encoded the minor coat phage protein III (g3p) located at the end of the phage particle. The fused antibody fragments are displayed on the virion surface and particles with the fragments can be selected by adsorbtion on insolubilized antigen (panning) (Huse, 1995). The selected particles are used after elution to reinfect bacterial cells (Table 11). The repeated rounds of adsorbtion and infection lead to enrichment factors of more than a millionfold. The panning mimics the antigen-driven process in the course of immunization of animals (Winter *et al.*, 1994; Burton and Barbas, 1994). Bacterial proteases can cleave the bond between the g3p protein and antibody fragments, which results in the production of soluble antibody fragments by infected bacterial cells. To release the soluble Fabs and scFvs, an excision of the g3p gene is made or an amber stop codon between the antibody gene and the g3p gene is engineered (Hoogenboom *et al.*, 1991).

As a source of the genetic information for development of combinatorial libraries, different tissues containing antibody synthesizing cells are used. Bone marrow, spleen, lymph nodes, tonsils, and peripheral blood are the tissues of choice to isolate mRNA for construction of recombinant antibody libraries. Us-

TABLE 11 Cloning Monoclonal Fab From Combinatorial Libraries on the Surface of Filamentous Phage

mRNA from lymphoid cells
⇓
cDNA
(or semisynthetic antibody genes)
⇓
PCR
Amplification, library construction
⇓
Cloning H and L chains in phage(mid) vectors for display on phage surface
⇓
Infection of bacteria
⇓
Panning against antigen
⇓ ⇓
Nonbinders Antigen binders
⇓
Reinfection of bacteria
⇓
Panning against antigen
⇓
Conversion to soluble Fab expressing phagemid
⇓
Monoclonal soluble Fab

ing recombinant libraries, various human antibody fragments were isolated from immunized and nonimmunized donors. In a synthetic approach, some or all of the CDRs of V_H and V_L are randomized partially or completely using oligonucleotide-directed mutagenesis or the PCR method.

Variable genes of nonimmunized donors were effectively used for isolation of antibody fragments with various antigen-binding activities bypassing immunization. Peripheral blood lymphocytes served as the source of mRNA for preparation of natural scFv libraries. The scFv fragments were selected with binding activity against turkey lysozyme, and hapten 2-phenyloxazol-5-one (Marks *et al.*, 1991) and for a number of self-antigens, such as thyroglobulin, tumor necrosis factor-α, carcinoembrionic antigen, MUC1-peptide, CD4, protein p53, immunoglobulin-binding protein, elongation factor (EF)-1α, blood group antigens, and some others (Marks *et al.*, 1993; Nissim *et al.*, 1994). However, the affinities of these scFcs were not high and were similar to those of the antibodies synthesized during primary immune response.

To improve antibody affinity several approaches were suggested, such as the genetic optimization of the combining site using codon-based mutagenesis (Glaser *et al.*, 1995), mutations at random *in vitro* (Hawkins *et al.*, 1992; Gram *et al.*, 1992), and increasing the functional affinity (avidity) of scFcs by their polymerization. Other methods include constructing semisynthetic libraries (Rosenblum and Barbas, 1995) and shuffling heavy and light chains (Marks *et al.*, 1992). A method is suggested for selection of phage antibodies by their affinity for antigen or by their kinetics of dissociation from antigen (Hawkins *et al.*, 1992). A mutant antibody was selected by this method with a fourfold improved affinity for a hapten.

Synthetic or semisynthetic libraries have a potential diversity greater than that of animals because they can contain specificities absent in the actual antibody repertoires. Using the framework of Fab with antitetanus activity, a new specificity was created after the heavy chain CDR3 loop was substituted with 16 random residues (Barbas *et al.*, 1992a). From the library of this semysynthetic Fab, antifluorescein Fabs were obtained with an affinity of $K_d = 0.1\ \mu M$. The scFv fragments against two haptens were created after the heavy chain CDR3 of the variable genes from human naïve repertoire was substituted with oligonucleotides five or eight residues long (Hoogenboom and Winter, 1992).

The combinatorial library approach provides an opportunity to isolate antibodies, which is very difficult or impossible to obtain by convenient methods. The important group of antibodies isolated by phage combinatorial libraries is the group of antibodies to viruses. Antiviral antibodies that are contained in minute amounts in human gammaglobulin commercial preparations are very effective in prevention and treatment of a wide variety of virus infections. However, there is a constant need for preparations with a higher content of antiviral antibodies to use a lesser amount of gammaglobulins per dose and to increase the efficacy of the therapy. Human monoclonal Fab fragments, which possess binding activities against several virus antigens, including human immunodeficiency virus (HIV) and respiratory syncytial, hepatitis B, herpes simplex, and measles viruses and cytomegalovirus, were derived from combinatorial libraries (Barbas *et al.*, 1992b,c). The source of the genetic information for such libraries was usually donors who have high titers of antiviral antibodies (Parren and Burton, 1997). This method is especially effective for production of fragments against highly variable viruses such as HIV because the combinatorial approach allows one to obtain antibodies to several viral epitopes. Isolated antiviral antibody fragments retain binding as well as virus-neutralizing activities.

In the first phage-display experiments, antibody fragments with moderate affinities were isolated from repertoires of 10^8 immunoglobulin genes, which is close to the number of B lymphocytes in the mouse. However, the chances of finding antibodies with higher affinities from the larger libraries are greater. Indeed, from a highly diverse large synthetic repertoire (6.5×10^{10}), the Fab antibody fragments were isolated with binding affinities in the nanomolar range (Griffiths *et al.*, 1994; Hoogenboom, 1997). Binding specificities isolated from this library include antibody Fab fragments against several haptens, 14 foreign antigens, and 17 autoantigens. From the pool of V gene segments from 43 nonimmunized donors, a repertoire was constructed of 1.4×10^{10} scFv fragments (Vaughan *et al.*, 1996). All measured binding affinities show a $K_d$value of less than 10 nM (i.e., like the K_d values usually associated with antibodies from a secondary response). Some of antihapten antibodies have even subnanomolar binding affinities (0.3–0.8 nM). These experiments are proving that large phage libraries are a reliable source of human monoclonal antibody fragments with high affinity .

B. Transgenic Animals

Another approach for the preparation of human monoclonal antibodies is generation of mouse strains transgenic for human immunoglobulin genes (Brüggemann and Neuberger, 1996). Miniloci, yeast artificial chromosomes, or phage P1 vectors are used to transfer immunoglobulin genes. Miniloci transgenic constructs contain a limited number of variable genes of heavy and light chains, J segment clusters, and several D gene segments, as well as sequences with transcription-enhancer regulatory elements (Fig. 39). Much larger contiguous segments of human immunoglobulin loci can be incorporated in yeast artificial chromosomes. One of the biggest transgenes of this type contains megabase-sized fragments in nearly germline configuration, including approximately 66 $V_H$genes, 32 $V_\varkappa$ genes, and μ, δ, $\gamma 2$, and $C_\varkappa$ constant region genes (Mendez *et al.*, 1997).

After exogenous human transgenes are integrated into mouse germ line, human immunoglobulin genes are rearranged and produce a functional primary antibody repertoire. The transgenic animals can be crossed with mice carrying disrupted endogenous loci (heavy- and light-chain-knockout mice). Practically all B cells of such recombinant mice primarily produce fully human antibodies. The level of human immunoglobulins in serum is high (up to 1 mg/ml) and after immunization the level of human IgG increases up to 2.5 mg/ml (Mendez *et al.*, 1997).

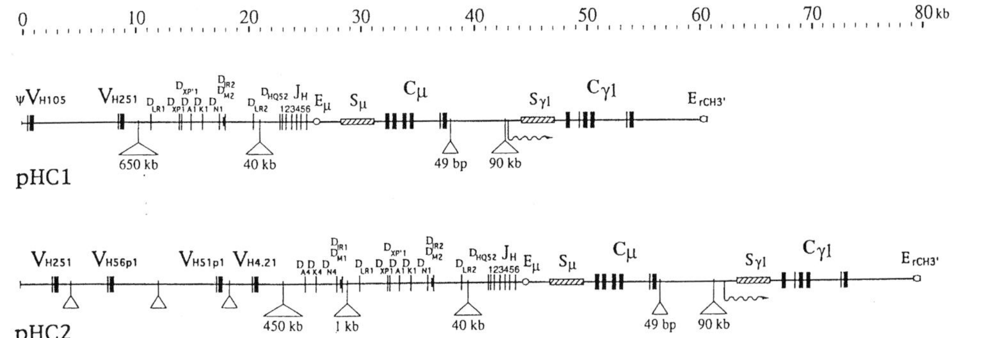

FIGURE 39 Human heavy chain minilocus transgenes expressed in mice. Open triangles indicate discontinuities between the structure of the transgene and the chromosomal structure of the intact heavy chain locus. S, switch region; E, enhancer. (Taylor *et al.*, 1994. Reprinted with permission from Oxford University.)

It was shown that the human transgenes undergo in such mice not only rearrangements, but also somatic mutation and class switching (Taylor *et al.*, 1994). The immunization of mice bearing human transgenes results in the production of a diverse human antibody repertoire (Green *et al.*, 1994; Wagner *et al.*, 1994). Stable hybridoma lines with good growth and secretion characteristics were established with B cells of the transgenic mice using conventional hybridoma technology. The avidity of human antibodies synthesized by such hybridomas can be very high. For example, the monoclonal anti-CD4 antibodies (IgG1$\varkappa$) had K_a values that range from 2.3×10^9 to 1.1×10^{10} M^{-1} and could inhibit a mixed leukocyte response (Fishwild *et al.*, 1996). In contrast to some other methods, the transgenic technology approach permits one to obtain the intact functional antibody molecules and not only their Fab or scFv fragments. The transgene technology could also be used to introduce other human loci participating in the immune response, such as the MHC complex or the T-cell receptor locus in the murine genome. Such transgenic animals would be invaluable for immunological studies.

V. ENGINEERING OF IMMUNOGLOBULINS WITH NOVEL PROPERTIES

To generate a molecule with combined properties of two different molecules, they can simply be linked chemically together (chemical fusion) or a part of one molecule can be inserted by genetic methods into the sequence of the second molecule (genetic fusion).

A. Fusion Proteins

Modern methods of genetic engineering are used successfully to construct fusion proteins, which have some antibody features as well as other characteristics that depend on added new sequences. These sequences can replace variable or constant antibody regions with subsequent loss either of the antigen-binding activity or effector functions. Various proteins were fused with immunoglobulin chains without the loss of biological activity (Fig. 40).

A group of fusion proteins was created by combination of the ligands for cell surface receptors with a human IgG3 molecule that has an extended hinge region (Sensel *et al.*, 1997). The fusion proteins retain the ability to combine both with an antigen and with respective cell receptors. The designed proteins were used to exploit cell receptors on the blood–brain barrier for delivery of antibody molecules to the brain. The native immunoglobulins cannot penetrate this

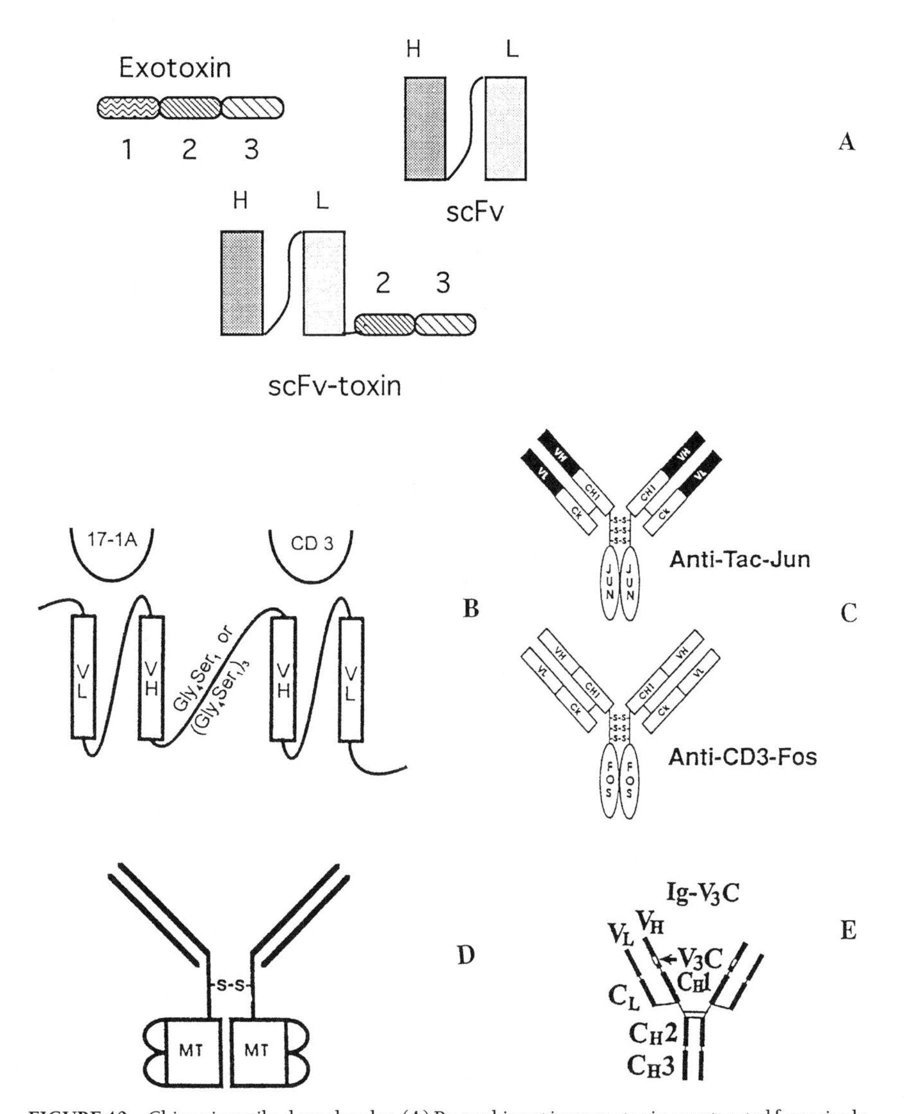

FIGURE 40 Chimeric antibody molecules. (A) Recombinant immunotoxin constructed from single chain Fv (scFv) fragment and inactive truncated *Pseudomonas* exotoxin. The N-terminal domain (1) mediating binding to mammalian cells is removed and the translocation (2) and toxic (3) domains are fused to scFv of a cancer-specific antibody. (B) Bispecific Fv fragment with the specificities directed to the CD3 antigen of T cells and to the epithelial 17–1A antigen. (Mack *et al.*, 1995.) (C) Chimeric antibody molecules used for the construction of a bispecific antibody that binds to receptor interleukin-2 and to CD3. (Kostelny *et al.*, 1992.) F(ab′)$_2$ fragments were fused with leucine zipper regions of the transcription factors Fos and Jun to form heterodimers. (D) Recombinant F(ab′)$_2$–metallothionein (MT) molecule. (Das *et al.*, 1992.) (E) Chimeric IgG molecule carrying a determinant from the V$_3$C loop of HIV-1. (Zaghouhani *et al.*, 1995.) A peptide of 19 amino acid residues was inserted into the variable region of the heavy chain. The chimeric IgG is able to react with anti-V$_3$C antibodies.

barrier but the IgG3–transferrin fusion protein was found in brain parenchyma after intravenous injection (Shin *et al.*, 1995). Such fusion proteins could be used for targeting within the brain or for the drug delivery into the brain. Another variant of drug-delivery molecules is a mouse-human himeric IgG3 antibody fused with hen avidin (Shin *et al.*, 1997). The fusion protein retains its antibody activity and coupled avidin molecules do not lose their ability to react with biotin. These avidin fusion antibodies can be used to deliver biotinylated ligands (drugs or peptides) to targets with antigenic properties specific for the antibodies used for fusion.

The antibody–cytokine fusion proteins can target cytokines to tumors and stimulate tumor lysis. An engineered protein consisting of an antiganglioside GD2 antibody and human interleukin-2 suppresses dissemination and growth of human neuroblastoma in an experimental model (Sabzevari *et al.*, 1994).

Several other examples also point to the potential usefulness of fusion proteins for medicine. The $F(ab')_2$ fragment fused with metallothionein, a metal-binding protein, could be effective for delivery of radioactive elements to tumor targets (Fig. 40D). (Das *et al.*, 1992). A special type of fusion proteins are the "antigenized" antibodies, which contain short foreign motifs in their variable regions (Fig. 40E). Viral or parasitic epitopes expressed in the heavy chain variable regions are highly immunogenic. They are recognized by T cells and are able to stimulate production of specific antibodies (Zaghouani *et al.*, 1995; Bona *et al.*, 1997). Their half-life *in vivo* is longer that that of the corresponding synthetic peptides.

A short motif Arg-Gly-Asp, which was found in adhesive proteins, is involved in cell adhesion, particularly in interactions of tumor cells with the extracellular matrix. The antibody molecules with the $(\text{Arg-Gly-Asp})_3$ sequence introduced in the CDR3 loop of the heavy chain were able to interact with cells of several tumor lines (Billeta *et al.*, 1997). Such fusion molecules are potential inhibitors of tumor metastasis.

B. Immunotoxins

Immunotoxins are molecules that can bind to specific target cells and selectively kill them. The specificity of the reaction with target cells is dependent on the antibody part and the cytotoxic effect is due to the toxin moiety (Vitetta and Uhr, 1985; Pastan *et al.*, 1992). Various toxins have been used for this purpose but *Pseudomonas* exotoxin A, plant toxin ricin, and diphtheria toxin are employed most ofen. As targets for immunotoxins, various surface antigens of tumor cells are applied, including receptors (e.g., interleukin and growth factor receptors), differential antigens, and some others.

The first immunotoxins were complexes of antibody molecules (or their Fab fragments) and purified toxins attached to each other by chemical linkages. The conjugates with whole IgG antibodies have long circulation times in the blood (typically 4–8 hours) and they have shown in clinical trials inhibitory activity against leukemias and lymphomas (Vitetta, 1994) as well as against solid tumors (Pai *et al.*, 1996).

More recent constructs are recombinant molecules made from single chain Fv antibody fragments and truncated toxins, which lack the cell-binding domain (Fig. 40A) (Pastan *et al.*, 1992; Brinkman, 1996). They penetrate cells of solid tumors more easily. However, small immunotoxins have short survival times (half-life up to 30 minutes). In some scFv fragments, the peptide link between V_L and V_H can interfere with binding and they are not always stable. Instead, a disulfide bond between residues in the conserved framework, which were mutated to cysteines, was used for the preparation of more stable Fvs (dsFvs) (Brinkman *et al.*, 1993).

For wider clinical usage of immunotoxins some important problems must be solved. The immunogenicity of immunotoxins, which is dependent mainly on their toxin part, is one of them. Another problem that has to be overcome is the toxicity of immunotoxins. Ricin-based immunotoxins can cause a capillary leak syndrome and immunotoxins with *Pseudomonas* exotoxin A can damage the liver.

VI. POLYMERIZATION OF IgG MOLECULES AND THEIR FRAGMENTS

The functional affinity (avidity) to antigen and to effector molecules of monomeric immunoglobulins is significantly less than that of polymeric IgM molecules (Karush, 1978). As one can expect, the polymerization of IgG antibody molecules results in increasing their functional affinity. The effect of the multivalency is particularly evident in the activation of the complement cascade. The ability of monomeric IgG molecules to activate complement is weak, but it increases significantly after aggregation due to the reaction with multivalent antigen. The chemical complexes of IgG are able to activate complement in the absence of antigen. The affinity of rabbit IgG chemical polymers to C1 and C1q complement components is dependent on the size of the polymers: dimers are less active than trimers and tetramers are most active (Wright *et al.*, 1980). Genetically engineered "tail-to-tail" dimers of human IgG1 are 200-fold more efficient at antibody-dependent complement-mediated cytolysis of antigen-bearing red blood cells (Shopes, 1992).

The fusion of the 18-amino acid tailpiece of μ chains (with penultimate cysteine) to the carboxy-termini of human γ chains results in production of polymeric IgG molecules by a mouse myeloma cell line (Smith *et al.*, 1995; Sensel *et al.*, 1997). The secreted IgG polymers are a heterogeneous population and are composed of oligomers of different sizes, up to six subunits per molecule. The J chain is not found in the IgG polymers. It is likely that for the incorporation of J chain some other determinants are required in addition to the tailpiece of μ chains. The complement activity of polymeric IgG1, IgG2, and IgG3 molecules is significantly enhanced as a result of polymerization. These polymers bind the C1q complement component even more effectively than IgM molecules. The IgG4 polymers have only a weak complement-binding activity. However, they direct complement-directed lysis of antigen-coated red blood cells as do other polymeric IgG isotypes. This result is unexpected because the IgG4 monomers are devoided of any complement activity. The capacity of IgG to bind to the Fc$\hat{\gamma}$RI high-affinity receptor does not change after polymerization. The interaction of IgG monomers with the Fc$\hat{\gamma}$RII low-affinity receptor is hard to detect but polymeric IgG of all four isotypes binds well to this receptor. Thus, the IgG polymers have a unique combination of features characteristic to both IgG and IgM molecules and could be useful for immunological studies as well as for therapy.^

The multimerization of antibody fragments to improve their stability and increase their functional affinity can be achieved either by chemical or by genetic methods. The chemical Fv dimers have a higher stability and prolonged serum half-life than a single Fv (Cumber *et al.*, 1992). The Fv tetramers were constructed using the genetic fusion of single chain Fv with a self-assembling polypeptide based on the tetramerization domain of the p53 transcription factor (Fig. 41) (Rheinnecker *et al.*, 1996). The Fv fragment was taken from an IgM monoclonal antibody against tumor-associated carbohydrate Lewis Y. The flexible hinge region of human IgG3 was inserted between scFv and the p53 protein. The flexibility of the complex permits the simultaneous reaction of all Fv fragments with cell surface antigens and the functional affinity of this self-assembled tetrameric scFv complex is much higher than that of Fv monomers or dimers. The main part of the scFv tetramers is of human origin, which reduces their immunogenic activity.

An original multivalent binding protein ("peptabody") was constructed by fusing a short peptide, a specific ligand for the B-cell lymphoma surface immunoglobulin idiotype, with two other peptides, a hinge region from camel IgG, and the coiled-coil assembly domain of the cartilage oligomeric matrix protein (Terskikh *et al.*, 1997). The peptabody protein expressed in *E. coli* is a stable homopentamer, which is bound to surface lymphoma idiotype with a high avidity (Fig. 42). It dissociates under denaturation and reducing conditions and reassociates as a pentamer with the initial binding activity that permits it to create hybrid molecules with two different specificities. Peptabodies can be displayed on the surface of filamentous phages and this technology could be used for rapid isolation of new pentabody molecules from random peptide libraries.

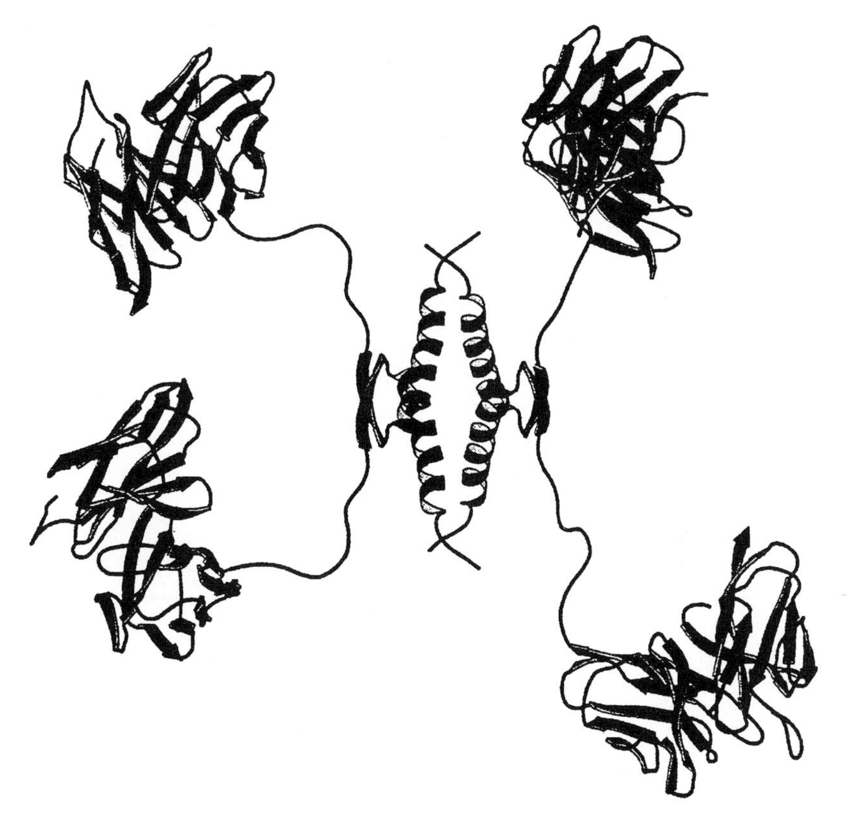

FIGURE 41 The molecular model of the tetrameric Fv construct.The variable domains (at the periphery) are connected by a linker, which is similar to the hinge region of IgG, to the tetrameization domain of transcription factor p53 (in the center). (Rheinnecker *et al.*, 1996. Reprinted with permission.)

VII. BISPECIFIC ANTIBODIES

Antibody molecules with two different antigenic specificities (bispecific antibodies) are useful tools for studies in cell biology and are a new type of immunotherapy agent (Fanger *et al.*, 1992; Plückthun and Pack, 1997). They may be prepared either by chemical methods or by biological techniques.

The first bispecific molecules have been obtained by mixing reduced $F(ab')_2$ fragments or IgG half-molecules from two different antibodies, followed by reoxidation of the hinge cysteines (Nisonoff *et al.*, 1975). Some of reconstituted bivalent fragments or IgG molecules possess both specificities of the parent

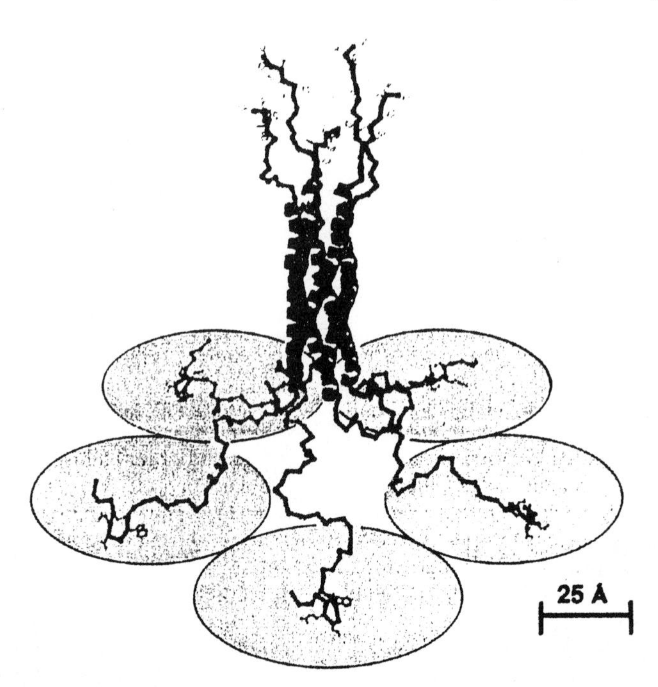

FIGURE 42 A model of three-dimensional structure of peptabody. The upper part of the structure shows six histidine residues at each C terminus. Five shaded circles (radius of 40 Å) under the structure schematically represent receptor molecules (Terskikh *et al.*, 1997. Reprinted with permission.)

antibody molecules. The SH-groups of one of the Fab′ partners can be modified by phenylenedimaleimide to provide maleimide groups. Then the second Fab′ with free SH-groups is added and the maleimide groups are crosslinked with the SH-groups, with the formation of the bispecific $F(ab')_2$ (Glennie *et al.*, 1987). The Fab′ fragments can also be linked by crosslinking agents (Brennan *et al.*, 1985; Snider, 1992; Wu *et al.*, 1993; Cook and Wood, 1994). For formation of bispecific $F(ab')_2$, leucine zipper peptides were used (Kostelny *et al.*, 1992). The peptides were derived from the Fos and Jun proteins and linked to the Fab′ portions of two different monoclonal antibodies, respectively, by gene fusion (Fig. 40C). As the zipper peptides preferentially form heterodimers, the Fab′–Jun and Fab′–Fos fragments dimerize with formation of bispecific $F(ab'\text{-}zipper)_2$ heterodimers.

Fusing two hybridoma cells results in hybrid hybridomas (quadromas) that form hybrid antibodies (Milstein and Cuello, 1983). This elegant biological method allows are to obtain bispecific antibodies without chemical treatments, which could lead to partial denaturation and a loss of antibody activity. One of the limitations of the method is the small amount of bispecific antibodies of desired specificity that can be obtained from quadromas and the relatively complex purification procedure. Despite this, quadromas were used to obtain different bispecific antibodies for many uses, including immunological studies and immunoassays (Takahashi *et al.*, 1991) .

Bispecific single chain Fv fragments (scFv) can be assembled from mono-scFvs with different specificities either by chemical procedures or at the gene level in bacterial (Gruber *et al.*, 1994) or mammalian (Mack *et al.*, 1995; Jost *et al.*, 1996) cells. Two scFvs are joined, usually by amino acid linkers (Fig. 40B). A special type of bivalent antibody scFv fragments is a diabody (Perisic *et al.*, 1994). Each scFv fragment of a diabody is formed by a V_H of one antibody and a V_L of the second one. The N-end of V_H and C-end of V_L are joined by a five-residue linker (Fig. 43). Such short linkers prevent the pairing of domains of the same chain and instead two chains form bivalent fragments. The antigen-combining sites are at opposite ends of a diabody molecule and they can react simultaneously with two antigens, even ones of large dimensions. Thousands of different diabody molecules can be generated and analyzed using phage-display technology (McGuinness *et al.*, 1996). The method can be applied to generation of human diabodies.

Bispecific antibodies are considered promising agents for immunotherapy and tools for immunological and cell biology studies (Fanger *et al.*, 1992; Holliger and Winter, 1993). One of the important applications of bispecific antibodies is recruiting cytotoxic cells for tumor cell lysis. To this end, bispecific antibodies are prepared with a specificity for T-cell antigens, in most cases for the CD3 complex, and for a tumor–associated antigen. Other types of effector cells, like monocytes and macrophages, possess Fc receptors and one of binding sites of bispecific antibodies can bear a specificity to these trigger molecules (Karpovsky *et al.*, 1984). Different surface molecules of cancer cells were used as target antigens, such as protein $p185^{HER2}$ of human breast tumor cells (Carter *et al.*, 1992), IgM expressed by mouse lymphoma cells (Brissink *et al.*, 1991; Demanet *et al.*, 1991), the glycoprotein 17–1A expressed by epithelial cells, particularly by colorectal cancer cells (Mack *et al.*, 1995), transferrin cell receptor (Jost *et al.*, 1996), or a hapten covalently linked to human cancer cells (Gruber *et al.*, 1994).

Bispecific antibodies were used for *in vivo* experiments and in clinical trials for treatment of cancer, as transporters of enzymes, toxins, and therapeutic drugs to specific targets like cancer cells, and for immunization as an adjuvant (Snider *et al.*, 1990). Several applications of bispecific antibodies for cell-target therapy were described, such as targeting of plasminogen to fibrin-containing thrombi and of immune complexes to cell receptors. Two other examples of bispecific antibodies have been recently described. One is a diabody with specificity to human C1q complement component and to lysozyme that is able to induce lysis of lysozyme–coated erythrocytes (Kontermann *et al.*, 1997). Other diabodies are directed against a target antigen and against conserved epitopes of

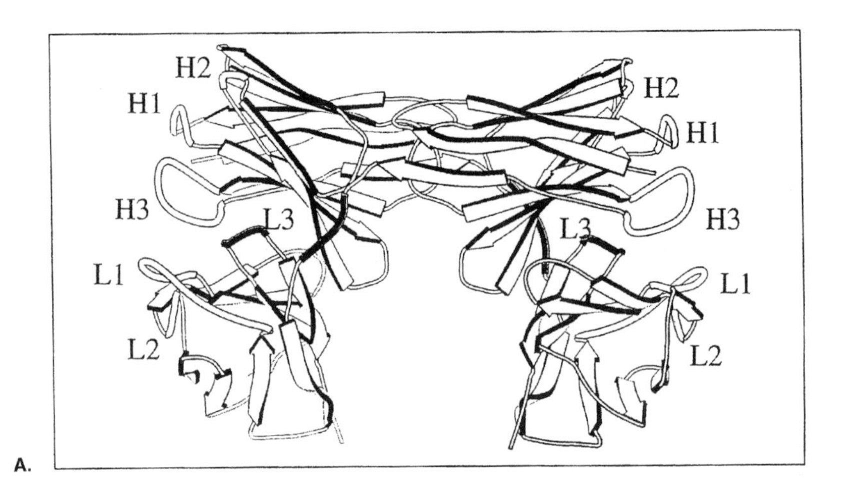

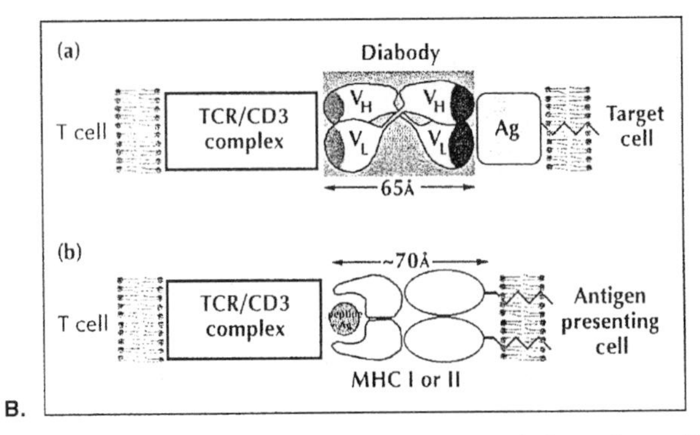

FIGURE 43 (A) Bivalent antibody molecule (diabody). (B) Crosslinking of a T cell and a target cell through a diabody (a) and through an MHC complex (b). (Perisic *et al.*, 1994. Reprinted with permission.)

immunoglobulin light chains. They are able to recruit different antibody effector functions by retargeting of serum immunoglobulins (Holliger *et al.*, 1997).

REFERENCES

Barbas C.F., Bain J.D., Hoekstra D.M., and Lerner R.A. (1992a). Semysynthetic combinatorial antibody libraries: a chemical solution to the diversity problem. *Proc. Natl. Ac. Sci. USA* **89**, 4457–4461.

Barbas C.F., Björling E., Chiodi F., Dunlop N., Cababa D., Jones T.M., Zebedee S.L., Persson M.A.A., Nara P.L., Norrby E., and Burton D.R. (1992b). Recombinant human Fab fragments neutralize human type 1 immunodeficiency virus in vitro. *Proc. Natl. Ac. Sci. USA* **89**, 9339–9343.

Barbas C.F., Crowe J.E., Cababa D., Jones T.M., Zebedee S.L., Murphy B.R., Chanock R.M., and Burton D.R. (1992c). Human monoclonal Fab fragments derived from a combinatorial library bind to respiratory syncytial virus F glycoprotein and neutralize infectivity. *Proc. Natl. Ac. Sci. USA* **89**, 10164–10168.

Behring E., and Kitasato S. (1890). Ueber das Zustandekommen der Diphtherie-Immunität und der Tetanus-Immunität bei Thiere. *Dtsch. Med. Wochenschr.* **49**, 1113–1114 (reproduced in *Mol. Immunol.*, 1991, **28**, 1319–1320).

Better M., Chang C.P., Robinson R.R., and Horwitz A.H. (1988). *Escherichia coli* secretion of an active chimeric antibody fragment. *Science* **240**, 1041–1043.

Billeta R., Lanza P., and Zanetti M. (1997). Ligand function of antigenized antibodies expressing the RGD motif. In: *Antibody Engineering* (Ed. Capra J.D.). pp.159–178. Karger, Basel.

Bird R.E., Hardman K.D., Jacobson J.W., Johnson S., Kaufman B.M., Lee S.-M., Lee T., Pope S.H., Riordan G.S., and Whitlow M. (1988). Single-chain antigen-binding proteins. *Science* **242**, 423–426.

Bona C.A., Bot A., and Brumeanu T.-D. (1997). Immunogenicity of viral epitopes expressed on genetically and enzymatically engineered immunoglobulins. In: *Antibody Engineering* (Ed. Capra J.D.). pp.179–206. Karger, Basel.

Boulianne G., Hozumi N., and Shulman M.J. (1984). Production of functional chimaeric mouse/human antibody. *Nature* **312**, 643–646.

Brennan M., Davison P.F., and Paulus H. (1985). Preparation of bispecific antibodies by chemical recombination of monoclonal immunoglobulin G_1 fragments. *Science* **229**, 81–83.

Brinkman U. (1996). Recombinant immunotoxins: protein engineering for cancer therapy. *Mol. Med. Today* **2**, 439–446.

Brinkman U., Reiter Y., Jung S.-H., Lee B., and Pastan I. (1993). A recombinant immunotoxin containing a disulfide-stabilized Fv fragment. *Proc. Natl. Ac. Sci. USA* **90**, 7538–7542.

Brissinck J., Demanet C., Moser M., Leo O., and Thielemans K. (1991). Treatment of mice bearing BCL_1 lymphoma with bispecific antibodies. *J. Immunol.* **147**, 4019–4026.

Brüggemann M., and Neuberger M.S. (1996). Strategies for expressing human antibody repertoires in transgenic mice. *Immunol. Today* **17**, 391–397.

Burton D.R., and Barbas C.F. (1994). Human antibodies from combinatorial libraries. *Adv. Immunol.* **57**, 191–280.

Cambrosio A., and Keating P. (1995). *Exquisite Specificity. The Monoclonal Antibody Revolution.* Oxford University Press, Oxford.

Carter P., Kelley R.F., Rodrigues M.L., Snedecor B., Covarrubias M., Velligan M.D., Wong W.L.T., Rowland A.M., Kotts C.E., Carver M.E., Yang M., Bourell J.H., Shepard H.M., and Henner D. (1992). High level *Escherichia coli* expression and production of a bivalent humanized antibody fragment. *Bio/Technology* **10**, 163–168.

Chapel H.M. (1993). Intravenous immunoglobulin: current concepts and application. In: *New Concepts in Immunodeficiency Diseases* (Eds. Gupta S, and Griscelli C.). pp. 457–474. Wiley, Chichester, England.

Clackson T., Hoogenboom H.R., Griffiths A., and Winter G. (1991). Making antibody fragments using phage display libraries. *Nature* **352**, 624–628.

Cohn E.J., Strong L.E., Hughes W.L., Mulford D.J., Ashworth J.N., Melin M., and Taylor H.L. (1946). Preparation and properties of serum and plasma proteins. IV. *J. Am. Chem. Soc.* **68**, 459–475.

Cook A.G., and Wood P.J. (1994). Chemical synthesis of bispecific monoclonal antibodies: potential advantages in immunoassay systems. *J. Immunol. Meth.* **171**, 227–237.

Cumber A.J., Ward E.S., Winter G., Parnell G.D., and Wawrzynczak E.J. (1992). Comparative stability in vitro and in vivo of a recombinant mouse antibody FvCys fragment and a bisFvCys conjugate. *J. Immunol.* **149**, 120–126.

Das C., Kulkarni P.V., Constantinescu A., Antich P., Blattner F.R., and Tucker P.W. (1992). Recombinant antibody-metallothionein: design and evaluation for radioimmunoimaging. *Proc. Natl. Acad. Sci. USA* **89**, 9749–9753.

Demanet C., Brissinck J., van Mechelen M., Leo O., and Thielemans K. (1991). Treatment of murine B cell lymphoma with bispecific monoclonal antibodies (anti-idiotype × anti-CD3). *J. Immunol.* **147**, 1091–1097.

Emery S.C., and Harris W.J. (1995). Strategies for humanizing antibodies. In: *Antibody Engineering*, 2nd ed. (Ed. Borrebaeck C.A.K.). pp. 159–183, Oxford University Press, Oxford.

Fanger M.W., Morganelli P.M., and Guyre P.M. (1992). Bispecific antibodies. *Crit. Rev. Immunol.* **12**, 101–124.

Fishwild D.M., O'Donnell S.L., Bengoechea T., Hudson D.V., Harding F., Bernhard S.L., Jones D., Kay R.M., Higgins K.M., Schramm S.R., and Lonberg N. (1996). High-avidity human IgGκ monoclonal antibodies from a novel strain of minilocus transgenic mice. *Nature Biotechnol.* **14**, 845–851.

Foote J., and Winter G. (1992). Antibody framework residues affecting the conformation of the hypervariable loops. *J. Mol. Biol.* **224**, 487–499.

Glaser S., Kristensson K., Chilton T., and Huse W. (1995). Engineering the antibody combining site by codon-based mutagenesis in a filamentous phage display system. In: *Antibody Engineering*. 2nd ed. (Ed. Borrebaeck C.A.K.). pp. 117–131. Oxford University Press, Oxford.

Glennie M.J., McBride H.M., Worth A.T., and Stevenson G.T. (1987). Preparation and performance of bispecific F(ab'γ)$_2$ antibody containing thioether-linked Fab'γ fragments. *J. Immunol.* **139**, 2367–2375.

Gram H., Marconi L.-A., Barbas C.F., Collet T.A., Lerner R.A., and Kang A.S. (1992). *In vitro* selection and affinity maturation of antibodies from a naive combinatorial immunoglobulin library. *Proc. Natl. Ac. Sci.USA* **89**, 3576–3580.

Green L.L., Hardy M.C., Maynard-Currie C.E., Tsuda H., Louie D.M., Mendez M.J., Abderrahim H., Noguchi M., Smith D.H., Zeng Y., David N.E., Sasai H., Garza D., Brenner D.G., Hales J.F., McGuinness R.P., Capon D.J,, Klapholz S., and Jakobovits A. (1994). Antigen-specific human monoclonal antibodies from mice engineered with human Ig heavy and light chain YACs. *Nature Genet.* 7, 13–21.

Griffiths A.D., Williams S.C., Hartley O., Tomlinson I.M., Waterhouse P., Crosby W.L., Kontermann R.E., Jones P.T., Low N.M., Allison T.J., Prospero T.M., Hoogenboom H.R., Nissim A., Cox J.P.L., Harrison J.L., Zaccolo M., Gherardi E., and Winter G. (1994). Isolation of high affinity human antibodies directly from large synthetic repertoires. *EMBO J.* **13**, 3245–3260.

Gronski P., Seiler F.R., and Schwick H.G. (1991). Discovery of antitoxins and development of antibody preparations for clinical uses from 1890 to 1990. *Mol.Immunol* **28**, 1321–1332.

Gruber M., Schodin B.A., Wilson E.R., and Kranz D.M. (1994). Efficient tumor cell lysis mediated by a bispecific single chain antibody expressed in *Escherichia coli*. *J. Immunol.* **152**, 5368–5374.

Güssow D., and Seemann G. (1991). Humanization of monoclonal antibodies. *Meth. Enzym.* **203**, 99–121.

Hasemann C., and Capra J.D. (1990). High-level production of a functional immunoglobulin heterodimer in a baculovirus expression system. *Proc. Natl. Acad. Sci. USA* **87**, 3942–3946.

Hawkins R.E., Russell S.J., and Winter G. (1992). Selection of phage antibodies by binding affinity. Mimicking affinity maturation. *J. Mol. Biol.* **226**, 889–896.

Hawkins R.E., Russell S.J., Baier M., and Winter G. (1993). The contribution of contact and non-contact residues of antibody in the affinity of binding to antigen. The interaction of mutant D1.3 antibodies with lysozyme. *J. Mol. Biol.* **234**, 958–964.

Hiatt A., Cafferkey R., and Bowdish K. (1989). Production of antibodies in transgenic plants. *Nature* **342**, 76–78.

Hochman J., Inbar D., and Givol D. (1973). An active antibody fragment (Fv) composed of the variable portions of heavy and light chains. *Biochemistry* **12**, 1130–1135.

Holliger P., and Winter G. (1993). Engineering bispecific antibodies. *Curr. Opin. Biotechnol.* **4**, 446–449.

Holliger P., Wing M., Pound J.D., Bohlen H., and Winter G. (1997). Retargeting serum immunoglobulin with bispecific diabodies. *Nature Biotechnol.* **15**, 632– 636.

Hoogenboom H.R. (1997). Designing and optimizing library selection strategies for generating high-affinity antibodies. *Trends Biotechnol.* **15**, 62–70.

Hoogenboom H.R., Griffits A.D., Johnson K.S., Chiswell D.J., Hudson P., and Winter G. (1991). Multi-subunit proteins on the surface of filamentous phage: methodologies for displaying antibody (Fab) heavy and lght chains. *Nucleic Acids Res.* **19**, 4133–4137.

Hoogenboom H.R., and Winter G. (1992). By-passing immunization. Human antibodies from synthetic repertoires of germline V_H gene segments rearranged *in vitro*. *J. Mol. Biol.* **227**, 381–388.

Huse W. (1995). Combinatorial antibody expression libraries in filamentous phage. In: *Antibody Engineering,* 2nd ed. (Ed. Borrebaeck C.A.K.). pp. 103–119, Oxford University Press, Oxford.

Huston J.S., Levinson D., Mudgett-Hunter M., Tai M.-S., Novotny J., Margolies M.N., Ridge R.J., Bruccoleri R.E., Haber E., Crea R., and Oppermann H. (1988). Protein engineering of antibody combining sites: recovery of specific activity in an anti-digoxin single-chain Fv analogue produced in *Escherichia coli*. *Proc. Natl. Acad. Sci. USA* **85**, 5879–5883.

Huston J.S., George A.J.T., Tai M.-S., McCartney J.E., Jin D., Segal D.M., Keck P., and Oppermann H. (1995). Single-chain Fv design and production by preparative folding. In: *Antibody Engineering,* 2nd ed. (Ed. Borrebaeck C.A.K.). pp. 185–223. Oxford University Press, Oxford.

Jost C.R., Titus J.A., Kurucz I., and Segal D.M. (1996). A single-chain bispecific Fv_2 molecule produced in mammalian cells redirects lysis by activated CTL. *Mol. Immunol.* **33**, 211–219.

Kang A.S., Barbas C.F., Janda K.D., Benkovic S.J., and Lerner R.A. (1991). Linkage of recognition and replication functions by assembling combinatorial antibody Fab libraries along phage surfaces. *Proc. Natl. Acad. Sci. USA* **88**, 4363–4366.

Kantha S.S. (1991). A centennial review: the 1890 tetanus antitoxin paper of von Behring and Kitasato and the related developments. *Keio J. Med.* **40**, 35–39.

Karpovsky B., Titus J.A., Stephany D.A., and Segal D.M. (1984). Production of target-specific effector cells using hetero-cross-linked aggregates containing anti-target cell and anti-Fcγ receptor antibodies. *J. Exp. Med.* **160**, 1686–1701.

Karush F. (1978). The affinity of antibody: range, variability, and the role of multivalence. *Compr. Immunol.* **5**,85–116.

Kelley R.F. (1996). Engineering therapeutic antibodies. In: *Proteins Engineering. Principles and Practice* (Eds. Cleland J.L., and Craik C.S.). pp. 399–434. Wiley-Liss, New York.

Kettleborough C.A., Saldanha J., Heath V.J., Morrison C.J., and Bending M.M. (1991). Humanization of a mouse monoclonal antibody by CDR-grafting: the importance of framework residues on loop conformation. *Protein Engin.* **4**, 773–783.

Kipriyanov S.M., Little M., Kropshofer H., Breitling F., Gotter S., and Dübel S. (1996). Affinity enhancement of a recombinant antibody: formation of complexes with multiple valency by a single-chain Fv fragment-core streptavidin fusion. *Protein Engin.* **9**, 203–211.

Knight D.M., Wagner C., Jordan R., McAleer M.F., DeRita R., Fass D.N., Coller B.S., Weisman H.F., and Ghrayeb J. (1995). The immunogenicity of the 7E3 murine monoclonal Fab antibody fragment variable region is dramatically reduced in humans by substitution of human for murine constant regions. *Mol. Immunol.* **32**, 1271–1281.

Kontermann R.E., Wing M.G., and Winter G. (1997). Complement recruitment using bispecific diabodies. *Nature Biotechnol.* **15**, 629–631.

Kostelny S.A., Cole M.S., and Tso J.Y. (1992). Formation of a bispecific antibody by the use of leucine zippers. *J. Immunol.* **148**, 1547–1553.

Mach J.-P. (1995). Monoclonal antibodies. In: *Oxford Textbook of Oncology* (Eds. Peckham M., Pinedo H., and Veronesi U.). Vol. 1, pp. 81–103. Oxford University Press, Oxford.

Mack M., Riethmüller G., and Kufer P. (1995). A small bispecific antibody construct expressed as a functional single-chain molecule with high tumor cell cytotoxicity. *Proc. Natl. Acad. Sci. USA* **92**, 7021–7025.

Magliani W., Conti S., De Bernardis F., Gerloni M., Bertoloti D., Mozzoni P., Cassone A., and Polonelli L. (1997). Therapeutic potential of antiidotypic single chain antibodies with yeast killer toxin activity. *Nature Biotechnol.*15, 155–158.

Marks J.D., Hoogenboom H.R., Bonnert T.P., McCafferty J., Griffiths A.D., and Winter G. (1991). By-passing immunization. Human antibodies from V-gene libraries displayed on phage. *J. Mol. Biol.* **222**, 581–597.

Marks J.D., Griffiths A.D., Malmquist M., Calckson T.P., Bye J.M., and Winter G. (1992). By-passing immunization. Building high affinity human antibodies by chain shuffling. *Bio/Technology* **10**, 779–783

Marks J.D., Ouwehand W.H., Bye J.M., Finnern R., Gorick B.D., Voak D., Thorpe S., Highes-Jones N.C., and Winter G. (1993). Human antibody fragments specific for human blood group antigens from a phage display library. *Bio/Technology* **11**, 1145–1149.

Mayforth R.D. (1993). *Designing Antibodies.* Academic Press, San Diego.

McCafferty J., Griffiths A.D., Winter G., and Chiswell D.J. (1990). Phage antibodies: filamentous phage displaying antibody variable domains. *Nature* **348**, 552–554.

McGuinnes B.T., Walter G., FitzGerald K., Schuler P., Mahoney W., Duncan A.R., and Hoogenboom H.R. (1996). Phage diabody repertoires for selection of large numbers of bispecific antibody fragments. *Nature Biotechn.* **14**, 1149–1154.

Mendez M. J., Green L.L., Corvalan J.R.F., Jia X.-C., Maynard-Currie C.E., Yang X.-d., Gallo M.L., Louie D.M., Lee D.V., Erickson K.L., Luna J., Roy C.M.-N., Abderrahim H., Kirschenbaum F., Noguchi M., Smith D.H., Fukushima A., Hales J.F., Finer M.H., Davis C.G., Zsebo K.M., and Jakobovits A. (1997). Functional transplant of megabase human immunoglobulin loci recapitulates human antibody response in mice. *Nature Genet.*, **15**, 146–156.

Milstein C., and Cuello A.C. (1983). Hybrid hybridomas and their use in immunohistochemistry. *Nature* **305**, 537–540.

Neuberger M.S., Williams G.T., Mitchell E.B, Jouhal S.S., Flanagan J. G., and Rabbits T.H. (1985). A hapten-specific chimaeric IgE antibody with human physiological effector function. *Nature* **314**, 268–270.

Nisonoff A., Hopper J.E., and Spring S.B. (1975). *The Antibody Molecule.* Academic Press, San Diego.

Nissim A., Hoogenboom H.R., Tomlinson I.M., Flynn G., Midgley C., Lane D., and Winter G. (1994). Antibody fragments from a 'single pot' phage display library as immunochemical reagents. *EMBO J.* **13**, 692–698.

Padlan E.A. (1991). A possible procedure for reducing the immunogenicity of antibody variable domains while preserving their ligand-binding properties. *Mol. Immunol.* **28**, 489–498.

Pai L.H., Wittes R., Setser A., Willingham M.C., and Pastan I. (1996). Treatment of advanced solid tumors with immunotoxin LMB-1: an antibody linked to *Pseudomonas* exotoxin. *Nature Med.*, **2**, 350–353.

Parren P.W.H.I., and Burton D.R. (1997) Antibodies against HIV-1 from phage display libraries: mapping of an immune response and progress towards antiviral immunotherapy. In: *Antibody Engineering* (Ed. Capra J.D.). pp.18–56. Karger, Basel.

Pastan I., Chaudhary V., and FitzGerald D.J. (1992). Recombinant toxins as novel therapeutic agents. *Annu. Rev. Biochem.* **61**, 331–354..

Pedersen J.T., Henry A.H., Searle S.J., Guild B.C., Roguska M., and Rees A.R. (1994). Comparison of surface accessible residues in human and murine Fv domains: implication for humanization of murine antibodies. *J. Mol. Biol.* **235**, 959–973.

Perisic O., Webb P.A., Holliger P., Winter G., and Williams R.L. (1994). Crystal structure of a diabody, a bivalent antibody fragment. *Structure* **2**, 1217–1226.

Plückthun A., and Pack P. (1997). New protein engineering approaches to multivalent and bispecific antibody fragments. *Immunotechnology* **3**, 83–105.

Putlitz J.zu, Kubasek W.L., Duchêne M., Marget M., von Specht B.-U., and Domday H. (1990). Antibody production in baculovirus infected insect cells. *Bio/Technology* **8**, 651–654.

Raag R., and Whitlow M. (1995). Single-chain Fvs. *FASEB J.* **9**, 73–80.

Rheinnecker M., Hardt C., Ilag L.L., Kufer P., Gruber R., Hoess A., Lupas A., Rottenberger C., Plückthun A., and Pack P. (1996). Multivalent antibody fragments with high functional affinity for a tumor-associated carbohydrate antigen. *J. Immunol.* **157**, 2989–2997.

Riechmannn L., Clark M., Waldmann H., and Winter G. (1988). Reshaping human antibodies for therapy. *Nature* **332**, 323–327.

Roguska M.A., Pedersen J.T., Keddy C.A., Henry A.H, Searle S.J., Lambert J.M., Goldmacher V.S., Blättler W.A., Rees A.R., and Guild B.C. (1994). Huamnization of murine monoclonal antibodies through variable domain resurfacing. *Proc. Natl. Acad. Sci. USA* **91**,969–973.

Rosen F. S. (1993). Putative mechanisms of the effect of intravenous γ-globulin. *Clin. Immunol. Immunopathol.* **67**, S41–S43.

Rosenblum J.S., and Barbas C.F. (1995). Synthetic antibodies. In: *Antibody Engineering,* 2nd ed. (Ed. Borrebaeck C.A.K.). pp. 89–116, Oxford University Press, Oxford.

Ruiz P., Haasner D., and Wiles M.V. (1997). Basic priniples of polymerase chain reaction (PCR) using limited amounts of starting material. In: *Immunology Methods Manual* (Ed. Lefkovits I.). pp. 285–305 Academic Press, San Diego.

Sabzevari H., Gillies S.D., Mueller B.M., Pancook J.D., and Reisfeld R.A. (1994). A recombinant antibody-interleukin 2 fusion protein suppresses growth of hepatic human neuroblastoma metastases in severe combined immunodeficiency mice. *Proc. Natl. Acad. Sci. USA* **91**, 9626–9630.

Sastry L., Alting-Mees M., Huse W.D., Short J.M, Sorge J.A., Hay B.N., Janda K.D., Benkovic S.J., and Lerner R.A. (1989). Cloning of the immunological repertoire in *Escherichia coli* for generetion of monoclonal catalytic antibodies: construction of a heavy chain variable region-specific cDNA library. *Proc. Natl. Acad. Sci. USA* **86**, 5728–5732.

Sensel M.G., Coloma M.J., Harvill E.T., Shin S.-U., Smith R.I.F., and Morrison S.L. (1997). Engineering novel antibody molecules. In: *Antibody Engineering* (Ed. Capra J.D.). pp.129–158. Karger, Basel.

Sharma S., and Rose D.R. (1997). Preparation, purification and crystallization of antibody Fabs and single-chain Fv domains. In: *Immunology Methods Manual* (Ed. Lefkovits I.). pp. 15–37. Academic Press, San Diego.

Shin S.-U., Friden P., Moran M., Olson T., Kang Y.-S., Pardridge W.M., and Morrison S.L. (1995). Transferrin-antibody fusion proteins are effective in brain targeting. *Proc. Natl. Acad. Sci. USA* **92**, 2820–2824.

Shin S.-U., Wu D., Ramanathan R., Pardridge W.M., and Morrison S.L. (1997). Functional and pharmacokinetic properties of antibody-avidin fusion proteins *J. Immunol.* **158**, 4797–4804.

Shopes B. (1992). A genetically engineered human IgG mutant with enhanced cytolytic activity. *J. Immunol.* **148**, 2918–2922.

Skerra A., and Plückthun A. (1988). Assembly of a functional immunoglobulin F_v fragment in *Escherichia coli. Science* **240**, 1038–1041.

Smith R.I.F., Coloma M.J., and Morrison S.L. (1995). Addition of a μ-tailpiece to IgG results in polymeic antibodies with enhanced effector functions including complement-mediated cytolysis by IgG4. *J. Immunol.* **154**, 2226–2236.

Snider D.P. (1992). Immunization with antigen bound to bispecific antibody induces antibody that is restricted in epitope specificity and contains antiidiotype. *J. Immunol.* **148**, 1163–1170.

Snider D.P., Kaubisch A., and Segal D.M. (1990). Enhanced antigen immunogenicity induced by bispecific antibodies. *J. Exp. Med.* **171**, 1957–1963

Takahashi M., Fuller S.A., and Winston S. (1991). Design and production of bispecific monoclonal antibodies by hybrid hybridomas for use in immunoassay. *Meth. Enzymol.* **203**, 312–327.

Taylor L.D., Carmack C.E., Huszar D., Higgins K.M., Mashayekh R., Sequar G., Schramm S.R., Kuo C.-C., O'Donnell S.L., Kay R.M., Woodhouse C.S., and Lonberg N. (1994). Human immunoglobulin transgenes undergo rearrangement, somatic mutation and class switching in mice that lack endogenous IgM. *Int. Immunol.* **6**, 579–591.

Terskikh A., Le Doussal J.-M., Crameri R., Fisch I., Mach J.-P., and Kajava A.V. (1997) "Peptabody": a new type of high avidity binding protein. *Proc. Natl. Acad. Sci.USA* **94**, 1663–1668.

Traunecker A. (1997). Expression of recombinant proteins in myeloma cell lines. In: *Immunology Methods Manual* (Ed. Lefkowits I.). Vol. 1, pp. 61–79. Academic Press, San Diego.

Vaughan T.J., Williams A.J., Pritchard K., Osbourn J.K., Pope A.R., Earnshaw J.C., McCafferty J., Hodits R.A., Wilton J., and Johnson K.S. (1996). Human antibodies with sub-nanomolar affinities isolated from a large non-immunized phage display library. *Nature Biotechnol.* **14**, 309–314.

Vitetta E.S. (1994). From the basic science of B cells to biological missiles at the bedside. *J. Immunol.* **153**, 1407–1420.

Vitetta E.S., and Uhr J.W. (1985). Immunotoxins. *Annu. Rev. Immunol.* **3**, 197–212.

Wagner S.D., Popov A.V., Davies S.L., Xian J., Neuberger M.S., and Brüggemann M. (1994). The diversity of antigen-specific monoclonal antibodies from transgenic mice bearing immunoglobulin gene transloci. *Eur. J. Immunol.* **24**, 2672–2681.

Ward E.S., Güssow D., Griffiths A.D., Jones P.T., and Winter G. (1989). Binding activities of a repertoire of single immunoglobulin variable domains secreted from *Escherichia coli*. *Nature* 341, 544–546.

Winter G., Griffiths A.D., Hawkins R.E., and Hoogenboom H.R. (1994). Making antibodies by phage display technology. *Annu Rev. Immunol.* **12**, 433–455.

Wright J.K., Tschopp J., Jaton J.-C., and Engel J. (1980). Dimeric, trimeric and tetrameric complexes of immunoglobulin G fix complement. *Biochem. J.***187**, 775–780.

Wu S., Sadegh-Nasseri S., and Ashwell J.D. (1993). Use of bispecific heteroconjugated antibodies (anti-T cell antigen receptor × anti-MHC class II) to study activation of T cells with a full length or truncated antigen receptor ε-chain. *J. Immunol.* **150**, 2211–2221.

Xiang J., Sha Y., Jia Z., Prasad L., and Delbaere L.T.J. (1995). Framework residues 71 and 93 of the chimeric B72.3 antibody are major determinants of the confirmation of heavy chain hypervariable loops. *J. Mol. Biol.* **253**, 385–390.

Zaghouani H., Anderson S.A., Sperber K.E., Daian C., Kennedy R.C., Mayer L., and Bona C. (1995). Induction of antibodies to the human immunodeficiency virus type 1 by immunization of baboons with immunoglobulin molecules carrying the principal neutralizing determinant of the envelope protein. *Proc. Natl. Acad. Sci.USA* **92**, 631–635.

PART II

Functional Aspects

CHAPTER 4

Antigen-Combining Site

I. GENERAL CHARACTERISTICS

Immunoglobulins can recognize a myriad of different antigen substances and form specific complexes with them through their antigen-combining sites. The reaction between antibody and antigen is basically a recognition process: dramatic conformational changes are usually not observed in both reactants as a result of the interaction, in contrast to enzyme-substrate reactions. The antibody–antigen complex formation was long ago considered the best model for studying the structural basis of the ability of protein molecules to recognize specifically various ligands, including other proteins.

In the past decades, it was recognized that complex genetic mechanisms are involved in the generation of information necessary for building antigen-combining sites. They include the rearrangement of three heavy (H) chain and two light (L) chain variable (V) gene segments, imperfect joining of these segments, and somatic mutations in the formed variable genes. The detailed structural knowledge is important for understanding how the genetic information is expressed on the level of the three-dimensional structure of specific binding sites in general. Using modern methods of genetic engineering it is possible to change the local structure of proteins and in this way to modify their specific activity, including the ability of antibodies to react with antigens. To this end, the knowledge of the architecture of the antigen-combining sites is of primary importance.

The most detailed information on the topology of antigen-combining sites was obtained by x-ray crystallographic studies, which were performed in many laboratories after the initial fundamental studies by the groups of David Davies, Alain Edmundson, and Roberto Poljak and their coworkers. These studies verified observations made through other methods, including nuclear magnetic resonance (NMR) and genetic manipulations.

The antigen-combining site is located on the tip of the antibody Fab portion and is composed of six complementarity-determining regions (CDRs) or hypervariable loops, three from the heavy chain (H1, H2, and H3), and three from the light chain (L1, L2, and L3). The loops are connected by relatively invariant stretches of the β-sheets (framework regions, FRs). Over the past years much of the data on the configuration of the antigen-combining site and on amino acid residues that participate in contacts with various antigens were established by x-ray crystallographic studies of antigen–antibody complexes (Davies *et al.*, 1990; Braden and Poljak, 1995; Padlan, 1996). Basic information on these studies is summarized in Tables, 12 to 14, which present separately the results of experiments with the complexes of the murine antibody Fab fragments with low molecular weight substances (Table 12), peptides, steroids, and oligosaccharides (Table 13), and protein antigens (Table 14).

Two important problems must be considered when the x-ray crystallographic data are discussed. The first one is related to the precision of structural x-ray models. Hypervariable loops where contact residues are concentrated are the most flexible parts of the variable domains and therefore determining their position is not an easy task. The second problem is related to the fact that individual atoms of protein molecules are in constant fluctuation in picosecond range. Loops and other large structural elements are also in constant movement on a nanosecond time-scale. Therefore, a model of a protein molecule built on the basis of x-ray crystallographic data provides only an average picture (Padlan, 1992).

A. Canonical Conformations of Hypervariable Loops

The hypervariable loops, with the possible exception of the most variable one, H3, have only a few main chain conformations, known as "canonical structures." These canonical structures are determined by the length of the loops and by the conserved amino acid residues. According to these criteria, kappa L1 and kappa

TABLE 12 Complexes of the Fab Fragments with Low Molecular Weight Substances Studied by X-Ray Crystallography

Fab from	Resolution (Å)	R value	Antigen	Buried Fab surface (Å²)	Fab contact residues	Reference
Mouse McPC603 IgA,ϰ	2.7	0.225	Phosphocholine	156	4 (H), 4 (L)	Satow *et al.*, 1986
Mouse 4-4-20 IgG2a,ϰ	2.7	0.215	Fluorescein	336	6 (H), 5 (L)	Herron *et al.*, 1989
Mouse BV04-01 IgG2b,ϰ	2.66	0.191	d(pT_3)	502	13 (total)	Herron *et al.*, 1991
Mouse AN02 IgG1,ϰ	2.9	0.195	DNP-spin label	334	2 (H), 5 (L)	Brünger *et al.*, 1991
Mouse NQ10/12.5 IgG1,ϰ	3.0	0.19	2-phenyl-oxasalone	230	7 (H), 7 (L)	Alzari *et al.*, 1990
Mouse 26-10 IgG2a,ϰ	2.5	0.174	Digoxin	383	8 (H), 2 (L)	Jeffrey *et al.*, 1993
Mouse 1F7 IgG1,ϰ	3.0		*Endo*-oxabicyclic dicarboxilic acid	265	7 (H), 1(L)	Haynes *et al.*, 1994
Mouse 40-50 IgG2b,ϰ	2.7		Oubain	411	4 (H), 1 (L)	Jeffrey *et al.*, 1995a
Mouse 17E8 IgG2b,ϰ	2.5	0.199	Norleucine phosphonate		4 (H), 3 (L)	Zhou *et al.*, 1994
Mouse NC6.8 IgG2b,ϰ	2.2	0.214	Trisubstituted guanidine	219–260	7 (H), 5 (L)	Guddat *et al.*, 1994
Mouse N1G9 IgG1,λ	2.4	0.196	(4-hydroxy-3- -nitrophenyl) acetate	184	7 (H), 3 (L)	Mizutani *et al.*, 1995
Mouse 88C6/12 IgG1,λ	3.0	0.190	4-hydroxy-3-nitrophenyl-ε-n-caproic acid	191	10 (H), 3 (L)	Yuhasz *et al.*, 1995
Mouse 1F7 IgG1,ϰ	3.0	0.220	*Endo*-oxabicyclic dicarboxilic acid	265	7 (H), 1 (L)	Haynes *et al.*, 1994

TABLE 13 Complexes of Murine Fab with Peptides, Steroids, and Oligosaccharides Studied by X-Ray Crystallography

Fab from	Resolution (Å)	R value	Complex with	Buried Fab surface (Å²)	Fab contact residues	Reference
B1312 IgG1,ϰ	2.8	0.220	Peptide from myohemerythrin, 12 residues	540	12 (H), 5 (L)	Stanfield *et al.*, 1990
Mab 131 IgG1,ϰ	3.0	0.250	Angiotensin	725	7 (H), 7 (L)	Garcia *et al.*, 1992
TE 33	3.0	0.386	Peptide from cholera toxin	545	10 (H), 9 (L)	Shoham *et al.*, 1991
17/9,IgG2a,ϰ	2.9 and 3.1	0.200 and 0.220	Peptide from IVN, 9 residues	468	11 (H), 6 (L)	Rini *et al.*, 1992
50.1, IgG2a	2.8	0.180	Peptide from HIV-1 16 residues	530	11 (H), 8 (L)	Rini *et al.*, 1993
C3	3.0		Peptide from poliovirus type 1, 17 residues	468		Wein *et al.*, 1995
26/9 IgG2a,ϰ	2.8	0.230	Peptide from IVN, epitope-6 residues	481	12 (H), 13 (L)	Churchill *et al.*, 1994
8F5 IgG2a,ϰ	2.5	0.180	Peptide from capsid of rhinovirus, 15 residues	945	14 (H), 12 (L)	Tormo *et al.*, 1994
R454511 IgG1,ϰ	2.65	0.185	Cyclosporin A	577	12 (H), 4 (L)	Altschuh *et al.*, 1992
DB3 IgG1,ϰ	2.7	0.210	Progesteron and its analogs	270–353	11 (H), 4–7 (L)	Arevalo *et al.*, 1993,a,b; 1994
BR96 IgG3,ϰ and himeric human IgG1ϰ	2.8 and 2.6	0.197 and 0.238	Nanoate methyl ester derivative of Lewis Y	422	12 (H), 2 (L) and 11 (H), 3 (L)	Jeffrey *et al.*, 1995b
Se155-4 IgG1λ	2.05	0.185	Branched dodescasaccharide from Sallmonella	304	8 (H), 4 (L)	Cygler *et al.*, 1991

IVN, influenza virus hemaggluinin.

TABLE 14 Complexes of Murine Fab and Fv with Protein Antigens Studied by X-Ray Crystallography

Antibody	Resolution (Å)	R value	Complex with	Buried Fab surface (Å^2)	Fab contact residues	Reference
D 1.3 IgG1κ, Fab	2.5	0.184	hen lysozyme	680	7 (H), 6 (L)	Fischman *et al.*, 1991
HyHEL-10 IgG1κ, Fab	3.0	0.246	hen lysozyme	720	12 (H), 8 (L)	Padlan *et al.*, 1989
HyHel-5 IgG1κ, Fab	2.54	0.245	hen lysozyme	750	10 (H), 7 (L)	Sheriff *et al.*, 1987
D1.3 IgG1κ, Fv	1.8	0.158	hen lysozyme	675	12 (H), 10 (L)	Bhat *et al.*, 1994
D1.3 IgG1κ, Fv W92D	1.8	0.208	hen lysozyme	604	12 (H), 9 (L)	Ysern *et al.*, 1994
D44.1 IgG1κ, Fab	2.5	0.210	hen lysozyme	632	9 (H), 3 (L)	Braden *et al.*, 1994
HyHEL-5 IgG1κ, Fab	2.65	0.183	bob white lysozyme	760		Chacko *et al.*, 1995
D11.15 Fab, Fv	2.4	0.214	pheasant lysozyme	648	8 (H), 4 (L)	Chitarra *et al.*, 1993
F9.13.7, Fab	3.0	0.140	guinea fowl lysozyme	770	12 (H), 4 (L)	Lescar *et al.*, 1995
NC41 IgG2a,κ, Fab	2.5	0.191	IVN avian	916	9 (H), 8 (L)	Tulip *et al.*, 1992a
NC41 IgG2a,κ, Fab	2.5	0.212	IVN mutant 1368R			Tulip *et al.*, 1992b
NC41 IgG2a,κ, Fab	2.5	0.165	IVN mutant N32 9D			Tulip *et al.*, 1992b
HC19 IgG1, Fab	3.3	0.169	IVH			Bizebard *et al.*, 1995
D1.3 IgG1κ, Fv	1.9	0.194	anti-idiotype Fv E5.2	912	18 total	Fields *et al.*, 1995
D1.3 IgG1κ, Fab	2.5	0.179	anti-idiotype Fab E225	800	4 (H), 9 (L)	Bentley *et al.*, 1990
730 1.4 IgG2a, Fab	2.9	0.210	anti-idiotype Fab409.5.3	860	19 total	Ban *et al.*, 1994
YsT9.1 IgG2bκ Fab	2.8	0.174	anti-adiotype FabT91AJ5	730	8 (H), 7 (L)	Evans *et al.*, 1994
Jel 42, Fab	2.8	0.193	PCP HPr from E.coli	690	15 (H), 5 (L)	Prasad *et al.*, 1993
28, Fab	3.0	0.180	HIV-1 RT-ds DNA			Jacob-Molina *et al.*, 1993
NC10 IgG2aκ, Fab	2.5	0.213	IVN,whale	689		Malby *et al.*, 1994
N10 IgG1κ, Fab	2.9	0.196	Staphylococcal nuclease	793	9 (H), 11 (L)	Bossart-Whitaker *et al.*, 1995
RF-AN	3.2	0.225	Fc from IgG4	640	7 (H), 2 (L)	Corper *et al.*, 1997

IVN, influenza virus neurominidase; RT, reverse transcriptase; PcP, phosphocarrier protein; IVH, influenza virus haemagglutinin.

L3 loops are devided into six structural classes each, lambda L1 into four classes, lambda L3 into two classes, H1 and H2 loops into three and four classes respectively. L2 loops have no class variation (Table 15; Fig. 44) (Chothia *et al.*, 1989, 1992; Webster and Rees, 1995; Searle *et al.*, 1995; Al-Lazikani *et al.*, 1997). Among the conserved residues are glycine, asparagine, and proline, which determine the special folding patterns of the peptide backbone, as well as several others amino acids that participate in specific packing interactions, such as hydrogen bonding or have preferred torsion angles. The canonical structures, which are selected during the primary antibody response, are conserved during the process

TABLE 15 Canonical Structures of Human CDR Loops

V region	CDR	Residues	Loop class	Loop length	Key residues
$V_{\varkappa}$[a]	$L_{\varkappa}1$	26-32	1 (murine)	10	2,25,29,33,71
		24-34[d]	2	7	
			3	13	
			4	12	
			5 (murine)	13	
			6	8	
	$L_{\varkappa}2$	50-52	1	3	48,64
		50-56[d]			
	$L_{\varkappa}3$	91-96	1	6	90,95,97
		89-97[d]	2	6	90.94,97
			3	5	90
			4	4	90,97
			5	7	90,96,97
			6		
V_{λ}[b]	$L_{\lambda}1$	25-32	1	10	
		25-32[d]	2	11	
			3	11	
			4	9	
	$L_{\lambda}3$	90-97	1	8	
			2	8	
V_H[c]	H1	26-32	1	9	24,26,27,29,34,94
		31-35b[d]	2	8	
			3	9	
	H2	52-56	1	5	55,71
		50-65[d]	2	6	52a,55,71
			3	6	54,71
			4	8	54,55,71

[a]Tomlinson *et al.*, 1995
[b]Al-Lazikani *et al.*, 1997
[c]Chothia *et al.*,1992
[d]Kabat and Wu, 1971

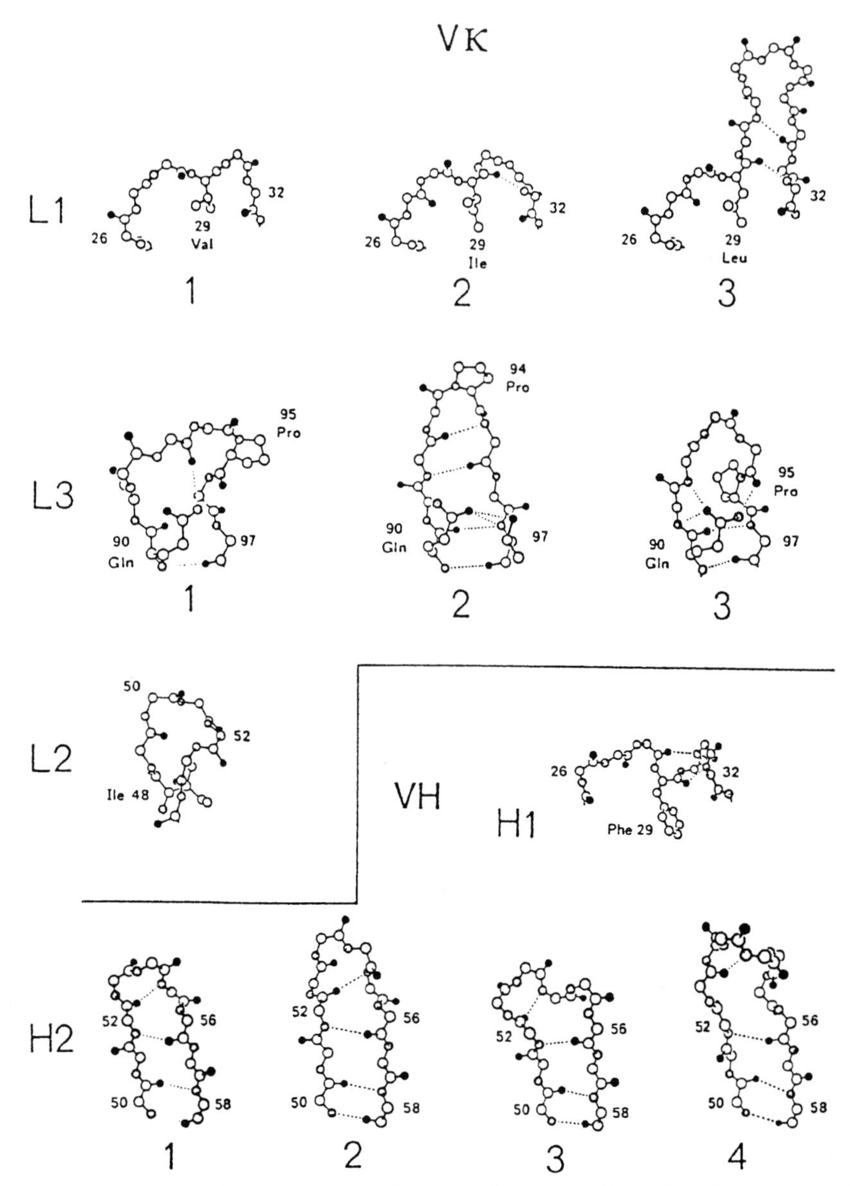

FIGURE 44 Canonical structures for the hypervariable regions of $V_{\varkappa}$ and V_H domains. In each drawing, the accessible surface is at the top and the framework region below. The main-chain conformation and some of the side chains that determine this conformation are shown (Chothia *et al.*, 1989. Reprinted with permission from Macmillan Magazines, Ltd.)

of affinity maturation. The residues supporting the conformation of loops are unmutated or undergo conservative mutations (Tomlinson *et al.*, 1995).

In antibody molecules, the canonical structures occur in various combinations, thus creating an enormous variety of antigen-combining sites. However, a very small number of all possible combinations are preferentially used for construction of the main part of the antibody–combining site repertoire and about 87% of the studied sequences belong to only 10 canonical structure classes (Vargas-Madrazo *et al.*, 1995). The loops with related canonical structures could comprise the combining sites of antibodies with different specificities. For example, the F9.13.7 and D11.14 antibodies have a very similar structure of H1 and H2 loops but the residues of these loops are responsible for contacts with the large parts of different lysozyme epitopes (Lescar *et al.*, 1995). Some sets of canonical structures are preferentially specific for small molecules (haptens) and they have long H2 and L1 loops. Other sets with shorter H2 and L1 loops are specific for the epitopes of large antigenic molecules such as proteins. There are also multispecific sets without such correlation (Fig. 45).

Whether four, five, or six hypervariable loops participate in antigen binding depends on the antigen specificity of the antibody (Wilson and Stanfield, 1993). For example, in antibody complexes with small antigens, such as some haptens and peptides, the second loop of the L chain (L2) is not usually engaged in contact with antigens, whereas residues of all six loops make contact with epitopes of large antigens, such as lysozyme. However, this rule is not a general one. For example, the antigen-combining site of the antibody against spin-labeled dinitrophenyl group (AN02) is built from all the L loops and H3 as well, and in the structure of anti-neurominidase antibody (NC41), the L1 loop is not used at all for its specific site.

Many sequence variations in CDRs only modify the surface of the combining site. This is illustrated in Figure 46, where the hypervariable regions of the Fab fragment from the murine monoclonal antibody to fluorescein 4-4-20 is shown, superimposed on the canonical structures taken from other Fabs. The structure of the 4-4-20 Fab fragment was determined at high resolution, but after the main canonical structures were identified and described (Herron *et al.*, 1989). The four loops of the 4-4-20 Fab have the main chain conformations, which closely coincide with previously analyzed structures. The fifth loop, L1, is also closely related but not identical to those of other known structures.

The concept that a few canonical structures exist for each hypervariable loop is most important for modeling the conformations of immunoglobulin hypervariable regions on the basis of their primary structure. More recently, several Fab structures studied at high resolution were compared and the similarity of common canonical CDR loop conformations was confirmed (Bajorath *et al.*, 1995). However, significant conformational differences between equivalent loops were found, which was mainly due to the deviations of residues located

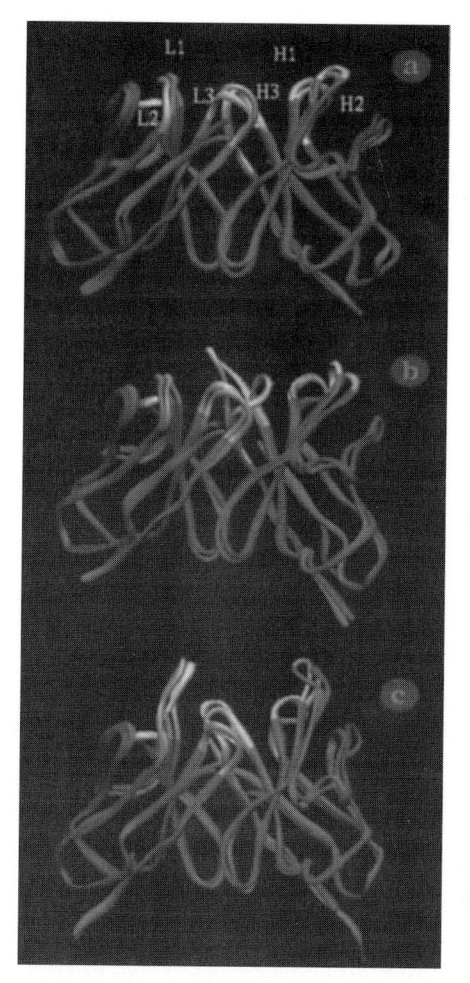

FIGURE 45 Representation of the backbone of antibodies (a) preferentially specific for proteins and (b) multispecific, and (c) preferentially specific for small molecules. (Vargas-Madrazo *et al.*, 1995. Reprinted with permission.)

on the border between conserved framework β-strands and variable loops. Hence, the modification of the variable region framework can also change the configuration of the antigen-combining site and as a consequence its antigen specificity.

In early experiments on recombination of heavy and light chains, it was found that the specificity of recombinant antibody molecules is dependent on the source of heavy chains (Franěk and Nezlin, 1963). In most cases, more

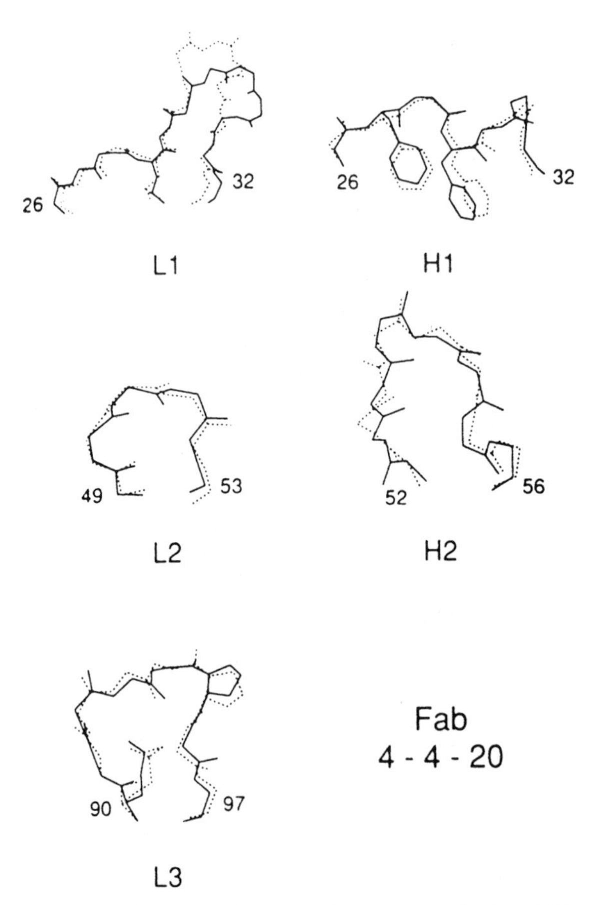

FIGURE 46 The conformations of the CDRs of a murine antibody 4-4-20 (continuous lines). They are superimposed on hypervariable regions (dotted lines) from other immunoglobulins with the same canonical structure. Four of the 4-4-20 CDRs (L2, L3, H1, and H2) have the main-chain conformations that coincide with those of previously known structures. (Chothia *et al.*, 1989. Reprinted with permission from Macmillan Magazines, Ltd.)

amino acid residues of the heavy chains are involved in the antigen binding than those of the light chains, although some exceptions are known, such as the AN02 antibody (Tables 12–14, Fig. 47). For example, an epitope of guinea fowl lysozyme consists of 13 residues and only 2 of them are in contact with the loops of the light chains, whereas other epitope residues are in contact with the heavy chain loops (Lescar *et al.*, 1995). In complexes of the Fv4155 antibody fragment with steroid hormones, the light chains contribute only about 20% of the Fv–ligand interface (Trinh *et al.*, 1997).

The H3 loop is exceptionally important for specificity of the antigen-combining site (Kabat and Wu, 1991). It has a highly variable length, from 2 to

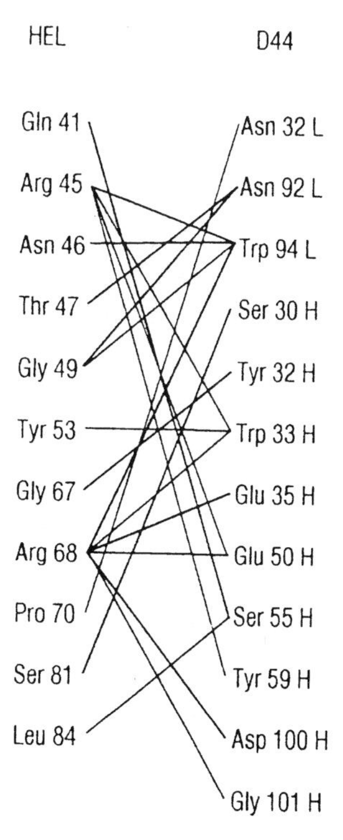

FIGURE 47 Contacting residues of the D44.1 anti-lysozyme antibody (right) and hen lysozyme (left). (Braden *et al.*, 1994. Reprinted with permission.)

26 residues (Fig. 48), and sequence. Its conformation also varied from compact involuted loops (Marquart *et al.*, 1980; Fan *et al.*, 1992) or sharp hairpin turns (Brady *et al.*, 1992) to long extended loops protruding from the surface (He *et al.*, 1992; Kodandapani *et al.*, 1995). H3 can completely block the combining site. In the Fv fragment isolated from a human IgM (Pot), the H3 loop forms an unusual compact structure and prevents access to the combining site by filling all the space between the variable domains (Fan *et al.*, 1992). This IgM molecule cannot bind any of the many tested peptides.

B. Correlation of Binding Site surface with Type of Antigen

Very schematically, the combining sites can be outlined as cavities, grooves, or planar surfaces (Rees *et al.*, 1994). There is a correlation between the shape of the combining site and type of antigen recognized (MacCallum *et al.*, 1996). In antibodies specific for haptens, the combining site resembles pockets or clefts

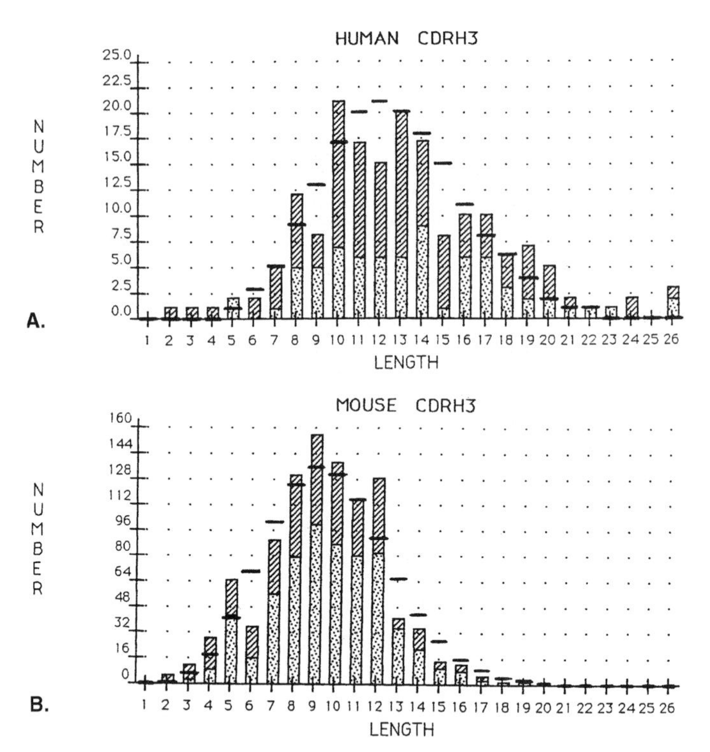

FIGURE 48 Length variation of CDR3 sequences from (A) human and (B) mouse. The stippled portions indicate the numbers of sequence with known antibody specificities. Entire bars indicate total numbers of complete sequences. A Poisson distribution with mean length of 8.7 residues is indicated by horizontal lines. (Wu *et al.*, 1993. Reprinted with permission.)

(Segal *et al.*, 1974). Such a shape for the binding site provides more contacts between hapten and the antibody. For example, fluorescein is bound in a relatively deep slot lined by residues of both heavy and light chains of the corresponding antibody 4-4-20 (Herron *et al.*, 1989). The carbohydrate-binding site is a long, wide groove lined with tyrosine residues (Cygler, 1994).

The peptide-binding sites are generally described as concave pockets of different size. They are formed by most or all CDR loops (Stanfield and Wilson, 1993; Lescar *et al.*, 1997). The binding site of the antibody against octapeptide, angiotensin II, is very deep and narrow, and this kind of binding site significantly increases the antibody contact area (Garcia *et al.*, 1992). In most cases, structural reorganizations of different types are necessary for the formation of a binding site to accommodate a binding part of an antigen. They are termed "induced fit."

The binding sites for larger molecules, such as intact proteins, present a more extensive and flatter surface (Fig. 49). The surfaces of antibody and protein

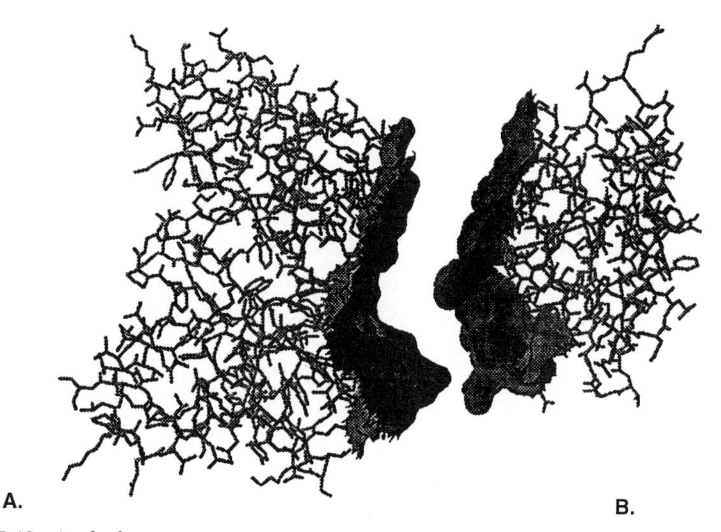

FIGURE 49 Antibody recognition of protein antigens. The Fab fragment of (A) antibody against hen lysozyme (HyHEL-10) and (B) hen lysozyme. (Saul and Alzari, 1996. Reprinted with permission from Humana Press.)

antigens are highly complementary in conformation and in charge, and protuberances from one of them protrude into depressions of the other. For example, the interacting surfaces of the HyHEL-10 and F9.13.7 antibodies are in general flat but they both have large protrusions built by H2 or H3 loops, respectively, which fit into the substrate-binding groove of lysozyme (Padlan *et al.*, 1989; Lescar *et al.*, 1995). The interface of the complex of staphylococcal nuclease and Fab N10 has an unusual U shape. This nontypical form of the interface could be due to an H3 loop being shorter than in other antibodies (Bossart-Whitaker *et al.*, 1995). The N10-contacting residues that build a U-shaped ridge surround a depression with a central protrusion. The corresponding nuclease epitope also has a U-like shape (Fig. 50, see color plate).

In the interface of the antibody—hen lysozyme complex, buried hydrophobic "holes" are found that are not filled with water molecules and have no van der Waals contacts. Several small pockets not filled by the antigen were found in the interface of the complex of Fab′ with the myohemerythrin peptide (Stanfield *et al.*, 1990). Single pockets were detected in the interface of the antibody–ouabain complex (Jeffrey *et al.*, 1995b) and of the complex of influenza virus hemagglutinin with the antibody HC19 (Bizebard *et al.*, 1995). Such holes may reduce the affinity of the antibodies.

The atomic packing density at the antigen—antibody interfaces is looser than in the interior of a protein globule. The areas of Fab coming in contact with antigens (buried surface inaccessible to solvent) range from 156–500 Å^2 for the complexes with low molecular weight substances (Table 12) to 632–916 Å^2 for the complexes forming with protein antigens (Table 14). The interface of the NC41 antibody with the influenza virus neuroaminidase is the most extensive (916 Å^2). There is no direct correlation between the areas "buried" by antigens and the affinity of antibodies. The number of "buried" atoms is larger than the calculated number of atoms, which are in direct pairwise contact (Sheriff, 1993b). For example, in the complex of the NC41 antibody Fab fragment with influenza virus neuraminidase, 17 residues of Fab make contact with the antigen and 36 residues form the buried surface area (Tulip *et al.*, 1992a). This difference is smaller for complexes with haptens, where the antigen surface is largely covered by antibodies.

In general, the antigen–antibody interfaces have poorer shape complementarity than can be found in other protein–protein interactions, such as contacts between protein oligomers. The complementarity of protein–protein interfaces can be characterized with a parameter S_c (Lawrence and Colman, 1993). The value of S_c for protein oligomeric interfaces is equal to 0.70–0.76 but for antibody–antigen interfaces S_c = 0.64–068. This poorer surface complementarity of the latter interfaces could be explained in several ways: by differences in amino acid contact residues with the other types of complexes, by the existence of an induced fit in many antibody-combining sites after reacting with antigens, and by large variations in the chemical structure of antigens.

C. Bonds between Antigen and Antibody

The linkages between antigenic epitopes and the antibody-combining sites include salt links (ionic bonds), hydrogen bonds, and van der Waals interactions. The relative contribution of each of the linkages varies among the antigen–antibody complexes. The number of van der Waals contacts between lysozyme and two antibodies against this antigen (HyHEL-5 and HyHEL-10) is 74 and 111, respectively, that of contacts between a peptide antigen and the antibody B1312 is 65, and between the antibody N1G9 and a hapten is 61.

The information on ionic and hydrogen bonds in antigen-antibody contacts is given in Table 16. Usually about 10–15 hydrogen bonds and 1–3 salt links are involved in contacts between protein and peptide epitopes and the antibody-combining sites. The complex of a large hapten, digoxin, a rigid steroid molecule, with the anti-digoxin antibody, is formed without any hydrogen or ionic bonds (Jeffrey *et al.*, 1993). In this complex, most of the contacts with antigens are from side-chain atoms of contacting residues, although some of their main-chain atoms also participate in linkages.

TABLE 16 Participation of Salt Linkages and Hydrogen Bonds in Contacts between the Antibody Fab Fragments and Various Antigens[a]

Fab from	Antigen	Number of	
		Salt links	Hydrogen bonds
HyHEL 5	HEL	3	16
HyHEL 10	HEL	1	21
D1.3	HEL	0	14
D1.3	Anti-idiotype Fab	1	9
Ys T9.1	Anti-idiotype Fab	0	12
F9.13.7	Guinea fowl lysozyme	3	12
D11.15	Pheasant lysozyme	2	5
NC41	IV neurominidase	3	12
N10	Staphylococcal nuclease	2	7
Jel 42	Phosphocarrier protein	0	11
B1312	Peptide from myohemerythrin	1	12
17/9	Peptide from IV hemagglutinin	1	12
50.1	Peptide from HIV-1	1	5
26–10	Digoxin	0	0
N1G9	(4-hydroxy-3-nitrophenyl) acetate	2	5
RF-AN	Fc from IgG4	1	3

[a]References in Tables 1–3.
HEL, hen egg lysozyme; IV, influenza virus.

D. Residues Responsible for Contacts with Antigens

Amino acid residues that determine antibody specificity belong to the most variable part of CDRs (Kabat and Wu, 1971). According to the criterion based on the analysis of three-dimensional structures of several antigen–antibody complexes, these residues belong to the regions 27d-34, 50-55 and 89-96 of light chains and 31–35b, 50–58, and 96–101 of heavy chains (Padlan *et al.*, 1995). CDR residues, which are located centrally within the combining site, are more often in contact with antigen. Some less central CDR residues participate in contacts only with large antigens (MacCallum *et al.*, 1996). Framework residues can also form contacts with antigens.

The number of CDR residues in contact with antigens is not very high (Tables 12–14). Nevertheless, the antigen-combining sites can accommodate a large number of antigens with various chemical structures. Therefore, the residues, which are more often in contact with antigen, must have properties favorable for antigen binding. The large residues can form many van der Waals and electrostatic interactions and the residues with flexible side chains can efficiently adjust various antigens. The amino acid residues with amphipathic properties could easily accommodate to hydrophobic and hydrophilic environment after formation of complexes with antigens (Padlan, 1990; Mian *et al.*, 1991).

Aromatic amino acids, tyrosine and tryptophan, fit these requirements. The analysis of residues participating in contacts with antigens reveals that both residues occur more frequently in CDRs than in the framework regions (Padlan, 1990, 1994). Moreover, these residues, when located in the framework, are completely or mostly buried, whereas in the CDRs they are significantly more often exposed to solvent (Padlan, 1990). Tryptophan and tyrosine were found regularly among the Fab contact residues (Table 17; Figs. 47 and 51). In some complexes, aromatic residues of the combining site participate in the majority of interatomic contacts with antigens, as in complexes of D1.3 with lysozyme (Fischman *et al.*, 1991), of Se155-4 with dodecasaccaride (Cygler *et al.*, 1991), of BR96 with nonoate methyl ester derivative of Lewis Y antigen (Jeffrey *et al.*, 1995a), of 4-4-20 with fluorescein (Herron *et al.*, 1989), and of 26-10 with digoxin (Jeffrey *et al.*, 1993).

Another amino acid residue that often occurs in the CDRs is asparagine (eight times more frequently than in the framework). This residue is probably more important for stabilization of the combining site structure than for direct contacts with antigens (Padlan, 1990). Apolar aliphatic residues, such as alanine, valine, isoleucine, and leucine, are absent in the contact areas. Only 5–6 amino acid residues in a combining site contribute actively to the binding energetic, forming barely a part of the total contact area (Novotny *et al.*, 1989). The affinity studies of the mutant antilysozyme Fv D1.3 fragment confirmed these findings. It was found that a band formed by five contact residues in the center of the antigen-combining site of this Fv is much more important for binding affin-

TABLE 17 Participation of Tyrosine and Tryptophane Residues of the Antigen-Combining Site in Contacts with Protein Antigens[a]

Antibody	Total contact residues	Tyrosine residues	Tryptophane residues
D1.3	13	4	2
HyHEL-10	17	3	2
D44.1	12	2	2
D11.15	12	4	2
NC41	17	3	2
Jel 42	20	7	0
RF-AN	9	1	1

[a]Specificity of antibodies in Table 14.

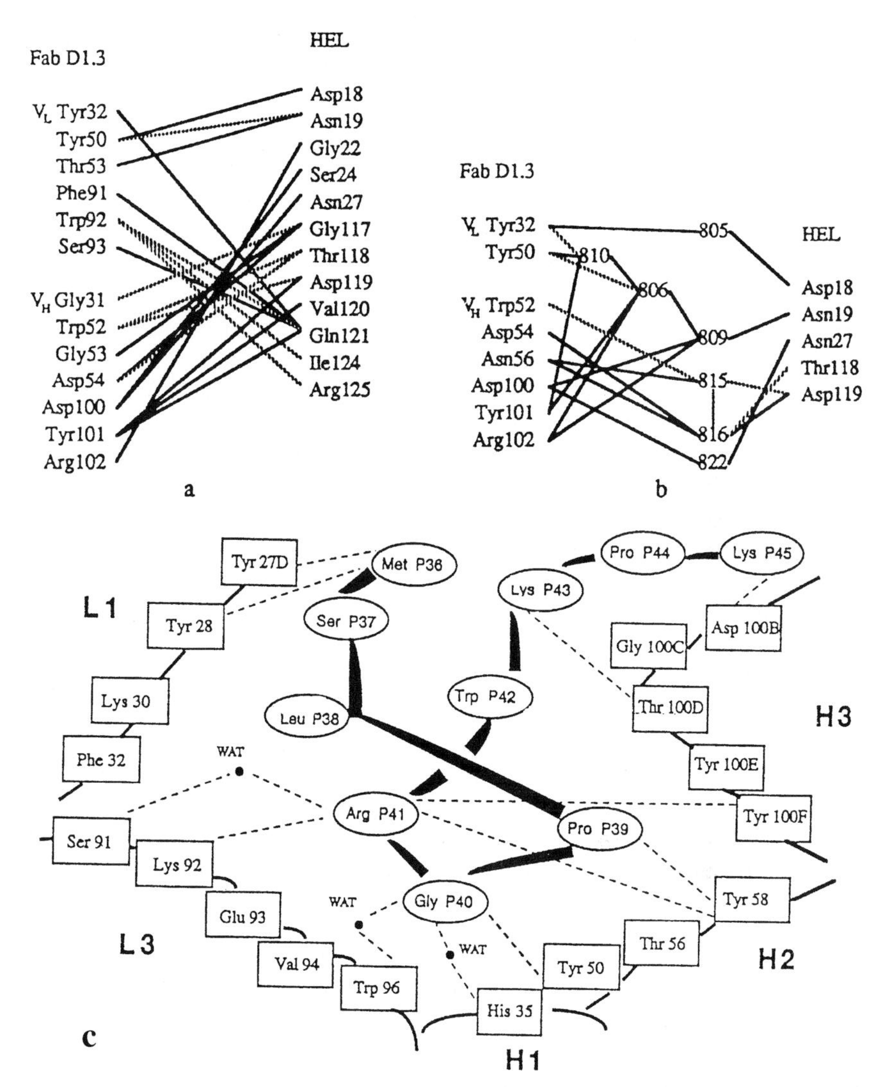

FIGURE 51 Contacts between hen lysozyme (HEL) and the D1.3 antibody Fab. (a) Direct contacts between Fab and antigen residues; (b) contacts mediated by water molecules. Residues or water molecules, which form at least one hydrogen bond are joined by a solid line. Dotted lines indicate residues that make only van der Waals contacts. (Fischman *et al.*, 1991.) (c) The Fab-peptide complex. The peptide composed of residues 36–46 of the HIV-1 protease is located in the combining site (view from above). All Fab-contacting residues are shown. Hydrogen bonds are indicated as broken lines. Three water molecules at the interface are shown as filled circles (Lescar *et al.*, 1997. Reprinted with permission.)

ity than residues at the periphery (Hawkins *et al.*, 1993). The relative importance of a particular residue in antigen binding is as a rule proportional to the number of its atomic contacts with ligand, as documented by x-ray structural studies.

In many complexes, a limited number of various framework residues (usually one or two for a complex) also has contacts with antigen (Evans *et al.*, 1994; Tulip *et al.*, 1992a; Bentley *et al.*, 1990). Five framework residues participate in the contacts between the 8F5 antibody and a viral peptide (Tormo *et al.*, 1994) and four in contacts between the NQ10/12.5 antibody and 2-phenyloxazalone (Alzari *et al.*, 1990).

It is now well established that framework residues, which do not participate in direct contacts with antigen, can influence the position and conformation of CDRs and modulate the affinity of binding (Chothia *et al.*, 1989; Tramonatono *et al.*, 1990). Some spontaneous mutants of monoclonal antibodies have decreased affinity to the antigen. For example, a variant of the anti-digoxin antibody has reduced affinity due to a mutation Arg → Ser at the framework position H-94 (Panka *et al.*, 1988). By contrast, several mutations in the framework region of the anti-lysozyme antibody D1.3 improved the binding affinity (Foote and Winter, 1992; Hawkins *et al.*, 1993). Structural analysis of mutants of the antibody B72.3 against the tumor-associated glycoprotein was performed on the basis of the B72.3 crystallographic structure (Xiang *et al.*, 1995). It was found that some heavy chain framework residues may directly induce changes in the CDR loop conformation through atomic interactions with CDR residues. Therefore, Ala-H71 provides room for packing CDR2/CDR1. The residues Lys-H73 and Lys-H93 contribute a salt-bridge to Asp-H55 in CDR2 and a hydrogen bond to the carbonyl group at H-96 in CDR3, respectively. If Ala-H71 was mutated to Phe or Lys-H93 was substituted to Ileu, the binding affinity was significantly decreased. A critical importance of FR1 residues for specific antibody activity is proved for all cold agglutinins with specificity to I/i carbohydrate antigen of erythrocytes (Li *et al.*, 1996).

E. Water Molecules Participate in Antigen-Antibody Interactions

Initially, it was thought that water molecules are excluded almost completely from the antigen–antibody interfaces. However, according to some recent x-ray structural data, received at high resolution, water molecules can take part in the antigen–antibody interactions (Fig. 51). In the complexes of antibodies D1.3, D44.1, and HyHEL-5 with lysozyme (structures at 2.5 Å and 2.65 Å resolutions), it was found that five, three, and four water molecules form a network of contacts between antibody and antigen surfaces (Fig. 51b) (Fischman *et al.*, 1991; Braden *et al.*, 1994; Chacko *et al.*, 1995).

More detailed data were obtained by the analysis of the structure of the lysozyme, the Fv D1.3 complex at 1.8 Å resolution (Bhat *et al.*, 1994). Forty-eight water molecules were located at the antigen–antibody interface. Some of them make hydrogen bonds with amino acid residues and others generate a complex network, which link antigen with antibody, contributing to the stability of the complex. Some water molecules bridge both surfaces directly but in other interactions two or even three water molecules are involved. A comparison between free and antigen-complexed Fv structures made in this study shows that the number of water molecules associated with an antigen-combining site is not diminished after the formation of the complex with antigen. Additional water molecules participate in bridging antibody and antigen after the formation of a complex. After the replacement of the contact residue Trp-L92 by aspartic acid in the D1.3 Fv fragment, two water molecules occupy space of the large tryptophan side chain. They extend the system of water molecules observed in the nonmutated Fv–lysozyme complex (Ysern *et al.*, 1994). Trapped water molecules also can fill the "empty" places, which are due to imperfect match between a combining site and an antigenic determinant.

Water molecules also play an important role in forming specific complexes between antibodies and other type of ligands such as haptens, peptides (Fig. 51c), and carbohydrates. They participate in hydrogen bonds, fill empty places between the antibody and antigen surfaces, and help to improve complementarity of the reactive surfaces (Bhat *et al.*, 1994; Cygler, 1994; Guddat *et al.*, 1994; Lescar *et al.*, 1997).

II. CONFORMATIONAL CHANGES LINKED WITH ANTIGEN BINDING

During the past decades, many efforts were made to detect conformational changes in antibody molecule after the formation of complexes with antigen that could be transmitted to the Fc part. It was hypothesized that such allosteric changes of the B-cell antigen receptor could initiate a series of events known as effector functions. Various approaches, including sophisticated physicochemical methods, were used to detect changes in the conformation of antibodies after ligand binding, which can trigger the Fc effector functions (Metzger, 1978). However, the results obtained in these early experiments were contradictory; this was partly due to the difficulty in interpreting the obtained data in structural terms.

In the past several years, much precise structural information has been accumulated, mainly by the x-ray crystallographic studies. The comparison of the structures of the free and antigen-bound Fab and Fv reveals several types of changes in the structure of the antibody-combining site and its vicinity, as a result of the complex formation with antigens. These changes can generally be characterized as a structural adaptation to antigen (Colman, 1988) or "induced fit." They include movements of the peptide backbones, changes in the orientation of side chains, and a relative displacement of the V_L and V_H domains.

Well-defined conformational changes, which can be described as orchestral movements of a number of residues toward the ligand, are noticed in studies of complexes with different antigens (i.e., low molecular weight substances, peptide, and proteins). Some changes are found in the vicinity of the binding site and others are more global, such as movements of the CDR loops and whole domains.

A. Induced Fit

The rearrangements of the H3 loop were registered in several systems. After binding of a nonomer peptide from influenza virus hemagglutinin by the 17/9 Fab fragment, the two strands of the H3 loop rotate around the long axis (Fig. 52, see color plate) and three H3 residues change their orientation significantly (Rini *et al.*, 1992; Schulze-Gahmen *et al.*, 1993). The residue Asn-H100a rotates to the distal site of the loop away from the binding site and thus releases the space for the antigen residue Tyr-105. The residues Asp-H99 and Glu-H100 jump from one side of the loop to the other. The shifts in C_α position are significant: for Asp-H99, Glu-H100, and Asn-H100a they were found at 3.9, 2.0, and 4.6 Å, respectively. Other residues near the binding site also change their position. The residues His-L34 and Tyr-H97, which are near each other in the unliganded Fab, are separated by a distance of about 8 Å after peptide binding, and the peptide leucine residue is now inserted between these two residues. As a result of all these movements, the combining site of the 17/9 antibody is formed as an extended groove linked to a deep pocket, which comfortably accommodate the peptide antigen.

Notable conformational changes of H3 accompanies the formation of a complex of the BV04-01 Fab fragment with trinucleotide of deoxythymidylic acid (Herron *et al.*, 1991). As a nucleotide residue is inserted under residues H97-99 (Thr-Gly-Thr), the H3 segment displaces 2.5 Å toward the exterior of the combining site and the indole ring of Trp-H100a shifts 4 Å to stack with another nucleotide. Conformational rearrangements of the hypervariable loops of heavy chains are found after the complex formation of 15-mer virus peptide with Fab 8F5 (Tormo *et al.*, 1994). They were described as rigid body rotations, which retain the internal loop conformation. Some movements are significant, as for example the displacement of Tyr-H102 of about 7 Å. Dramatic changes in the distal tip of the H3 loop were found in the antigen-combining site of the HC19 antibody complexed with a domain isolated from the influenza virus hemagglutinin. Several atoms of the H3 loop residues change their positions about 10 Å (Bizebard *et al.*, 1995). These movements allow nearly half of the hydrogen bonds to form in the interface. The formation of the binding site for a large hapten, trisubstituted guanidine, also is accompanied by structural adjustments. The main residue participating in the adaptation to the hapten is Tyr–H96 (H3 loop) whose side chains move 4.5 Å to form linkages with the hapten (Guddat *et al.*, 1994). Significant structural changes are found in the position of loops of an antibody against peptide from HIV-1 protease after its reaction with this peptide antigen (Fig. 53) (Lescar *et al.*, 1997).

Comparison of the structure of free Fab′ from the antibody BR96 that recognizes Lewis Y carbohydrate with the structure of Lewis Y antigen–BR96 complex, reveals significant conformational changes of the combining site. The loops L1, L3, and H2 change their conformation and/or spatial position after combining with antigen. However, the orientation of V_H relative to V_L does not change significantly and the conformation of H3 is also not influenced by antigen binding despite the significant contacts between this loop and the antigen (Sheriff *et al.*, 1996).

NMR studies can describe antigen–antibody interactions in solution. Using two-dimensional transferred nuclear Overhauser effect difference spectroscopy, the interactions of a peptide from cholera toxin with monoclonal antibodies were studied. The difference spectra indicate that the conformational changes in the Fab of these antibodies upon peptide binding are confined to the combining site (Scherf *et al.*, 1995).

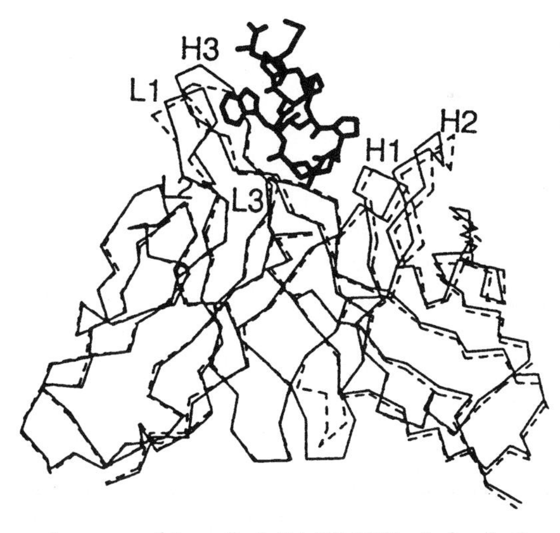

FIGURE 53 The α-carbon atoms of the antibody Fab F11.2.32 in the free (broken thin line) and complexed (continuous thin lines) forms. The conformational changes upon binding of the antigen are seen after superimposition of the models of the free and complexed forms of the Fab. The antigenic peptide (as in Figure 51c) is drawn as a thick line. (Lescar *et al.*, 1997. Reprinted with permission.)

B. Global Changes

In many complexes, changes in quaternary structure of the variable domains have been seen (Herron *et al.*, 1991; Tormo *et al.*, 1994; Stanfield *et al.*, 1993; Bhat *et al.*, 1990). To study the relative positions of variable domains after ligand binding, one of the domains from the complex is superimposed on the corresponding variable domain in the unliganded Fab. In this way variable domains are found in various complexes to be displaced from their positions in the unliganded Fab (Guddat *et al.*, 1994; 1995). This kind of movement of the domain pairs relative to each other is registered as a change in the elbow bend angle after antigen binding.

The biggest elbow angle change (from 184 to 153 degrees) was found in NC6.8 Fab after reaction with a large hapten, the trisubstituted guanidine compound (Guddat *et al.*, 1994). The significant structural alterations of Fab occurring in this system were also registered in another way. The hapten was added to crystals of the unliganded NC6.8 Fab. As a result, the crystals begin to dissolve and after a while, new crystals with a different shape appear. The formation of the ligand–Fab complex is accompanied by large structural changes, including flexing of heavy chain and extension of light chain, and the complex accommodates within a new crystal lattice with a different space group and unit cell dimensions. Another observation made in these experiments is concerned with the change of the position of Glu–H224: after the ligand binding it is displaced 19 Å. This glutamine residue, the carboxyl end of Fd, is located near the upper hinge region of the antibody molecule, and one could speculate that the conformational changes in Fab, as a consequence of the ligand binding, can be transmitted through the hinge to the Fc region. However, the found shift could also be associated with different crystal packing.

The large domain shift was found after formation of complexes between the 50.1 antibody Fab fragment and an HIV-1 antigenic peptide (Stanfield *et al.*, 1993). When V_L domains of the liganded and unliganded Fabs were superimposed, the rotations and translations that are necessary for V_H superposition were significant (16.1° and 2.7 Å, and 16.3° and 2.8 Å for two stidied complexes, respectively). Both variable domains in these complexes move away from each other to widen the combining site pocket to accomodate antigen.

Structural changes in Fab as a consequence of ligand binding do not occur in all studied complexes. For example, after binding of digoxin, a large and rigid hapten, no conformational changes were detected in 26-10 Fab (Jeffrey *et al.*, 1993). The specificity in digoxin binding depends mainly on the shape complementarity to this antigen, and no hydrogen bonds or salt links are generated between this hapten and the residues of the antigen-binding site.

III. COMPLEX OF V_H DOMAIN OF CAMELID HEAVY CHAIN ANTIBODIES WITH ANTIGEN

The recent interest in the camel V_H domain is linked with efforts to obtain minimal antigen-binding fragments. From early experiments, it is known that the fragments composed of V_L and V_H have the complete antigen-binding capacity of the intact antibody molecules (Givol, 1991). It was also found that the isolated heavy chains of anti-hapten antibodies retain some antigen-binding capacity in the absence of the light chains (Utsumi and Karush, 1964; Haber and Richards, 1966). But the attempts to obtain the smallest antigen-binding unit in a form of V_H fragments from normal antibody molecules ran into difficulties partially due to the poor solubility of isolated V_H.

The *Camelidae* possess antibodies composed only of the heavy chains. Their V_H domains have a good solubility due to several substitutions on the surface and in the same time they possess the antigen-binding activity of the whole molecule. Two recent x-ray studies were devoted to understanding the camelid V_H structure. In one of them, the crystal structure of V_H of camel anti-lysozyme heavy chain antibodies was studied in a complex with the antigen (Desmyter *et al.*, 1996) and in the other, the structure of the llama V_H domain from the antibodies against human chorionic gonadotropin was solved (Spinelli *et al.*, 1996).

The overall structure of the camel and llama V_H domains fits the known variable domain fold. The CDR1 deviates significantly from the canonical structures previously known for this loop, whereas CDR2 adopts canonical structure class 2. A portion of the unusually long camel V_H CDR3 forms an exposed, nonflexible loop that penetrates into the lysozyme cleft. This protruding loop donates about 70% of the V_H contacting surface and it reacts with lysozyme residues, which the anti-lysozyme antibodies from other species do not contact at all. The CDR3 loop folds back on the V_H framework blocking a part of residues normally involved in contacts with V_L. The llama V_H CDR3 is smaller. However, the llama V_H combining site is also not flat because the CDR1 loop is exposed to the solvent and there is a groove between CDR1 and CDR3.

IV. INTERACTION OF AN AUTOANTIBODY (RHEUMATOID FACTOR) WITH Fcγ

Autoantibodies with specificity to Fcγ (rheumatoid factors) are usually found in serum and synovia of patients suffering from rheumatoid arthritis. Rheumatoid factors form complexes with IgG in joints and can induce a tissue damage by activating complement cascade and inflammation.

According to the x-ray crystallographic study of the complex between human Fc and a rheumatoid factor (RF-AN), the autoantigen Fc epitope includes a region between $C_\gamma 2$ and $C_\gamma 3$ domains (Fig. 54, see color plate) (Corper *et al.*, 1997). Staphylococcal protein A binds to Fcγ at the same site and that is why protein A competes with many rheumatoid factors for binding with Fcγ. The residues of the CDR3 loop of the autoantibody heavy chain donate the main part of contacts with the Fc epitope (four contacts from a total of nine). This loop protrudes into the cleft between $C_H 2$ and $C_H 3$ domains of Fcγ.

The character of the interaction of the RF-AN Fab fragment with Fc differs from that found in other complexes of antibodies with protein antigens. First, the contact area is small and the buried molecular surface area contributed by the Fab fragment is just 640 Å^2. Second, only one side of the antigen combining site is involved in the interaction. No contacts are found between the Fc epitope and the two loops of the Fab light chains (CDR1 and CDR3). Thus, the large part of the autoantibody-combining site is unoccupied and may potentially recognize another antigen even simultaneously with the Fc fragment. A proline residue at position 56 of the autoantibody light chain participates in direct contacts with several Fc residues. This residue is a result of the somatic mutation (Ser → Pro) of the germ line progenitor of RF-AN. It was proposed that the autoantibody activity of rheumatoid factor RF-AN may originate by somatic mutation of an antibody originally directed against a foreign antigen (Corper *et al.*, 1997).

V. STRUCTURAL ASPECTS OF ANTIBODY SPECIFICITY

X-ray structural studies have contributed much to our understanding of the specificity of antigen–antibody reactions. Several systems were investigated. They include complexes of a single antibody with different but related antigens, complexes of a single epitope with different antibodies, and the effect of mutations on antibody–antigen reactions (Bentley *et al.*, 1994; Sheriff, 1993a,b).

A. Complexes of a Single Antibody with Several Related Antigens

Two group of ligands—low molecular weight substances and mutated variants of protein antigens were used to understand how structurally related antigens can accommodate the same binding site.

Detailed x-ray structural studies were performed on complexes of several related steroids competing with progesterone for binding with the antibody DB3

(Fig. 55) (Arevalo *et al.*, 1993a,b; 1994). Progesterone-11-α-hemisuccinate coupled with albumin was used as an immunogen. Despite significant structural differences, the studied steroid haptens can combine with DB3 with high affinities. This effect can be explained in general by the fortuitous shape complementarity of haptens with the portion of the binding site that is unoccupied with the immunizing hapten. The hapten-binding site of DB3 is not changed considerably as a result of complex formation with all these ligands. However, some adjustments are necessary for the accommodation of each ligand. In the unliganded DB3 the hapten-combining site is "closed" by the indole side chain of Trp–H100 and the rotation of this side chain is a prerequisite for the hapten binding (Fig. 56, see color plate). The P1 pocket of the DB3-binding site, a hydrophobic cavity formed mainly by four aromatic residues, is rigid and accommodates the D ring, which is similar for all ligands. The interaction with the D ring is most important for the specificity of binding of the studied group of the ligands. The P2 pocket is more flexible and has less ligand complementarity, which permits it to accommodate the ligand steroid skeleton in two different orientations without

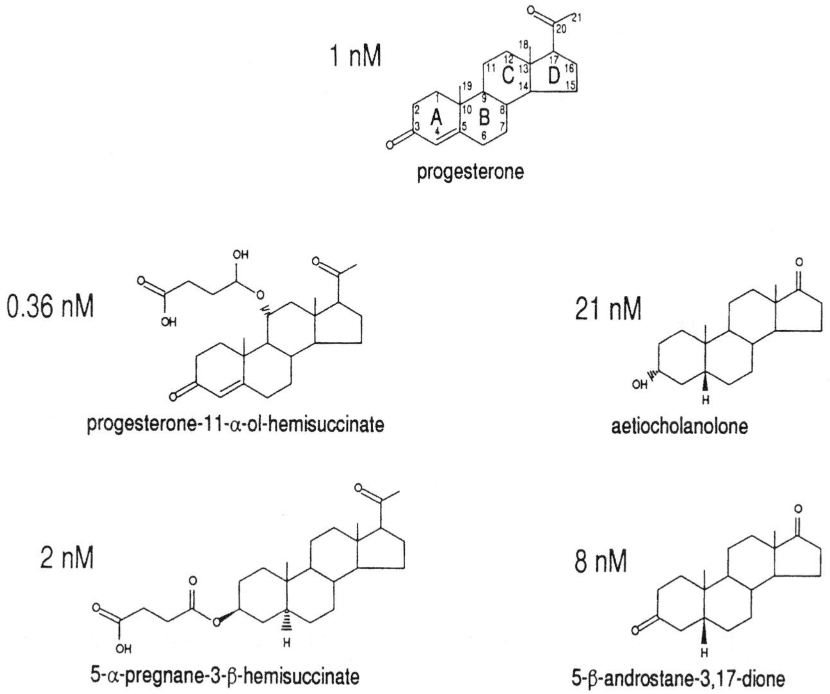

FIGURE 55 Structure of progesterone and the four analogs cross-reactive with antibody DB3 induced by immunization with progesterone-11-α-hemisuccinate. Comparative binding affinities (IC_{50}) are given (Arevalo *et al.*, 1994. Reprinted with permission.)

large modifications (Fig. 56b, see color plate). Another feature of the DB3-binding site important for understanding the steroid cross-reactivity is the ability of the P3 and P3′ pockets to accommodate variants of the steroid A ring.

Natural mutants of lysozyme isolated from white eggs of various bird species and mutants of influenza virus neurominidase were used for studying the specificity of antibodies against protein antigens. Antibody D1.3 reacts with high affinity only with lysozymes that have glutamine at position 121 (hen and bobwhite quail lysozymes). This residue is accommodated by a hydrophobic pocket surrounded by aromatic residues from both light and heavy chains and it makes multiple close contacts with a number of antibody residues (Fischmann *et al.*, 1991). However, the same D1.3 antibody reacted with very low affinity with other lysozymes, which have histidine in the position 121 instead Gln-121. The simplest explanation for this effect is that His-121 cannot fit well the tightly packed hydrophobic pocket, which is suitable for Gln-121.

The HyHEL-5 antibody reacts with another epitope of hen lysozyme. The residue Arg-68, which is located in the central area of the epitope, is the most important for the reaction (Fig. 57, see color plate) and the affinity of the reaction with the natural mutant Arg-68 → Lys (bobwhite quail lysozyme) is 10^3 less. This dramatic loss in affinity could be explained by the inclusion of an additional water molecule into the interface followed by substantial changes in the arrangement of hydrogen bonds (Chachko *et al.*, 1995). No significant transformations of the binding site surface were noted. Therefore, a change even of a small part of the interface can cause a significant loss of binding affinity.

The crystal structure of the NC41 antibody complex with avian influenza virus neurominidase was compared with complexes of NC41 with two cross-reactive mutants of the antigen, which have minor structural differences (Tulip *et al.*, 1992b). Despite the location of mutations at the interface, the affinity of mutants is reduced only slightly because in both cases amino acid substitutions are accommodated by relatively small rearrangements near the mutation sites. In one mutant, Ile-368 mutated for arginine residue and in another Asn-329 for asparagine. The Arg-368 side chain has to shift by about 2.9 Å to accommodate the NC41-combining site and this shift in the antigen molecule is accompanied by the movement of the His–L55 light chain residue about 1.3 Å. In the second complex, the side chain of Asp-329 has to rotate in such a way that the carboxylate group is placed on the periphery of the interface.

B. Complexes of a Single Epitope with Different Antibodies

Two antibodies against hen lysozyme—F9.13.7 and HyHEL-10—form complexes with a nearly identical epitope of guinea fowl and hen lysozymes, which is constructed from a similar number of amino acid residues. However, the combining sites of the antibodies are built from loops that have no sequence homology, and they react in a different way with heterologous antigens (Padlan *et al.*, 1989; Lescar *et al.*, 1995). According to x-ray crystallographic studies, side chains of some lysozyme residues are flexible enough to accommodate different combining sites. The lysozyme residue Arg-73 projects toward the combining site of both antibodies. However, in the complex with F9.13.7, this residue is mostly buried and occluded from the solvent by the stacking interactions with Tyr–L92 but in the complex of HyHEL-10 it is exposed to the solvent. In addition, the lysozyme Trp-62 residue has different orientations in the complexes. Furthermore, Arg-21 is located in the complex with HyHEL-10 in a pocket formed by the H2 and L3 antibody loops. However, the combining site of F.913.7 lacks this pocket and Arg-21 is exposed to the solvent beside having contacts with Tyr–H32.

Another example of such types of cross-reactions is the formation of complexes by the NC10 and NC41 antibodies with influenza virus neurominidase. Even though both antibodies react with almost the same site of the antigen, their combining site structure is different (Malby *et al.*, 1994). Although the sites have identical sequences of the first heavy chain CDR, these common residues do not participate in the contact with antigen and some structural adjustments in the interface are required for the interaction.

The D11.15 antibody against hen lysozyme has a broad specificity. It reacts with lysozymes isolated from other avian species and in some cases even with a higher affinity than with the immunogen (heteroclytic binding; Mäkelä, 1965). The crystal structure of the heteroclytic complex of D11.15 with pheasant egg lysozyme was determined (Chitarra *et al.*, 1993) and the broad specificity of the D11.15 antibody was explained by its binding to the epitope common to different lysozymes. The sequence variations in the heteroclytic antigens are located at the edge of the epitope and hence it is accommodated sterically by the D11.15-combining site.

C. Effect of Mutations on the Antigen-Combining Site

Comprehensive sequence studies of the antibody peptide chains have been done in the past to elucidate the role of different genetic mechanisms, leading to the appearance of high affinity antibodies during an immune response. The information obtained by x-ray crystallographic studies explains what the structural basis of changes of the antibody affinity are in the maturation of an immune response and what is the structural role of various mutations (Spinelli and Alzari, 1994).

Several experimental systems were used in these studies. In one of them, the three-dimensional structure of a complex of the Fab anti-phenyloxazolone (phOX) antibody NQ10/12.5 with the hapten was determined (Alzari *et al.*, 1990). This antibody originates during the secondary immune response due to several somatic mutations of germ line genes. Two positions of the light chains, L34 and L36, are frequently mutated with increased affinity and it was found that these residues are in close contact with the hapten. One of them, Phe-L36, participates in the formation of a hydrophobic pocket for the hapten. Two other residues, Tyr-L32 and Leu-L96 that contact with the hapten are highly conserved among the phOX antibody family. A characteristic feature of the heavy chains of this family is the presence of a short D segment. According to the three-dimensional data, such a D segment is not an obstacle for a channel existing between the hapten binding pocket and solvent. Longer D segments are not suitable because they probably could partially fill the pocket. The contacting residues of the heavy chains are not frequently mutated.

During affinity maturation of the immune response against (4-hydroxy-3-nitrophenyl)acetate (NP), the replacement of Trp-H33 for leucine increases the affinity of the reaction by 10-fold. In the complex between NP and the Fab from the primary response antibody N1G9, the residue Trp-H33 is in close contact with NP and with neighboring residues (Mizutani *et al.*, 1995). Its side chain is less suitable for the formation of the complementary-binding pocket than the smaller side chain of leucine and this replacement helps to create more suitable binding sites by easing the cramped contacts of the tryptophan residue. Some anti-NP antibody molecules, which appear during the maturation, also have a higher affinity for NP but without the Trp-H33 → Leu replacement. They have another replacement: Ser-H100 for Gly. The Ser-H100 residue has no contacts with the hapten but could change the location of the side chain of the neighboring Tyr-H97 residue, which interacts with Trp-H33 and in this way alter the conformation of the Trp-H33 side-chain. The secondary anti-NP antibody 88C6/12 has a leucine residue in position H33. According to the x-ray structural analysis of the complex of 88C6/12 with a hapten 4-hydroxy-3-nitrophenacetyl-ε-amino-caproic acid, the Leu-H33 residue plays an important role in binding the hapten (Yuhasz *et al.*, 1995). A tryptophan residue at this position is much less favorable for the reaction because it prevents the hapten nitrophenyl ring from fitting inside the binding cavity.

The anti-NP antibody 88C6/12 cross-reacts at a higher affinity with the iodo-derivative of NP acetate (NIP). This heteroclitic reaction could be explained by the weak electrostatic interaction of the iodine atom with nitrogen of the nearby-located Lys-H58 (Fig. 58, see color plate). However, the observed heteroclytic reaction could be not dependent on direct iodine–antibody interactions. The binding site of this antibody has a positively charged region, which interacts with the predominant resonance form of the nitrophenyl ring system.

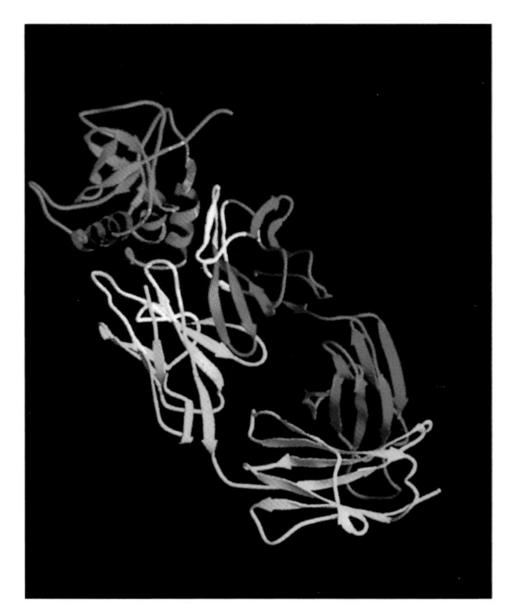

FIGURE 50 Complex of staphylococcal nuclease and the Fab fragment of the antibody N10. Nuclease is colored in cyan with the contact residues in red. The N10 heavy chain is in green, with the light chain in yellow. All CDR regions are colored in silver and the contact residues in magenta. (Bossart-Whitaker *et al.*, 1995. Reprinted with permission.)

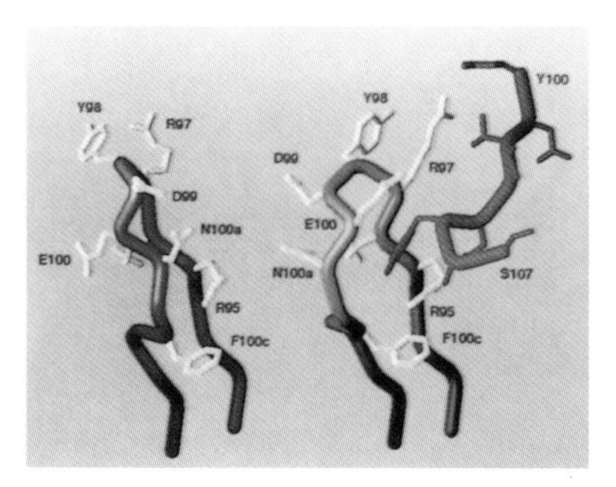

FIGURE 52 The conformation of the H3 loop of the 17/9 Fab anti-peptide antibody in the free (blue) and liganded (green) form. The influenza virus hemagglutinin peptide is shown in red. The H3 residues Asp-H99–Asn-H100a have substantially different conformations in the free and bound forms. (Rini *et al.*, 1992. Reprinted with permission from Amer. Ass. for the Adv. of Science AAAS.)

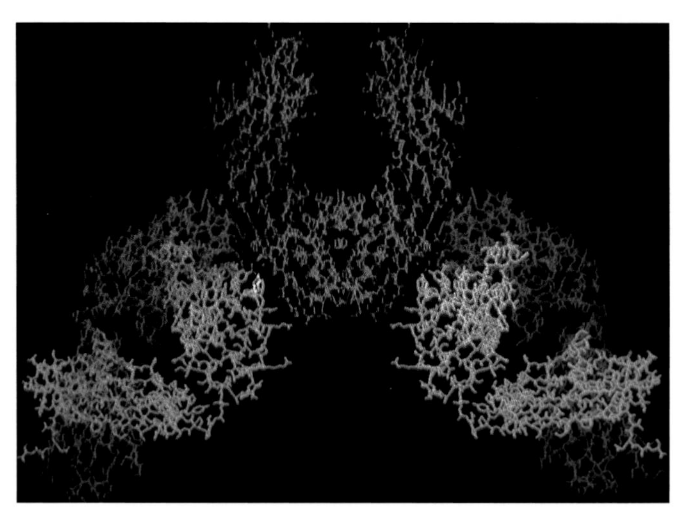

FIGURE 54 The complex between the Fab fragment of the rheumatoid factor (blue) and the Fc fragment of the IgG4 molecule (red). The light chains of the Fab fragment are shown in light blue and the heavy chain in dark blue. (Corper *et al.*, 1997.)

A

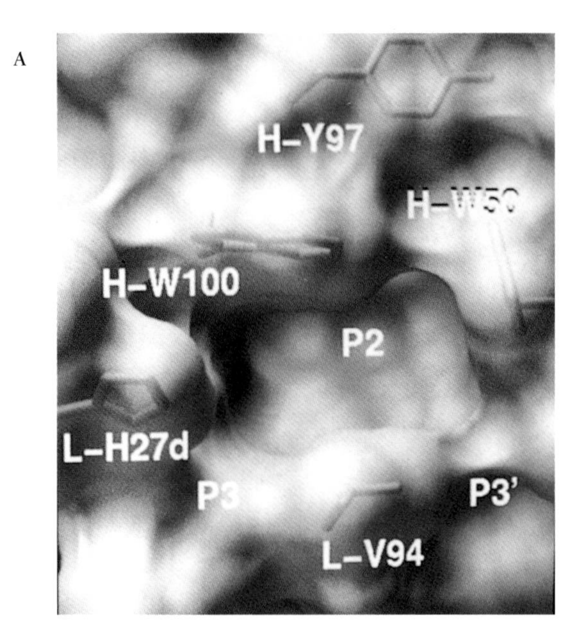

FIGURE 56 The (A) unliganded and (B) hapten bound combining sites of the DB3 anti-progesterone antibody. Four pockets comprising the site are shown (P1, P2, P3, and P4). Hapten carbon atoms are shown green and oxygen atoms red. (A) The unliganded combining site is partially blocked by indole of Trp-H100 that covers P1. (B) Two binding orientations of the progesterone hapten are shown by translucent and by solid tubes. (Arevalo *et al.*, 1994.)

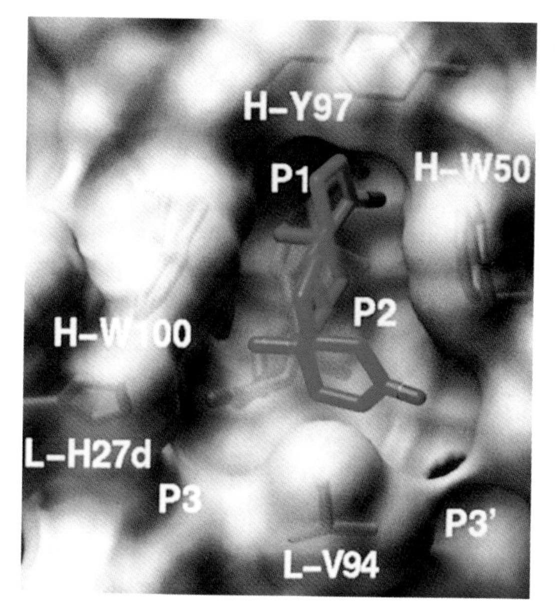

FIGURE 56 CONTINUED

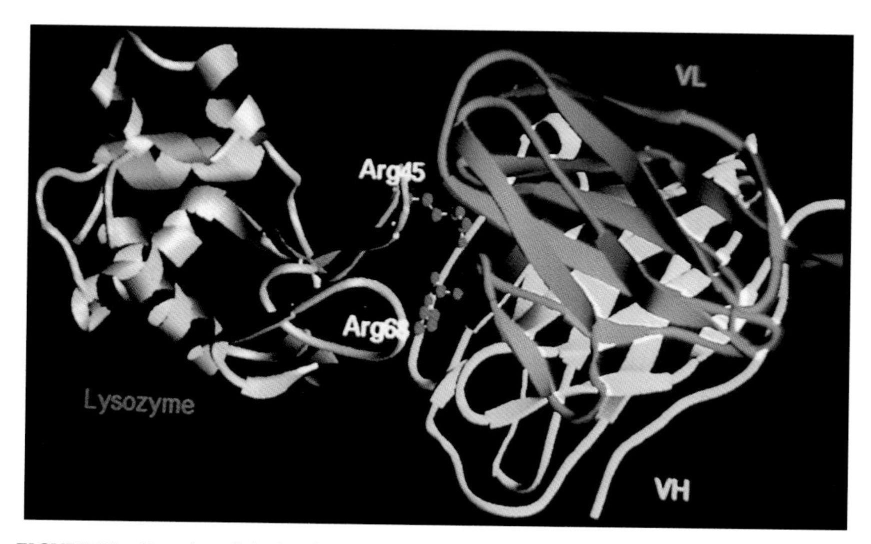

FIGURE 57 Complex of chicken lysozyme and the Fab fragment of HyHEL-5 antibody. The antigen is in bronze, the Fab light chain in blue, and the heavy chain in yellow. Two lysozyme arginine residues (68 and 45), both in red, can extend into the Fab groove. (Chacko *et al.*, 1995.)

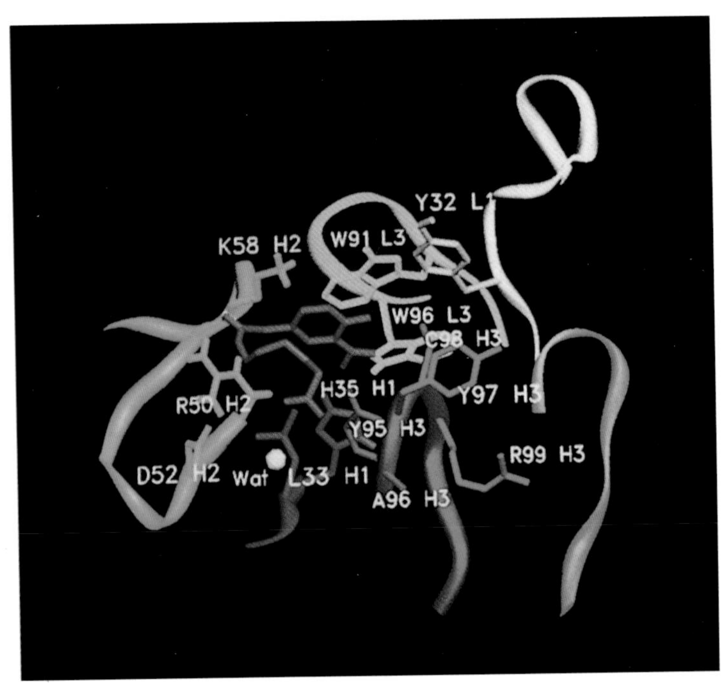

FIGURE 58 Contacts of the heavy and light chains of the 88C6/12 anti-nitrophenyl antibody with the hapten 4–hydroxy-3–nitrophenyl acetate (orange). The white ribbon is L1, green is L2, yellow is L3, violet is H1, turquoise is H2, and dark blue is H3. A white ball is the bound water molecule. (Yuhasz *et al.*, 1995.)

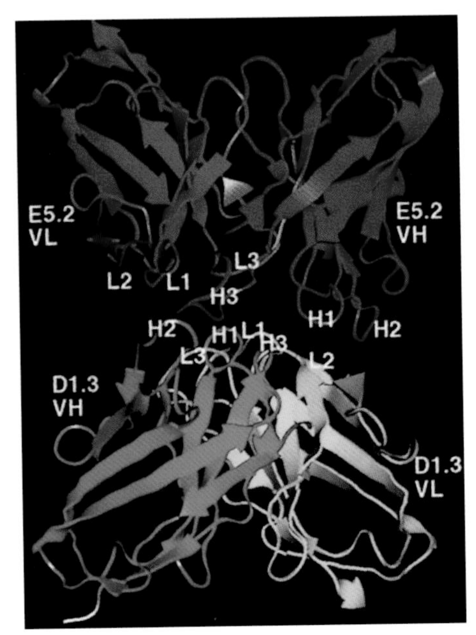

FIGURE 59 Idiotype–anti-idiotype complex (D1.3–E5.2) is formed by the contacts of all six CDR loops of each Fv fragment. The V_H H3 loop of E5.2 is more protruding. (Braden *et al.* 1996.)

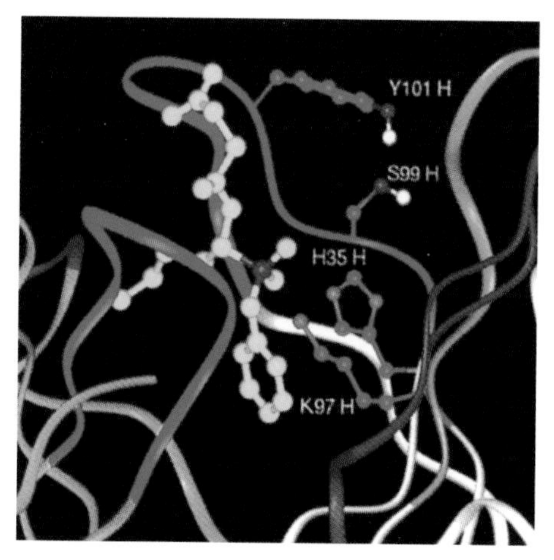

FIGURE 64 The contacts of norleucine phosphonate hapten (Figure 63, Scheme 1, 4) and the residues of the 17EB antibody-combining site. The light chains are shown in pink and the heavy chains in white. The CDR regions in both chains are colored as orange, red, and green for CDR1, CDR2, and CDR3, respectively. The hapten is shown in yellow and its phosphate atom in red. The side chains of the important amino acids are shown in ball-and-stick format and color-coded as C, green; O, red; N, blue; and H, white. (Zhou *et al.*, 1994.)

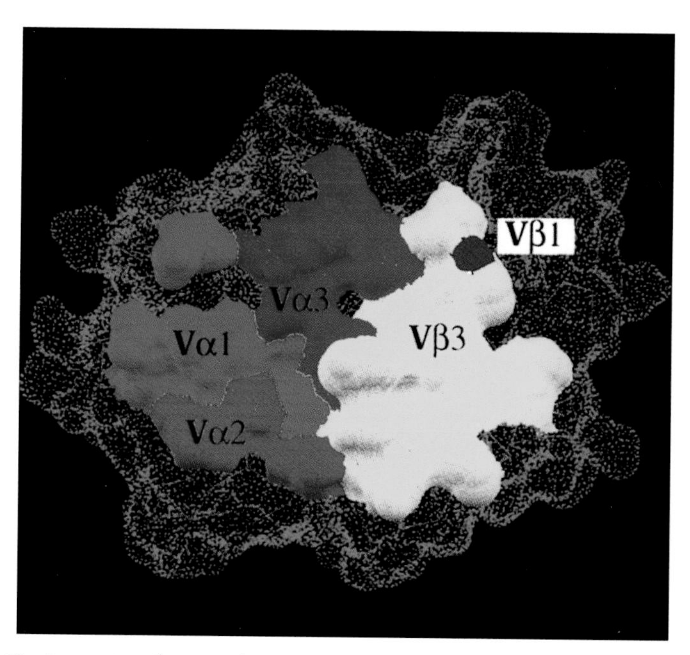

FIGURE 69 Interaction of TCR with Tax peptide and HLA-A2. MHC–peptide solvent-accessible surface buried by the CDRs of the T-cell receptor according to contact region. (Garboczi *et al.*, 1996.)

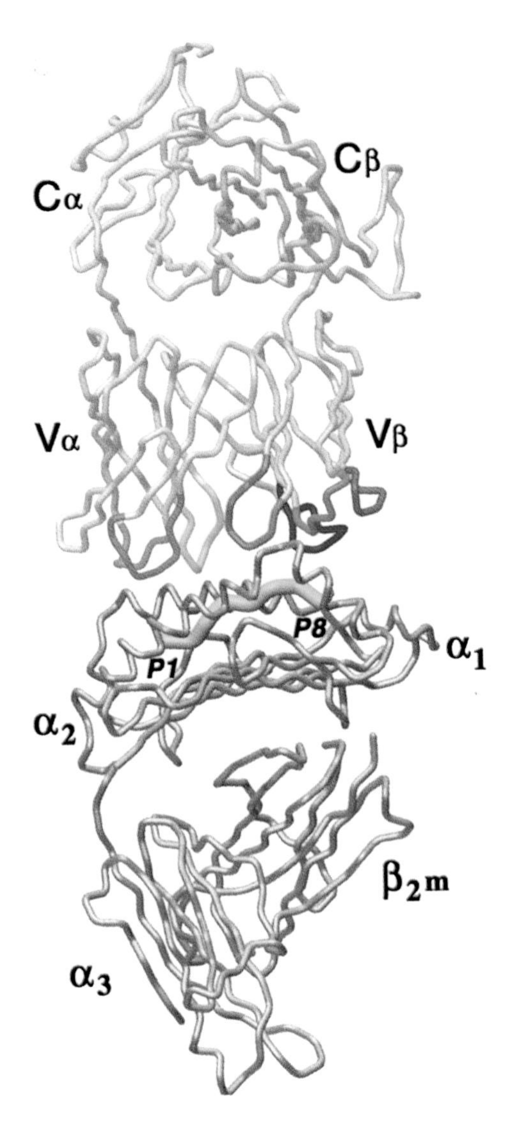

FIGURE 70 TCR–MHC–peptide complex. Octamer peptide in the MHC groove is presented as a yellow tube. The 2C T-cell receptor is above. Its α1 and α2 CDRs are colored pink, β1 and β2 CDRs blue, CDR3s yellow. HV4 (c″-d) loop of V_β is orange. (Garcia *et al.*, 1996.)

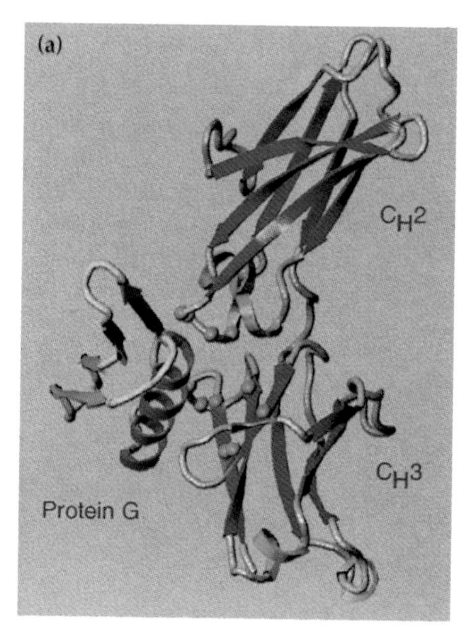

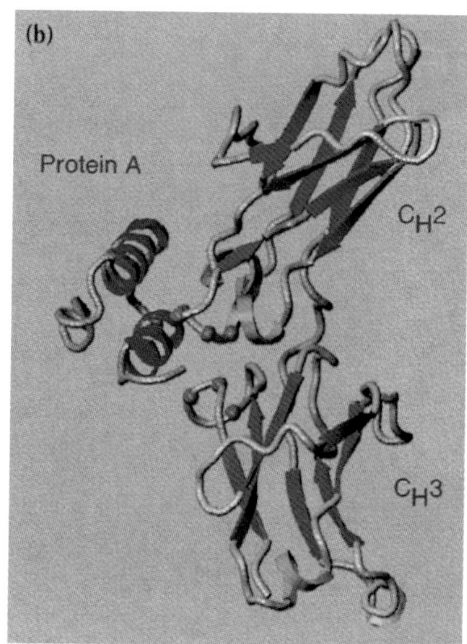

FIGURE 79 Models of (a) protein G–Fc and (b) protein A–Fc complexes. β-strands and loops in protein G are colored green and yellow, respectively, and α-helix is colored red. Helices and loops in protein A are colored light and dark violet, respectively. Cα atoms of the Fc residues that interact with protein G are marked in green and that interact with protein A are marked in pink. (Sauer-Eriksson *et al.*, 1995.)

Iodination stabilizes this resonance ring structure and in this way favors the interaction with the antibody-combining site (Yuhasz *et al.*, 1995). The iodinated hapten has more van der Waals interactions and additional hydrogen bonds in the interface.

In the course of the affinity maturation of the catalytic antibody 48G7 that has esterolytic activity, the nine replacement mutations were fixed, which increased affinity for the hapten (the transition state analog) by a factor of 10^4 (Patten *et al.*, 1996). The three-dimensional structure of 48G7 with the hapten was solved. It was found that none of the mutated residues has direct contacts with the bound hapten. Most likely, they participate in reorganization or stabilization of the combining site conformation. More recently, a crystal structure of the germ line antibody Fab fragment and its complex with the same hapten have also been solved (Wedemayer *et al.*, 1997a). The hapten binds to the mature 48G7 antibody by a lock-and-key mechanism without marked structural changes in the combining site. By contrast, the structure of the combining site of the germ line antibody changes significantly after binding, which results in a configuration with enhanced complementarity to the antigen. The main structural changes that happened in the germ line Fab after hapten binding already preexist in the mature Fab due to somatic mutations, which occur up to 15 Å from bound hapten.

The Fv and Fab fragments of the D1.3 antibody against hen lysozyme are good models for evaluating the results of site-directed mutagenesis because the three-dimensional structure of the fragments is known in detail. Using quenching of fluorescence as a result of the reaction with antigen, it was shown that the affinity of the D1.3 antibody can be increased severalfold by site-directed mutagenesis (Foote and Winter, 1992). All replaced residues are in framework regions. They underlay closely the binding site but do not participate in the direct contact with the antigen. Similar results were obtained with mutants of Fv D1.3: a fivefold improved affinity was obtained from a combination of mutations. However, none of the mutated residues was located in the combining site itself and they modulate the affinity for the antigen (Hawkins *et al.*, 1993). These observations are important for understanding how antibody affinity is improved by somatic mutations in the course of the immune response. The replacements of contact residues due to random mutations have little chance to improve the binding affinity of an antibody but more likely will damage it. For example, the replacement of the contact residue Trp-L92 of the D1.3 combining site by aspartic acid leads to a decrease by three order of magnitude in the equilibrium constant for the binding of Fv to lysozyme (Ysern *et al.*, 1994). By contrast, each mutation of residues, which are near the binding site and indirectly affect its structure, could slightly increase the binding affinity. The gradual accumulation of such mutations can lead to a significant improvement in the antibody-binding capacity (Hawkins *et al.*, 1993).

D. Structure of Idiotype–Anti-Idiotype Complexes

Individual antigenic determinants or idiotopes of an antibody molecule (Oudin, 1974) associate with the variable regions usually of both the heavy and light chains. A set of these determinants of an antibody molecule (Ab1) is designated as an "idiotype." Antibodies against idiotypes or anti-idiotypic antibodies (Ab2) can be classified as directed toward the parts of the variable region outside the antigen-combining site (Ab2α) or toward the antigen-combining site itself (Ab2β and Ab2γ). The Ab2β antibodies react with the Ab1 combining site in the same manner as the antigen and can carry an "internal image" of the original antigen. They block the reaction of Ab1 with an antigen. The Ab2γ antibodies are also directed against the combining site or a part of it but have no "internal image" of the antigen and may only partially block the Ab1–antigen reaction. The immunization with Ab2β elicits the formation of the Ab3 antibodies, which react with the antigen as Ab1. It was suggested by Niels Jerne that the immune system responds to antigens by formatting an interconnected regulatory network consisting of interacting Ab1 and Ab2 antibodies (Jerne, 1974).

The antibodies, which resembled antigens (Ab2β), have been found in many antigenic systems, including bacteria, parasites, proteins, carbohydrates, and small molecules (Thanavala and Pride, 1994). It was proposed to use anti-idiotypic antibodies as surrogate antigens to elicit the immune response, especially in cases where it is hard to isolate antigens, like viral antigens. The main and puzzling question is how the antigen-combining site with nearly the same general architecture in different antibodies can mimic so many different antigenic substances? Is the mimicry really structural or only functional? To answer this question several idiotype–anti-idiotype systems were studied using x-ray crystallography (Poljak, 1994; Pan *et al.*,1995).

The first group of experiments was performed on complexes of the Fab and Fv fragments of the anti-lysozyme monoclonal antibody D1.3 and anti-idiotype antibodies. The Fab fragments of D1.3 and of the anti-idiotypic antibody E225 are aligned approximately along their major lengths (Bentley *et al.*, 1990). The contact residues of both Fabs belong mainly to CDRs. However, three framework V_L residues of D1.3 Fab and a framework V_L residue of E225 Fab also participate in contacts. The D1.3 idiotope is built from 13 residues and 7 of them also participate in contacts with lysozyme. So, only a partial overlap was found between the idiotope and the antigen-combining site. The nature of interactions of these seven common residues in D1.3 Fab–lysozyme and D1.3 Fab-E225 Fab complexes is different: they make eight hydrogen bonds and three van der Waals contacts in the first complex and four hydrogen bonds, one salt bridge and three van der Waals interactions in the second complex. Hence, in this idiotype–anti-idiotype system the antigen molecular mimicry was not found.

Other types of results were obtained in experiments with complexes of Fv D1.3 and another idiotypic antibody, E5.2 (Fig. 59, see color plate) (Fields *et al.*, 1995; Braden *et al.*, 1996). From 18 D1.3 residues contacting the E5.2-combining site and 17 D1.3 residues contacting lysozyme, 13 are in contact with both E5.2 and lysozyme. Moreover, positions of D1.3-contacting atoms are very similar in both cases. Six of twelve hydrogen bonds in the interface of the D1.3–E5.2 complex are structurally equivalent to hydrogen bonds in the interface of the D1.3 complex with the antigen. The positions of many water molecules located in and near the interface of both complexes are similar. The observed antigen mimicry of the E5.2 anti-idiotype depends on the similarity of van der Waals interactions, hydrogen bonds, and solvent interactions. The existence of the mimicry was confirmed by the finding that in response to immunization by E5.2 some mice produce antibodies (Ab3), which can react with lysozyme, although in a different way than the original D1.3 antibody (Ab1). The lysozyme epitope specific to D1.3 is partly α-helical and it is hard to expect the strict structural homology between this epitope and the E5.2 antigen-combining site, which has no α-helices. In the case of D1.3 the antigenic mimicry by antibody CDRs is a functional one and does not depend directly on sequence homology.

The 730.1.4 antibody against the E2 peplomer of the feline infectious peritonitis virus can recognize a viral epitope even after denaturation of the latter and, therefore, it is specific to the unfolded sequence. The anti-idiotypic antibody 409.5.3, which was induced by immunization with 730.1.3, belongs to the Ab2β type anti-idiotypic antibodies and can mimic virus antigen. The crystal structure of the complex of Fabs 730.1.4 and 409.5.3 has been determined (Ban *et al.*, 1994, 1995). The Fab fragments are rotated by 61 degrees about the long axis with respect to one another and the heavy chain of one Fab interacts in the complex with the heavy chain of the second Fab. Light chains also interact with each other. This idiotype–anti-idiotype system has a unique feature: there is a sequence homology between L1 and H1 loops, on the one hand, and some regions of the antigen, on the other (Table 18). The homologous regions of the anti-idiotype 409.5.3 participate in many direct contacts with the original antibody 703.1.4. Therefore, it could be likely that in this system the homologous

TABLE 18 Amino Acid Homology Sequences of Anti-idiotypic 409.5.3 Antibody and Feline Infectious Perotonitis Virus

Virus protein	–Ile–Ser–Ser–Ser–Ile–Ser–
L1 loop	–Val–Ser–Ser–Ser–Ile–Ser–
Virus protein	–Gly–Phe–Ser–Phe–Asn–Asn–
H1 loop	–Gly–Phe–Thr–Phe–Asn–Asn–

Composed from data by Ban *et al.*, 1994.

loops of the anti-idiotype antibody mimic some epitopes of the virus antigen. However, this hypothesis must be verified by structural studies of the virus epitope and of its complex with 703.1.4.

A structural study was performed on the angiotensin II system that is composed of the whole idiotypic row of antibodies, Ab1, Ab2β, and Ab3. Antibodies Ab1 and Ab3 (numbered 110 and 131) have significant sequence homology, especially in the combining sites, and both of them can bind angiotensin II with a high affinity (Amzel *et al.*, 1994). This small antigen is an octapeptide that is very flexible in solution, but has a compact conformation after binding to Ab3 (Garcia *et al.*, 1992). The conformation of seven of the eight residues of the bound angiotensin resembles that of the L3 loop of the human immunoglobulin REI. It is likely that a CDR loop of Ab2β mimics angiotensin and therefore it can elicit the antibodies of the Ab3 type, which in turn can react with the antigen. The CDR antigen mimicry was found also in experiments with the reovirus type 3 hemagglutinin (Williams *et al.*, 1989). A sequence similarity between CDR2 loops of anti-idiotypic antibody (Ab2) and the antigen allows the synthesis of peptides corresponding to the homologous CDR regions, which elicited an antibody response to virus hemagglutinin.

A structural explanation why Ab2 is not always able to mimic an antigen was found in experiments, in which lipopolysaccharide (LPS) from *Brucella abortus* was used as an antigen (Evans *et al.*, 1994). The monoclonal antibody YsT9.1 against LPS (Ab1) has an antigen-binding pocket in the form of a deep groove complemented to the rod-like structure of the antigen. The crystal structure of the complex of the YsT9.1 Fab with the Fab of anti-idiotypic antibody was determined at high resolution (Fig. 60). In this complex, the anti-idiotypic antibody (Ab2) is not able to contact with the major portion of the antigen-binding groove of Ab1 and it is hard to expect that Ab2 can mimic the antigen. Indeed, this anti-idiotypic antibody is not able to induce LPS-specific Ab3-type antibodies and hence it belongs to the Ab2γ-type of the anti-idiotypic antibodies.

Thus according to crystallographic studies, the antigen mimicry can be realized in two different ways: by the existence of structural homology between some variable parts of the CDR loops of anti-idiotypic antibodies (Ab2) and a protein antigen, or by a functional mimicry when such homology is absent.

The unusual self-interactions of the Fab fragments from the Ab3 antibody were found in crystals (Ban *et al.*, 1996). The Fab fragments were isolated from anti-anti-idiotypic antibody (Ab3) specific for anti-idiotypic antibody (Ab2), which in turn mimics an epitope mapped by the antibody (Ab1) against human melanoma-associated antigen. The Fab fragments of Ab3 tightly interacts in crystals by their CDRs, resembling closely the idiotype–anti-idiotype interactions of Ab1-Ab2 type (Ban *et al.*, 1995). It was proposed that in the course of the idiotypic network formation such self-complementarity is increasing but at the same time the structural information on antigen is gradually lost.

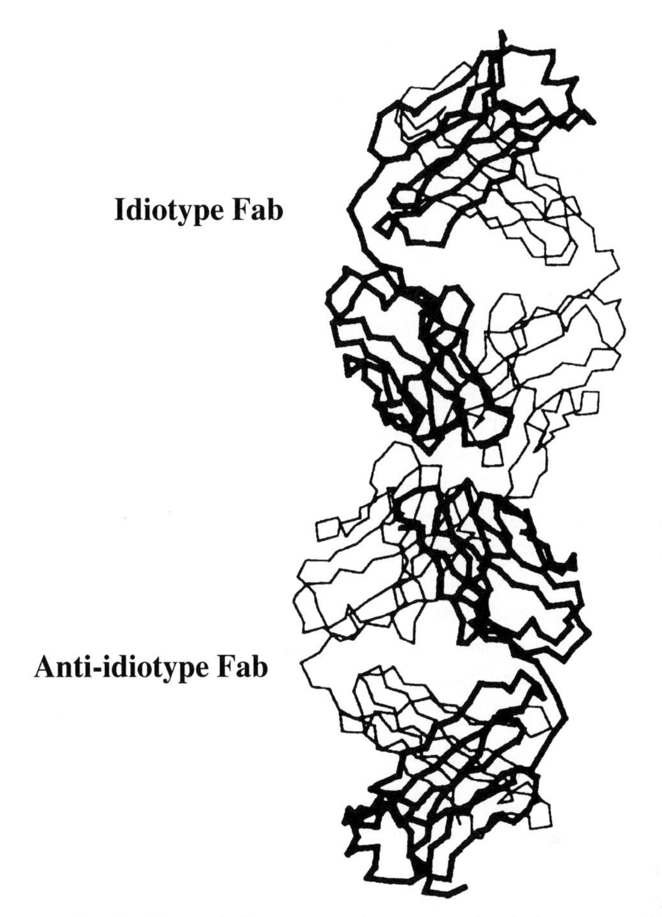

FIGURE 60 α-carbon backbone of idiotype–anti-idiotype complex (Fab YsT9.1.–Fab T91AJ5). The Fab fragments are bound head-to-head with contacts made almost exclusively through hypervariable loops. (Evans *et al.*, 1994. Reprinted with permission.)

VI. MODELING ANTIBODY-COMBINING SITES

The number of the known three-dimensional structures of antibodies is constantly growing. However, the determination of the three-dimensional structure of protein molecules by x-ray crystallography is still a time-consuming technique despite all recent innovations in the field. For these studies, well-diffracting crystals are necessary. Recently, methods for antibody crystallization have improved remarkably and nearly half the Fabs or even more can be crystallized (Stura *et al.*, 1993). However, growing such crystals is still a matter of art and they are not always obtainable, even if highly experienced specialists are involved in the study.

The information on the antibody three-dimensional structure is of primary importance to be able to solve many important problems. For example, the knowledge of the combining site structure at atomic resolution is required for practical applications such as antibody engineering, including the humanization of murine antibodies, for rationalized drug design, and to produce catalytic antibodies. The approximate models are also of great value for the initial steps in using diffractional data.

There are still no satisfactory methods for the prediction of a peptide chain folding only on the basis of the amino acid sequence. However, different antibody molecules have the same general topology, and a high degree of homology exists between large stretches of their peptide chains. Even among the variable CDRs, there are positions where identical amino acid residues are found for different antibody molecules. Such homology, as well as the availability of many crystal structures of antibodies, greatly simplifies modeling procedures.

The modeling can be used for various purposes: modeling single-site mutations, insertions, and deletions or modeling a single CDR as well as a whole set of CDRs of Fv fragments only from amino acid sequences (Martin *et al.*, 1991). Two approaches have been used in modeling the antibody-combining sites: knowledge-based procedures (or homology modeling), which are based on using information from known structures as templates, and another, *ab initio* modeling, based on conformational searches or molecular dynamics and conformational energy evaluations (Chothia *et al.*, 1986; Padlan and Kabat, 1991; Thornton, 1991; Padlan, 1992; Amzel, 1992; Rees *et al.*, 1994).

A. Knowledge-Based Methods

The knowledge-based methods involve: (1) searching the most similar canonical loop from the known crystal structures according to length, key residues within the loop, and residues near to both loop terminals; (2) replacing side chains according to the sequence that is modeled; and (3) finding the compatibility with other loops and final energy refinement to improve the geometry of the model and to relieve bad contacts (Chothia *et al.*, 1989; Lesk and Tramontano, 1991; Bajorath and Fine, 1992; Edwards *et al.*, 1992; Brünger, 1993). The knowledge-based modeling is relatively simple, and commonly known programs can be used for computer-aided molecular modeling. The homology approaches are highly dependent on how precisely the crystallographic structures used are described. Many of the known antibody Fab structures were characterized only to medium resolutions and higher resolution studies as well as additional refinement is necessary to receive more detailed pictures (Padlan, 1992;

Lascombe *et al.*, 1992). For example, many significant differences have been found between α-carbon positions of the initial and more recent models of Fab New (Poljak *et al.*, 1973; Saul and Poljak, 1992). In some positions, displacements in main-chain atoms as large as 5–6 Å have been observed (Fig. 61). It is also necessary to remember that the relative positions of V_H and V_L are not the same in different Fab fragments and this factor can affect the spatial position of CDR residues and the shape of the combining site. There is no distinct canonical structure for the most variable H3 loop and this fact also limits the usage of the knowledge-based approaches.

An example of the knowledge-based procedures is presented in Figure 62 as a flow chart for ABGEN, an automated approach for the Fv structure modeling (Mandal *et al.*, 1996). One component of this approach is ABalign, which aligns the test V-region sequence with the known sequences. The found protein with the highest homology is chosen as the scaffold for the modeling using the second component, ABbuild.

B. *ab* Initio Procedures

The *ab initio* methods involve modeling multiple loop conformations using conformational searching on the basis of escaping steric constraints and some other considerations. The next step is energy minimization, and a conformation with the lowest potential energy is selected as the final model (Bruccoleri

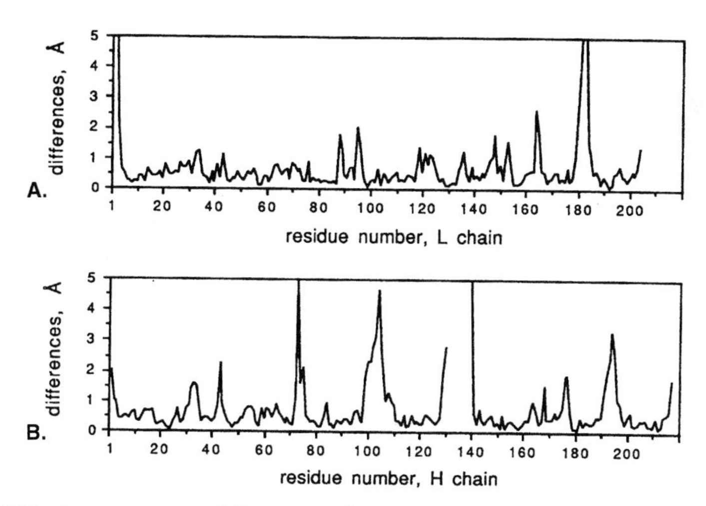

FIGURE 61 Root mean square differences (in Å) between α-carbon positions of the initial and final models of Fab New. (Saul and Poljak, 1992. Reprinted with permission.)

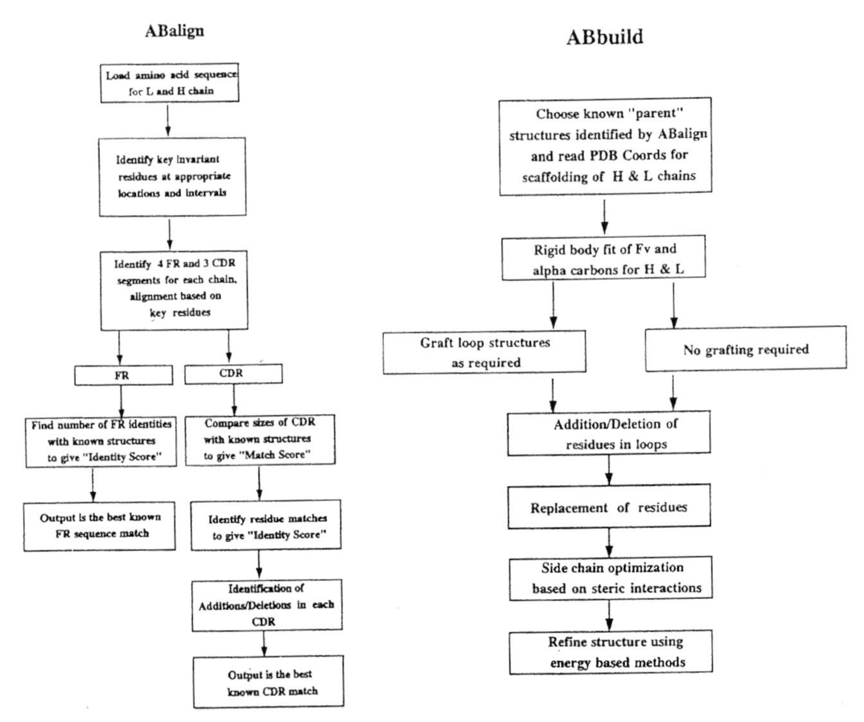

FIGURE 62 Flow chart for ABGEN modeling of antibody Fv regions. (Mandal *et al.*, 1996. Reprinted with permission.)

and Novotny, 1992; Gibrat *et al.*, 1992; Bajorath and Fine, 1992). The method is more suitable for small loops. The combined knowledge-based and *ab initio* methods were also applied (Pedersen *et al.*, 1992; Webster and Rees, 1995).

Modeling the framework part of immunoglobulin folds, which are highly conserved, is a more simple problem and it can be relatively easily performed by side-chain substitutions. The antibody CDR loops are extremely variable in their length, sequence, and conformation. Therefore, their modeling is more complicated. Despite all the difficulties, several groups using the above-mentioned approaches obtained relatively good results in modeling the five most conserved loops that have known canonical structures. Particularly complicated is the modeling of the highly variable H3 loop, especially if it is longer than 10 residues. However, modeling has been performed even for the long H3 using artificial neural networks (Reczko *et al.*, 1995).

To measure the accuracy of a model, the comparison studies with the x-ray structural data are performed. In several such studies, the general design of the antigen-binding site as well as the CDR loop conformation were correctly predicted and root mean square (rms) deviations were usually less than 1 Å on the backbone (Chothia *et al.*, 1985, 1989; Eigenbrot *et al.*, 1993; Essen and Skerra,

1994; Bajorath and Sheriff, 1996). The shape of the combining site is dependent not only on the conformation of the CDRs, but also on their spatial positions relative to the framework and to each other. The prediction accuracy is lower when not only the CDR conformation but also the CDR spatial positions are taken into account (Bajorath and Sheriff, 1996).

C. Application of Modeling

Despite all recent achievements in modeling, the precise construction of the antibody-combining site and its complexes with antigens is still hard to achieve. The conformational lability and the antigen-induced changes of the conformation of the antigen-combining site are the main problems. However, modeling can provide useful information concerning some important structural problems.

One of the effective applications of modeling is the analysis of mutations that affect the antigen-binding activity of antibodies. Such a study was performed on mutants of a monoclonal antibody against the cardiac glycosides digoxin and digitoxin (Novotny *et al.*, 1990). One of the mutant antibody variants has a single residue mutation Ser → Arg in position 94 of the heavy chain (H–94), which changed the binding specificity despite the fact that this position is located outside the antigen-combining site and is considered a part of the framework. According to computer analysis, arginine in position H–94 can form a hydrogen bond to side chains of several residues, including Asp–H101, Arg–L46, and Asp–L55 (a CDR residue). The formed net of hydrogen bonds results in rearrangements of these side chains and, as a consequence, changes in the combining site surface and the altered affinity as well.

Modeling is applied to select residues in the CDR regions that participate in contacts with antigen. The main criteria for choosing such positions are the location of a residue in the antigen-binding area and its surface accessibility. The chosen residues can be used as targets for mutagenesis. Such experiments were performed with a model of a monoclonal anti-digoxin antibody (Near *et al.*, 1993). Mutations of selected residues more frequently changed binding of the lactone ring of digoxin and it was concluded that this ring is a dominant part of this antigen. X-ray crystallographic studies of the complex of digoxin with the same antibody molecule confirm this prediction (Jeffrey *et al.*, 1993). Using the ABGEN procedure (Fig. 62), an accurate prediction of the NC6.8 antibody residues participating in contacts with the antigen, the guanidinium sweetener, was made (Viswanathan *et al.*, 1995). All nine key amino acid residues involved in antigen recognition and complexation were identified before the high-resolution x-ray structural data were obtained (Guddat *et al.*, 1994).

Modeling is an important element in the design of a new functional site. Using a template-based approach, a specific metal-binding site was created in the variable region of an antibody molecule (Roberts *et al.*, 1990). As a template, the known structure of the anti-fluorescein antibody (Herron *et al.*, 1989) was used. In the following experiments, three light chain variable residues were replaced on histidine residues. They formed a metal-binding site close to the fluorescein-binding site without disrupting the overall structure. Copper added to the mutant Fv caused significant quenching of fluorescence of the tryptophan residues located near the metal-binding site but this effect was much less expressed in the wild-type Fv. In another experiment, metal-binding sites were designed in the antigen-combining site of an catalytic antibody on the basis of a three-dimensional antibody model (Roberts and Getzoff, 1995; Wade *et al.*, 1993). Introducing metal ions and cofactors into the antibody-combining sites could be used for construction of novel catalytic antibodies with a desired specificity and high catalytic activity. Antibody molecules with metal ions in the combining sites could also be applied in affinity chromatography and for the development of biosensors.

The artificial antibody M41 against cystatin, a peptide protease inhibitor, has been successfully constructed (Schiweck and Skerra, 1997). The model is based on the coordinates of the anti-lysozyme antibody HyHEL-10. The binding site of HyHEL-10 was reshaped by replacing 19 amino acid residues and 10 residues were substituted in the framework. The backbone structure of HyHEL-10 was retained. The M41 antibody was produced as a Fab fragment in *Escherichia coli* and its crystal structure was solved at a high resolution. The comparison of the crystal structure with the computer model of M41 reveals a large similarity between both structures in the mutual arrangement of all CDRs and most of their backbone conformation.

Only a few of all the Fab fragments whose structures were solved by x-ray crystallography were studied in complexes with antigens. Modeling is now widely using to obtain information on the localization of antigens in the combining site and on contacting amino acid residues. In some studies, the known three-dimensional structure of antibodies or the low-resolution map of complexes was used and antigen was docked in an optimal position by computer modeling (Stigler *et al.*, 1995). Other studies employed computer modeling for predicting structures of both Fab and antigen epitope (Höhne *et al.*, 1993; Totrov and Abagyan, 1994). In experiments with virus antigens, the best binding epitope peptide is used for docking, and for verifying constructed models, the affinity of epitope variants are tested (Liu *et al.*, 1994). Models of antigen–antibody complexes could be used for predicting substitutions in the interface, which result in higher affinity interactions between both reactants.

The identification of residues for “humanization,” the procedure that reduces a potential antigenicity of rodent monoclonal antibodies for humans, is another important practical application of modeling. The main goal of this procedure is to transfer a minimum number of residues from a rodent antibody to human immunoglobulin that could provide the full binding activity of a constructed chimeric molecule (Mayforth, 1993). Modeling variable regions provides information on how rationally to find residues responsible for direct contacts with antigen (Padlan, 1991; 1992; Padlan *et al.*, 1995) and those framework residues, which could influence the overall conformation of the combining site (Foote and Winter, 1992).

If the conformation of a viral peptide located in the antibody-combining site is the same as in a virus protein, it is possible to model the antibody interactions with the whole virus particles. By superimposition of the structure of Fab from the antibody to human rhinovirus on viral capsid, it was found that Fabs can be placed without steric difficulty on 60 equivalent viral epitopes (Tormo *et al.*, 1995). The neighboring fragments locate nearly parallel without clashing, which points to the possibility of bivalent interactions of IgG antibody molecules with the viral protein. Indeed, it was found experimentally that thirty IgG molecules can bind to one virion particle and neutralize virus reproduction.

VII. STRUCTURAL ASPECTS OF CATALYTIC ANTIBODY ACTIVITY

A. Catalytic Antibodies Induced by Immunization

The hypothesis that antibodies can be used for the catalysis of chemical reactions was first suggested in 1969 (Jencks, 1969). However, this idea was realized only in 1986 when two laboratories used the transition state analogs of some chemical reactions for immunization (Pollack *et al.*, 1986; Tramontano *et al.*, 1986). It was shown that antibody molecules specific to these haptens are capable of catalyzing the hydrolysis of carbonates and esters.

The ensuing years yielded a growing number of catalytic antibodies, which catalyze various chemical reactions (Schultz and Lerner, 1995; Driggers and Schultz, 1996; Wade and Scanlan, 1997). The method for the screening of catalytic antibodies based on immunological identification of forming products was developed (Tawfik *et al.*, 1993). This assay, catalytic enzyme-linked immunosorbent assay (catELISA), is based on immobilizing a substrate to a solid support. After incubation with the catalyst, the solid-phase product is detected by a standard ELISA with antibodies that specifically bind the product but not the substrate. Recently, a method was suggested that allows direct chemical selection for catalysis from combinatorial antibody libraries. For this, a reaction was designed that leads to a covalent interaction between phage particles carrying catalytic antibodies and an insoluble matrix with coupled substrate (Janda *et al.*, 1997).

As a rule, haptens used for eliciting catalytic antibodies are synthetic analogs of the transition state of the catalyzed reaction and dissimilar to the reaction products. For example, the hydrolysis of esters and carbonates go through tetrahedral intermediates and the haptens used for immunization are their stable analogs. The binding of the catalytic antibodies to the transition state structure stabilizes them, decreases the activation energy, and thereby increases the reaction rate. It was found that human autoantibodies can also posses catalytic activity. For example, some of them can hydrolyze vasoactive intestinal peptide (VIP) or thyroglobulin and human autoantibodies with DNA-hydrolyzing activity were described.

The catalytic antibodies have very good substrate specificity but relatively low rate acceleration (typically 10^3–10^4 and not higher that 10^6). The strategy for obtaining more efficient catalytic antibodies includes improved hapten design and modification of the antibody-combining site either chemically or by site-specific mutagenesis. For the rational design of antibody catalysts, precise structural information on their combining site is of primary importance. X-ray crystal structures of several catalytic antibodies in complexes with transition state analogs were recently solved and other important structural studies were performed (Ulrich and Schultz, 1998).

The first three-dimensional structure of an catalytic antibody with a transition state analog was published in 1994 (Table 12) (Zhou *et al.*, 1994). The complex was formed by the antibody 17E8, which catalyzes the hydrolysis of norleucine and methionine phenyl esters (Fig. 63, Scheme 1) and norleucine phosphonate hapten 4. This hapten locates in a deep cleft between the light and the heavy chain CDR3s where the hapten phenyl ring is bound in a hydrophobic pocket. A proposed mechanism for hydrolysis by 17E8 is similar to the hydrolytic mechanism of serine proteases. The active site of these proteases have Ser–His–Asp catalytic triad that is responsible for the high rate acceleration characteristic for the whole family of these enzymes. The combining site of 17E8 has a dyad Ser-H99-His-H35, which locates in the proximity to the phosphorous atom of the bound hapten (Fig. 64, see color plate). A lysine residue (Lys-H97), which probably stabilizes the developing oxyanion in the transition state and is an analog of an oxyanion hole of proteases, is present in the vicinity of the hapten. The comparison of the structures of the 17E8–hapten 4 and trypsin–pancreatic trypsin inhibitor complexes was done. The positions of the serine and histidine side chains in both active sites are superimposable. These data point out that the 17E8-combining site resembles the active site of serine proteases in many respects. The structural information obtained suggests the sites for specific mutations that may enhance the 17E8 hydrolytic activity.

Antibody CNJ206 induced by a phosphonate hapten catalyzes the hydrolysis of *p*-nitrophenyl ester (Fig. 63, Scheme 2). The complex of Fab CNJ206 with the hapten 3 was crystallized and its three-dimensional structure was deter-

Scheme 1

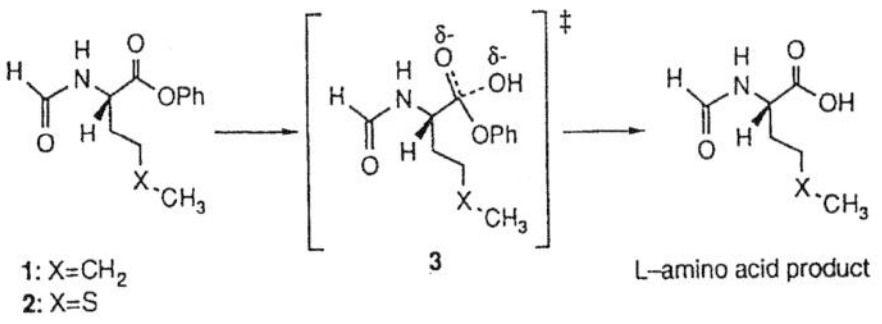

1: X=CH2
2: X=S

3

L-amino acid product

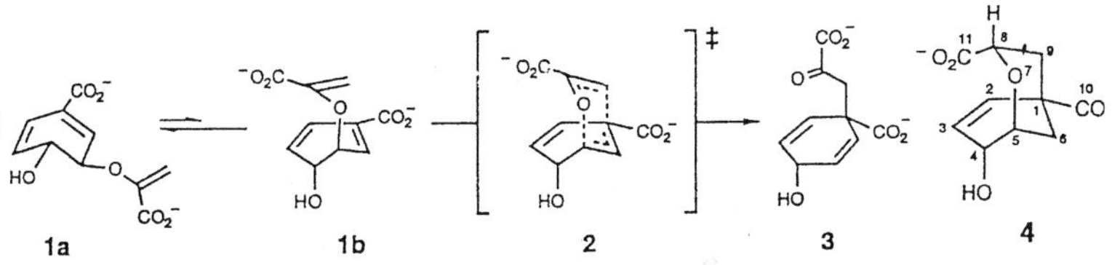

4

Scheme 2

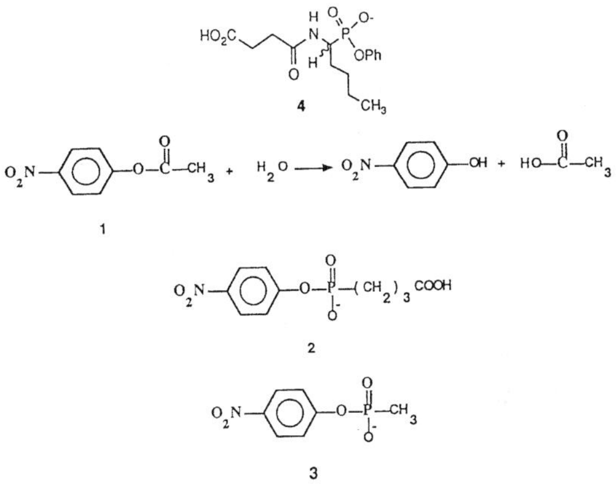

1

2

3

Scheme 3

1a 1b 2 3 4

Scheme 4

1: R=

2: R=

3

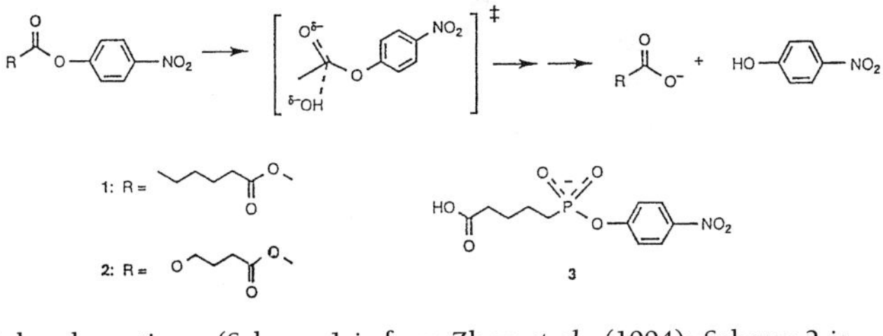

FIGURE 63 Antibody-catalyzed reactions. (Scheme 1 is from Zhou *et al.*, (1994); Scheme 2 is from Charbonnier *et al.*, (1995); Scheme 3 is from Haynes *et al.*, (1994), and Scheme 4 is from Patten *et al.*, (1996). All schemes reprinted with permission.

mined (Charbonnier *et al.*, 1995). The hapten is located in a deep cavity, which contains a large proportion of aromatic residues. Some structural similarity was noted between the combining sites of 17E8 and CNJ206. The bottoms of their hapten cavities are built from the same residues and they have the same conformation. From the 17E8 catalytic dyad Ser-His only His-H35 is present in the

CNJ206 combining site (Fig. 65). The hisitidine residues in both combining sites maintain very similar conformations. However, Ser-H99 of 17E8 is replaced by glycine in CNJ206. According to the structural analysis, the replacement of Gly-H99 by serine residue could be done without changing the conformation of the loop CDR3 of CNJ206. In both combining sites the phosphonate oxyanoin is stabilized by hydrogen bond donors. In 17E8 the ε-amino group of Lys-H97 participates in the stabilization and in CNJ206 peptide NH-groups of H100 and H101 (CDR3) and His-H35 are involved in this (Fig. 65).

The x-ray crystal structure of bacterial chorismatic mutase and the catalytic antibody 1F7 with chorismatic mutase activity have been solved as complexes with a transition state analog (Haynes *et al.*, 1994). The enzyme and 1F7 catalyzes the conversion of chorismate into prephenate (Fig. 63, Scheme 3) with the rate enhancements of about 2×10^6 and 10^2–10^4, respectively. The significant differences in positions of the transition state analog 4 in both complexes were found. The nature and number of interactions between the active

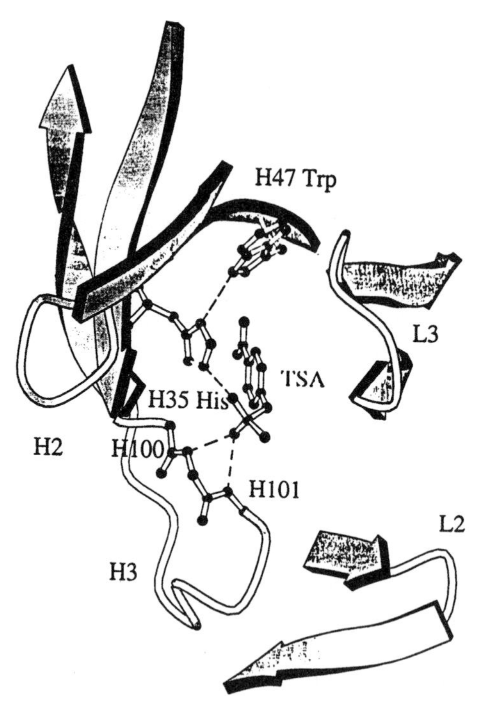

FIGURE 65 TSA hapten (Figure 63, Scheme 2, 3) in the combining site of the CNJ206 catalytic antibody. Hydrogen bonds involved in the stabilization of TSA are shown by broken line. (Charbonnier *et al.*, 1995. Reprinted with permission.)

site residues of the enzyme and 1F7 and the hapten **4** are also distinct. These differences are probably responsible for the lower activity of the catalytic antibody relative to the natural enzyme.

The antibody 48G7, which catalyses the hydrolytic reaction of p-nitrophenyl ester **1** and cabonate **2** (Fig. 63, Scheme 4), was induced by the transition state analog **3** (Patten *et al.*, 1996). The three-dimensional crystal structure of the 48G7 complex with the hapten **3** was solved at a high resolution (Wedermayr *et al.*, 1997a). No significant changes occur upon hapten binding. The hapten is bound with the aryl end buried deep in the binding site of a cavity type (10 Å deep and 5 Å wide) while the other end remains to be solvent-exposed. The phosphonate moiety is responsible for the majority of the binding energy. The same orientation is characteristic for complexes of other similar haptens with antibodies CNJ206 and 17E8.

The germ line variable genes of a 48G7 precursor were cloned, and expressed in *E. coli* cells. The affinity of the resulting germ line Fab fragment for the analog **3** is less than 10^{-4} that of the 48G7 antibody molecule and its catalytic activity was significantly lower when compared with 48G7. The primary structure of both antibodies was compared and it was found that during the affinity maturation process of 48G7 nine mutations occur. None of the mutated residues has direct contact with the hapten **3**. Hence, in the evolution of the catalytic activity of 48G7, the mutations that reorganized and stabilized the conformation of the active site were the most important ones (Wedermayer *et al.*, 1997b).

The x-ray structures of a family of three esterase-like antibodies (D2.3, D2.4, and D2.5) were analyzed at high resolution (Charbonnier *et al.*, 1997). The antibodies were identified by screening the entire hybridoma repertoire obtained in response to a phosphonate transition state analog. The D2 antibody molecules display significant structural differences in the combining sites. However, they bind the hapten in the same mode and the conformations of catalytic residues are similar. These results as well as the comparisons of several other hydrolytic antibodies studied suggest that the immune system responds to immunization by antigens with aryl phosphonates and phosphonamide groups by a common structural motif. The limited structural variations between the motif in the studied antibodies are responsible for the differences in catalytic efficiency (MacBeath and Hilvert, 1996). X-ray structures were also solved for unliganded D2.3 and its complexes with a substrate analog and with one of the reaction products. All obtained data provide a full description of the catalytic reaction pathway (Gigant *et al.*, 1997).

For the catalytic activity of antibody 6D9, the light chain residue His–L27 is critically important. The His-L27 → Ala antibody mutant has no catalytic activity and a large reduction in transition state binding, whereas the substrate binding does not change (Miyashita *et al.*, 1997).

B. Autoantibodies with Catalytic Activity

In the blood of patients with autoimmune disorders and normal individuals autoantibodies were found with specific proteolytic activity. The best studied are autoantibodies against vasoactive intestinal peptide (VIP), a 28-residue neuropeptide widely distributed in the nervous system. The anti-VIP antibodies with catalytic activity were isolated from sera of two asthma subjects and one normal individual (Paul *et al.*, 1989; 1991). One of these antibodies catalyzes hydrolysis of the Gln-16–Met-17 peptide bond in VIP, but two other anti-VIP antibodies cleave up to seven peptide bonds between different amino acid residues located mainly in the middle of VIP. The catalytic activity was specific and the antibodies did not hydrolyze polypeptides unrelated to VIP-like insulin in the primary structure.

In other experiments a monoclonal antibody (c23.5) against VIP was raised by immunization of mice by the peptide. This antibody hydrolyzes VIP or a synthetic peptide corresponding to the VIP residues 14–22 at peptide bonds between residues Lys-20–Lys-21 and Lys-21–Tyr-22 (Paul *et al.*, 1992). The isolated light chains of c23.5 also have proteolytic activity and cleave four peptide bonds located between the VIP residues 16 and 21. The heavy chains were catalytically inactive. By site directed mutagenesis, various residues in the c23.5 light chains were substituted by alanine residues (Gao *et al.*, 1995). The hydrolytic activity of the light chains was reduced by about 95% by substitution of Ser-27a (CDR1) or His-93 (CDR3) by alanine residues (or His-93 by arginine). According to a computer-generated model of the c23.5 light chains, the imidazole nitrogen of His-93 and the oxygen of Ser-27a are close enough (2.8 Å) to form a hydrogen bond, which facilitates nucleophylic attack on the peptide carbonyl group by Ser-27a (similarly to the mechanism of serine proteases).

Catalytic autoantibodies specific to thyroglobulin, which were isolated from sera of patients with Hashimoto's thyroiditis, are able to hydrolyze thyroglobulin. As a result of the proteolysis, several thyroglobulin fragments of different sizes were identified (Li *et al.*, 1995). Some data point to the location of antibody residues important for antigen binding near residues responsible for catalysis.

DNA hydrolyzing autoantibodies of IgM and IgG classes were found in sera of patients suffering from autoimmune diseases, particularly from systemic lupus erythmatosus (Shuster *et al.*, 1992). The incubation of plasmid DNA with these antibodies yields DNA degradation products that are different from those of deoxyribonuclease I and blood deoxyribonuclease. Kinetic properties of the Fab fragment isolated from an autoimmune antibody revealed a high affinity and significant catalytic efficiency of the DNA hydrolysis (Gololobov *et al.*, 1995).

It is quite possible that at least some of the catalytic autoantibodies are in fact anti-idiotypic antibodies, which mimic enzymes with corresponding activities. A confirmation of this possibility was received in experiments with an anti-idiotypic monoclonal antibody 9A8 with cholinesterase activity (Izadyar *et al.*, 1993). The 9A8 antibody was raised by immunization with another monoclonal antibody directed against the active site of acetylcholinesterase. The active site of this enzyme has a catalytic triad Glu,Ser,His. Using various inhibitors, two residues of this triad (Ser and His) were identified in the combining site of the 9A8 catalytic antibody. The cholinesterase activity of 9A8 resembles the activity of the enzyme in some respects but differs in others from both quantitative and qualitative points of view.

Catalytic autoantibodies can play a significant functional role. Anti-VIP antibodies of high affinity, which bind and hydrolyze VIP, were found in many asthma patients. This neuropeptide is a mediator of a system responsible for airway relaxation. The neutralization and cleavage of VIP by anti-VIP catalytic antibodies could be an important factor in the development of asthma (Paul, 1994). It has been known that light chains of immunoglobulins are synthesized in an excess of heavy chains (Nezlin and Kulpina, 1966). The catalytic activity of the free anti-VIP light chains could be an important factor in the development of asthma symptoms. High-affinity autoantibodies to thyroglobulin may remove this protein from serum and diminish in this way the autoimmune response (Li *et al.*, 1995). The high occurrence of catalytic antibodies in autoimmune strains of mice after immunization by a transition state analog raise a question about the possible link between autoimmunity and the appearance of catalytic antibodies (Tawfik *et al.*, 1995). It is probable that the immune system is able to suppress the formation of forbidden clones of antibody molecules, which may catalyze the modifications of self-antigens.

Specific antibodies against crystals can be regarded as a kind of catalytic antibody. After injections of crystals such as monosodium urate monohydrate, magnesium urate octahydrate, and allopurinol, rabbits developed antibodies that are able to catalyze the nucleation of the same type of crystals (Kam *et al.*, 1994). The enhancement of the crystal formation is highly specific and the cross-reactivity is low. Recently monoclonal anti-crystal antibodies were obtained and studies of their interactions with crystals are awaited.

REFERENCES

Al-Lazikani B., Lesk A.M., and Chothia C. (1997). Standard conformations for the canonical structures of immunoglobulines. *J. Mol. Biol.* **273**, 927–948.

Altschuh D., Vix O., Rees B., and Thierry J.-C. (1992). A conformation of cyclosporin A in aqueous environment revealed by the x-ray structure of a cyclosporin-Fab complex. *Science* **256**, 92–94.

Alzari P.M., Spinelli S., Mariuzza R.A., Boulot G., Poljak R.J., Jarvis J.M., and Milstein C. (1990). Three-dimensional structure determination of an anti-2-phenyloxazolone antibody: the role of somatic mutation and heavy/light chain pairing in the maturation of an immune response. *EMBO J.* **9**, 3807–3814.

Amzel L.M. (1992). Modeling the variable region of immunoglobulins. *Immunomethods* **1**, 91–95.
Amzel L.M., Garcia K.C., and Desiderio S. (1994). Do antiidiotypic antibodies mimic antigen? *Res. Immunol.* **145**, 53–55.
Arevalo J.H., Hassig C.A., Stura E.A., Sims M.J., Taussig M.J., and Wilson I.A. (1994). Structural analysis of antibody specificity. Detailed comparison of five Fab′-steroid complexes. *J. Mol. Biol.* **241**, 663–690.
Arevalo J.H., Stura E.A., Taussig M.J., and Wilson I.A. (1993a). Three-dimensional structure of an anti-steroid Fab′ and progesterone-Fab′ complex. *J. Mol. Biol.* **231**, 103–118.
Arevalo J.H., Taussig M.J., and Wilson I.A. (1993b). Molecular basis of crossreactivity and the limits of antibody-antigen complementarity. *Nature* **365**, 859–863.
Bajorath J., and Fine R.M. (1992). On the use of minimization from many randomly generated loop structures in modeling antibody combining sites. *Immunomethods* **1**, 137–146.
Bajorath J., Harris L., and Novotny J. (1995). Conformational similarity and systematic displacement of complementarity determining region loops in high resolution antibody x-ray structures. *J. Biol. Chem.* **270**, 22081–22084.
Bajorath J., and Sheriff S. (1996). Comparison of an antibody model with an x-ray structure: the variable fragment of BR96. *Proteins Struct. Funct. Genet.* **24**, 152–157
Ban N., Day J., Wang X., Ferrone S., and McPherson A. (1996). Crystal struture of an anti-anti-idiotype shows it to be self-complementary. *J. Mol. Biol.* **255**, 617–627.
Ban N., Escobar C., Garcia R., Hasel K., Day J., Greenwood A., and McPherson A. (1994). Crystal structure of an idiotype-anti-idiotype Fab complex. *Proc. Natl. Acad. Sci. USA* **91**, 1604–1608.
Ban N., Escobar C., Hasel K.W., Day J., Greenwood A., and McPherson A. (1995). Structure of an anti-idiotypic Fab against feline peritonitis virus-neutralizing antibody and a comparison with the complexed Fab. *FASEB J.* **9**, 107–114.
Bentley G.A., Boulot G., Riottot M.M., and Poljak R.J. (1990). Three-dimensional structure of an idiotope-anti-idiotope complex. *Nature* **348**, 254–257.
Bentley G.A., Boulot G., and Chitarra V. (1994). Cross-reactivity in antibody-antigen interactions. *Res. Immunol.* **145**, 45–48.
Bhat T.N., Bentley G.A., Fischmann T.O., Boulot G., and Poljak R.J. (1990). Small rearrangements in structures of Fv and Fab fragments of antibody D1.3 on antigen binding. *Nature* **347**, 483–485.
Bhat T.N., Bentley G.A., Boulot G., Greene M.I., Tello D., Dall'Acqua W., Souchon H., Schwarz F.P., Mariuzza R.A., and Poljak R.J. (1994). Bound water molecules and conformational stabilization help mediate an antigen-antibody association. *Proc. Natl. Acad. Sci. USA.* **91**, 1089–1093.
Bizebard T., Gigant B., Rigolet P., Rasmussen B., Diat O., Bösecke P., Wharton S.A., Skehel J.J., and Knossow M. (1995). Structure of influenza virus haemagglutinin complexed with a neutralizing antibody. *Nature* **376**, 92–94.
Bossart-Whitaker P., Chang C.Y., Novotny J., Benjamin D.C., and Sheriff S. (1995). The crystal structure of the antibody N10–staphylococcal nuclease complex at 2.9 Å resolution. *J. Mol. Biol.* **253**, 559–575.
Braden B.C., Souchon H., Eisele J.-L., Bentley G.A., Bhat T.N., Navaza J., and Poljak R.J. (1994). Three-dimensional structures of the free and the antigen-complexed Fab from monoclonal anti-lysozyme antibody D44.1. *J. Mol. Biol.* **243**, 767–781.
Braden B.C., and Poljak R.J. (1995). Structural features of the reactions between antibodies and protein antigens. *FASEB J.* **9**, 9–16.
Braden B.C., Fields B.A., Ysern X., Dall'Acqua W., Goldbaum F.A., Poljak R.J., and Mariuzza R. (1996). Crystal structure of an Fv-Fv idiotype–anti-idiotype complex at 1.9 Å resolution. *J. Mol. Biol.* **264**, 137–151.
Brady R.L., Edwards D.J., Hubbard R.E., Jiang J.-S., Lange G., Roberts S.M., Todd R.J., Adair J.R., Emtage J.S., King D.J., and Low D.C. (1992). Crystal structure of a chimeric Fab′ fragment of an antibody binding tumour cells. *J. Mol. Biol.* **227**, 253–264.

Bruccoleri R.E., and Novotny J. (1992). Antibody modeling using the conformational search program CONGEN. *Immunomethods* **1**, 96–106.

Brünger A.T. (1993). Structure determination of antibodies and antibody-antigen complexes by molecular replacement. *Immunomethods* **3**, 180–190.

Brünger A.T., Leahy D.J., Hynes T.R., and Fox R.O. (1991). 2.9 Å Resolution structure of an anti-dinitrophenyl-spin-label monoclonal antibody Fab fragment with bound hapten. *J. Mol. Biol.* **221**, 239–256.

Chacko S., Silverton E., Kam-Morgan L., Smith-Gill S., Cohen G., and Davies D. (1995). Structure of an antibody-lysozyme complex. Unexpected effect of a conservative mutation. *J. Mol. Biol.* **245**, 261–274.

Charbonnier J.-B., Carpenter E., Gigant B., Golinelli-Pimpaneau B., Eshhar Z., Green B.S., and Knossow M. (1995). Crystal structure of the complex of a catalytic antibody Fab fragment with a transition state analog: structural similarities in esterase-like catalytic antibodies. *Proc. Natl. Acad. Sci. USA.* **92**, 11721–11725.

Charbonnier J.-B., Golinelli-Pimpaneau B., Gigant B., Tawfik D.S., Chap R., Schindler D.G., Kim S.-H., Green B.S., Eshhar Z., and Knossow M. (1997). Structural convergence in the active sites of a family of catalytic antibodies. *Science* **275**, 1140–1142.

Chitarra V., Alzari P., Bentley G.A., Bhat T.N., Eiselé J.-L., Houdusse A., Lescar J., Souchon H., and Poljak R.J. (1993). Three-dimensional structure of a heteroclitic antigen-antibody cross-reaction complex. *Proc. Natl. Acad. Sci. USA* **90**, 7711–7715.

Chothia C., Lesk A.M., Gherardi E., Tomlinson I.M., Walter G., Marks J.D., Llewelyn M.B., and Winter G. (1992). Structural repertoire of the human V_H segments. *J. Mol. Biol.* **227**, 799–817

Chothia C., Lesk A.M., Levitt M., Amot A.G., Mariuzza R.A., Phillips S.E.V., and Poljak R. (1986). The predicted structure of immunoglobulin D1.3 and its comparison with the crystal structure. *Science* **233**, 755–758.

Chothia C., Lesk A.M., Tramontano A., Levitt M., Smith-Gill S.J., Air G., Sheriff S., Padlan E.A., Davies D., Tulip W.R., Colman P.M., Spinelli S., Alzari P.M., and Poljak R.J. (1989). Conformations of immunoglobulin hypervariable regions. *Nature* **342**, 877–883.

Chothia C., Novotny J., Bruccoleri R., and Karplus M. (1985). Domain association in immunoglobulin molecules. The packing of variable domains. *J. Mol. Biol.* **186**, 651–663.

Churchill M.E.A., Stura E.A., Pinilla C., Appel J.R., Houghten R.A., Kono D.H., Balderas R.S., Fieser G.G., Schulze-Gahmen U., and Wilson I.A. (1994). Crystal structure of a peptide complex of anti-influenza peptide antibody Fab 26/9. Comparison of two different antibodies bound to the same peptide antigen. *J. Mol. Biol.* **241**, 534–556.

Colman P.M., (1988). Structure of antibody-antigen complexes: implications for immune recognition. *Adv. Immunol.* **43**, 99–132.

Corper A.L., Sohi M.K., Bonagura V.R., Steinitz M., Jefferis R., Feinstein A., Beale D., Tausiig M.J., and Sutton B.J. (1997). Structure of human IgM rheumatoid factor Fab bound to its autoantigen IgG Fc reveals a novel topology of antibody-antigen interaction. *Nature Struct. Biol.* **4**, 374–381.

Cygler M. (1994). Recognition of carbohydrates by antibodies. *Res. Immunol.* **145**, 36–40.

Cygler M., Rose D.R., and Bundle D.R. (1991). Recognition of a cell-surface oligosaccharide of pathogenic *Salmonella* by an antibody Fab fragment. *Science* **253**, 442–445.

Davies D.R., Padlan E.A., and Sheriff S. (1990). Antibody-antigen complexes. *Ann. Rev. Biochem.* **59**, 439–473.

Desmyter A., Transue T.R., Ghahroudi M..A., Dao Thi M.-H., Poortmans F., Hamers R., Muyldermans S., and Wyns L. (1996). Crystal structure of a camel single-domain V_H antibody fragment in complex with lysozyme. *Nature Struct. Biol.* **3**, 803–811.

Driggers E.M., and Schultz P.G. (1996). Catalytic antibodies. *Adv. Prot. Chem.* **49**, 261–287.

Edwards D.J., Hubbard R.E., and Brady R.L. (1992). Homology modeling of antibody combining sites. *Immunomethods* **1**, 71–79.

Eigenbrot C., Randal M., Presta L., Carter P., and Kossiakoff A.A. (1993). X-ray structures of the antigen-binding domains from three variants of humanized anti-p185^{HER2} antibody 4D5 and comparison with molecular modeling. *J. Mol. Biol.* **229**, 969–995.

Essen L.-O., and Skerra A. (1994). The *de novo* design of an antibody combining site. Crystallographic analysis of the V_L domain confirms the structural model. *J. Mol. Biol.* **238**, 226–244.

Evans S.V., Rose D.R., To R., Young N.M., and Bundle D.R. (1994). Exploring the mimicry of polysaccharide antigens by anti-idiotypic antibodies. The crystallization, molecular replacement, and refinement to 2.8 Å resolution of an idiotope–anti-idiotope Fab complex and of the unliganded anti-idiotope Fab. *J. Mol. Biol.* **241**, 691–705.

Fan Z.-C., Shan L., Guddat L.W., He X.-M., Gray W.R., Raison R.L., and Edmundson A.B. (1992). Three-dimensional structure of an Fv from a human IgM immunoglobulin. *J. Mol. Biol.* **228**, 188–207.

Fields B.A., Goldbaum F.A., Ysern X., Poljak R.J., and Mariuzza R.A. (1995). Molecular basis of antigen mimicry by an anti-idiotope. *Nature* **374**, 739–742.

Fischmann T.O., Bentley G.A., Bhat T.N., Boulot G., Mariuzza R.A., Phillips S.E., Tello D., and Poljak R.J. (1991). Crystallographic refinement of the three-dimensional structure of the Fab D1.3-lysozyme complex at 2.5 Å resolution. *J. Biol. Chem.* **266**, 12915–12920.

Foote J., and Winter G. (1992). Antibody framework residues affecting the conformation of the hypervariable loops. *J. Mol. Biol.* **224**, 487–499.

Franěk F., and Nezlin R. (1963). Recovery of antibody combining activity by interaction of different peptide chains isolated from purified horse antitoxins. *Folia Microbiol.* **8**, 197–201.

Gao Q.-S., Sun M., Rees A.R., and Paul S. (1995). Site-directed mutagenesis of proteolytic antibody light chain. *J. Mol. Biol.* **253**, 658–664.

Garcia K.C., Ronco P.M., Verroust P.J., Brünger A.T., and Amzel L.M. (1992). Three-dimensional structure of an angiotensin II–Fab complex at 3 Å: hormone recognition by an anti-idiotypic antibody. *Science* **257**, 502–507.

Gibrat J.-F., Higo J., Collura V., and Garnier J. (1992). A simulated annealing method for modeling the antigen-combining site of antibodies. *Immunomethods* **1**, 107–125.

Gigant B., Charbonnier J.-B., Eshhar Z., and Green B.S. (1997). X-ray structures of a hydrolytic antibody and of complexes elucidate catalytic pathway from substrate binding and transition state stabilization through water attack and product release. *Proc. Natl. Acad. Sci. USA* **94**, 7857–7861.

Givol D. (1991). The minimal antigen-binding fragment of antibodies—Fv fragment. *Mol. Immunol.* **28**, 1379–1387.

Gololobov G.V., Chernova E.A, Schourov D.V., Smirnov I.V., Kudelina I.A., and Gabibov A.G. (1995). Cleavage of supercoiled plasmid DNA by autoantibody Fab fragment: application of the flow linear dichroism technique. *Proc. Natl. Acad. Sci. USA* **92**, 254–257.

Guddat L.W., Shan L., Anchin J.M., Linthicum D.S., and Edmundson A.B. (1994). Local and transmitted conformational changes on complexation of an anti-sweetener Fab. *J. Mol. Biol.* **236**, 247–274.

Guddat L.W., Shan L., Fan Z.-C., Andersen K.N., Rosauer R., Linthicum D.S., and Edmundson A.B. (1995). Intramolecular signaling upon complexation. *FASEB J.* **9**, 101–106.

Haber E., and Richards F.F. (1966). The specificity of antigenic recognition of antibody heavy chains. *Proc. Natl. Acad. Sci. USA* **166**, 176–187.

Hawkins R.E., Russell S.J., Baier M., and Winter G. (1993). The contribution of contact and non-contact residues of antibody in the affinity of binding to antigen. *J. Mol. Biol.* **234**, 958–964.

Haynes M.R., Stura E.A., Hilvert D., and Wilson I.A. (1994). Routes to catalysis: structure of a catalytic antibody and comparison with its natural counterpart. *Science* **263**, 646–652.

He X.M., Rüker F., Casale E., and Carter D.C. (1992). Structure of a human monoclonal antibody Fab fragment against gp41 of human immunodeficiency virus type 1. *Proc. Natl. Acad. Sci. USA* **89**, 7154–7158.

Herron J.N., He X.M., Mason M.L., Voss E.W., and Edmundson A.B. (1989). Three-dimensional structure of a fluorescein-Fab complex crystallized in 2–methyl-2,4–pentanediol. *Proteins Struct. Funct. Genet.* **5**, 271–280.

Herron J.N., He X.M., Ballard D.W., Blier P.R., Pace P.E., Bothwell A.L.M., Voss E.W., and Edmundson A.B. (1991). An autoantibody to single-stranded DNA: comparison of the three-dimensional structures of the unliganded Fab and a deoxynucleotide-Fab complex. *Proteins Struct. Funct. Genet.* **11**, 159–175.

Höhne W.E., Küttner G., Kieβig S., Hausdorf G., Grunow R., Winkler K., Wessner H., Gieβmann E., Stigler R., Schneider-Mergener J., von Baehr R., and Schomburg D. (1993). Structural base of the interaction of a monoclonal antibody against p24 of HIV-1 with its peptide epitope. *Mol. Immunol.* **30**, 1213–1221.

Izadyar L., Friboulet A., Remy M.H., Roseto A., and Thomas D. (1993). Monoclonal anti-idiotypic antibodies as functional internal images of enzyme active sites: production of a catalytic antibody with a cholinetserase activity. *Proc. Natl. Acad. Sci. USA* **90**, 8876–8880.

Jacobo-Molina A., Ding J., Nanni R.G., Clark A.D., Lu X., Tantillo C., Williams R.L., Kamer G., Ferris A.L., Clark P., Hizi A., Hughes S.H., and Arnold E. (1993). Crystal structure of human immunodeficiency virus type 1 reverse transcriptase complexed with double-stranded DNA at 3.0 Å resolution shows bent DNA. *Proc. Natl. Acad. Sci. USA* **90**, 6320–6324.

Janda K.D., Lo L.-C., Lo C.-H.L., Sim M.-M., Wang R., Wong C.-H., and Lerner R.A. (1997). Chemical selection for catalysis in combinatorial antibody libraries. *Science* **275**, 945–948.

Jeffrey P.D., Bajorath J., Chang C.Y., Yelton D., Hellström I., Hellström K.E., and Sheriff S. (1995a). The x-ray structure of an anti-tumour antibody in complex with antigen. *Nature Struct. Biol.* **2**, 446–477.

Jeffrey P.D., Schildbach J.F., Chang C.Y., Kussie P.H., Margolies M.N., and Sheriff S. (1995b). Structure and specificity of the anti-digoxin antibody 40–50. *J. Mol. Biol.* **248**, 344–360.

Jeffrey P.D., Strong R.K., Sieker L.C., Chang C.Y.Y., Campbell R.L., Petsko G.A., Haber E., Margolies M.N., and Sheriff S. (1993). 26-10 Fab-digoxin complex: affinity and specificity due to surface complementarity. *Proc. Natl. Acad. Sci. USA* **90**, 10310–10314.

Jencks W.P. (1969). *Catalysis in Chemistry and Enzymology.* p. 288. MacGraw-Hill, New York.

Jerne N.K. (1974). Towards a network theory of the immune system. *Ann. Immunol. (Inst.Pasteur)* **125C**, 373–389.

Kabat E.A., and Wu T.T. (1971). Attempts to locate complementarity determining residues in the variable positions of light and heavy chains. *Ann. N Y Acad. Sci.* **190**, 382–393.

Kabat E.A., and Wu T.T. (1991). Identical V region amino acid sequences and segments of sequences in antibodies of different specificities. Relative contributions of V_H and V_L genes, minigenes, and complementarity-determining regions of antibody-combining sites. *J. Immunol.* **147**, 1709–1719.

Kabat E.A., Wu T.T., Perry H.M., Gottesman K.S., and Foeller C. (1991). *Sequences of Proteins of Immunological Interest*, 5th ed. Public Health Service, National Institutes of Health, Washington, DC.

Kam M., Perl-Treves D., Sfez R., and Addadi L. (1994). Specificity in the recognition of crystals by antibodies. *J. Mol. Recogn.* **7**, 257–264.

Kodandapani R., Veerapandian B., Kunicki T.J., and Ely K.R. (1995). Crystal structure of the OPG2 Fab. An antireceptor antibody that mimicks an RGD cell adhesion site. *J. Biol. Chem.* **270**, 2268–2273.

Lascombe M.-B., Alzari P.M., Poljak R.J., and Nisonoff A. (1992). Three-dimensional structure of two crystal forms of FabR19.9 from a monoclonal anti-arsonate antibody. *Proc. Natl. Acad. Sci. USA* **89**, 9429–9433.

Lawrence M.C., and Colman P.M. (1993). Shape complementarity at protein/protein interfaces. *J. Mol. Biol.* **234**, 946–950.

Lescar J., Pellegrini M., Souchon H., Tello D., Poljak R.J., Peterson N., Greene M., and Alzari P.M. (1995). Crystal structure of a cross-reaction complex between Fab F9.13.7 and guinea fowl lysozyme. *J. Biol. Chem.* **270**, 18067–18076.

Lescar J., Stouracova R., Riottot M.-M., Chitarra V., Brynda J., Fabry M., Horeisi M., Sedlacek J., and Bentley G.A. (1997). Three-dimensional structure of an Fab-peptide complex: structural basis of HIV-1 protease inhibition by a monoclonal antibody. *J. Mol. Biol.* **267**, 1207–1222.

Lesk A.M., and Tramontano A. (1991). Antibody structure and structural predictions useful in guiding antibody engineering. In: *Antibody Engineering* (Ed. Borrebaeck C.A.K.). pp. 1–38. Freeman, New York.

Li L., Paul S., Tyutyulkova S., Kazatchkine M.D., and Kaver S. (1995). Catalytic activity of anti-thyroglobulin antibodies. *J. Immunol.* **154**, 3328–3332.

Li Y., Spellerberg M.B., Stevenson F.K., Capra J.D., and Potter K.N. (1996). The I binding specificity of human V_H4-34 (V_H4-21) encoded antibodies is determined by both V_H framework region 1 and complementarity determining region 3. *J. Mol. Biol.* **256**, 577–589.

Liu H., Smith T.J., Lee W., Mosser A.G., Rueckert R.R., Olson N.H., Cheng R.H., and Baker T.S. (1994). Structure determination of an Fab fragment that neutralizes human rhinovirus 14 and analysis of the Fab-virus complex. *J. Mol. Biol.* **240**, 127–137.

MacBeath G., and Hilvert D. (1996). Hydrolytic antibodies: variations on a theme. *Chem. Biol.* **3**, 433–445.

MacCallum R.M., Martin A.C.R., and Thornton J.M. (1996). Antigen-anibody interactions: contact analysis and binding site topography. *J. Mol. Biol.* **262**, 732–745.

Mäkelä O. (1965). Single lymph node cells producing heteroclitic bacteriophage antibody. *J. Immunol.* **95**, 378–386.

Malby R.L., Tulip W.R., Harley V.R., McKimm-Breschkin J.L., Laver W.G., Webster R.G., and Colman P.M. (1994). The structure of a complex between the NC10 antibody and influenza virus neuraminidase and comparison with the overlapping binding site of the NC41 antibody. *Structure* **2**, 733–746.

Mandal C., Kingery B.D., Anchin J.M., Subramaniam S., and Linthicum D.C. (1996). ABGEN: a knowledge-based automated approach for antibody structure modeling. *Nature Biotechnol.* **14**, 323–327.

Marquart M., Deisenhofer J., Huber R., and Palm W. (1980). Crystallographic refinement and atomic models of the intact immunoglobulin molecule Kol and its antigen-binding fragment at 3.0 Å and 1.9 Å resolution. *J. Mol. Biol.* **141**, 369–391.

Martin A.C.R., Cheetham J.C., and Rees A.R. (1991). Molecular modeling of antibody combining sites. *Meth. Enzymol.* **203**, 121–153.

Mayforth R.D. (1993). *Designing Antibodies.* Academic Press, San Diego.

Metzger H. (1978). Effect of antigen on antibodies: recent studies. *Contemp. Topics Molec. Immunol.* **7**, 191–224.

Mian I.S., Bradwell A.R., and Olson A.J. (1991). Structure, function and properties of antibody binding sites. *J. Mol. Biol.* **217**, 133–151.

Miyashita H., Hara T., Tanimura R., Fukuyama S., Cagnon C., Kohara A., and Fujii I. (1997). Site-directed mutagenesis of active site contact residues in a hydrolytic abzyme: evidence for an essential histidine involved in transition state stabilization. *J. Mol. Biol.* **267**, 1247–1257.

Mizutani R., Miura K., Nakayama T., Shimada I., Arata Y., and Satow Y. (1995). Three-dimensional structures of the Fab fragment of murine N1G9 antibody from the primary immune response and of its complex with (4-hydroxy-3-nitrophenyl)acetate. *J. Mol. Biol.* **254**, 208–222.

Near R.I., Mudgett-Hunter M., Novotny J., Bruccoleri R., and Ng S.C. (1993). Characterization of an anti-digoxin antibody binding site by site-directed *in vitro* mutagenesis. *Mol. Immunol.* **30**, 369–377.

Nezlin R., and Kulpina L. (1966). Microglobulin synthesized in cell culture of the lymph nodes and spleen of the rabbit. *Nature* **212**, 845.

Novotny J., Bruccoleri R.E., and Haber, E. (1990). Computer analysis of mutations that affect antibody specificity. *Proteins Sruct. Funct. Genet.* **7**, 93–98.

Novotny J., Bruccoleri R.E., and Saul F.A. (1989). On the attribution of binding energy in antigen-antibody complexes McPC 603, D1.3, and HyHEL-5. *Biochemistry* **28**, 4735–4749.

Oudin J. (1974). L'idiotypie des anticorps. *Ann. Immunol. (Inst. Pasteur)* **125C**, 309–337.

Padlan E.A. (1990). On the nature of antibody combining sites: unusual structural features that may confer on these sites an enhanced capacity for binding ligands. *Proteins Struct. Funct. Genet.* 7, 112–124.

Padlan E.A. (1991). A possible procedure for reducing the immunogenicity of antibody variable domains while preserving their ligand-binding properties. *Mol. Immunol.* **28**, 489–498.

Padlan E.A. (1992). Modeling of antibody combining sites: goals, expectations, and realities. *Immunomethods* **1**, 65–70.

Padlan E.A. (1994). Anatomy of the antibody molecule. *Mol. Immunol.* **31**, 169–217.

Padlan E.A. (1996). X-ray crystallography of antibodies. *Adv. Prot. Chem.* **49**, 57–133.

Padlan E.A., Abergel C., and Tipper J.P. (1995). Identification of specificity-determining residues in antibodies. *FASEB J.* **9**, 133–139.

Padlan E.A, and Kabat E.A. (1991). Modeling of antibody combining sites. *Meth. Enzymol.* **203**, 3–21.

Padlan E.A., Silverton E.W., Sheriff S., Cohen G.H., Smith-Gill S.J., and Davies D.R. (1989). Structure of an antibody-antigen complex: crystal structure of the HyHEL-10 Fab-lysozyme complex.. *Proc. Natl. Acad. Sci. USA* **86**, 5938–5942.

Pan Y., Yuhasz S.C., and Amzel L.M. (1995). Anti-idiotypic antibodies: biological function and structural studies. *FASEB J.* **9**, 43–49.

Panka D.J., Mudgett-Hunter M., Parks D.R., Peterson L.L., Herzenberg L.A., Haber E., and Margolies M.N. (1988). Variable region framework differences result in decreased or increased affinity of variant anti-digoxin antibodies. *Proc. Natl. Acad. Sci. USA* **85**, 3080–3084.

Patten P.A., Gray N.S., Yang P.L., Marks C.B., Wedemayer G.J., Boniface J.J., Stevens R.C., and Schultz P.G. (1996). The immunological evolution of catalysis. *Science* **271**, 1086–1091.

Paul S. (1994). Catalytic activity of anti-ground state antibodies, antibody subunits, and human autoantibodies. *Appl. Biochem. Biotechnol.* **47**, 241–253.

Paul S., Volle D.J., Beach CM., Johnson D.R., Powell M.J., and Massey R.J. (1989). Catalytic hydrolysis of vasoactive intestinal peptide by human antibody. *Science* **244**, 1158–1162.

Paul S., Sun M., Mody B., Eklund S.H., Beach C.M., Massey R.J., and Hamel F. (1991). Cleavage of vasoactive intestinal peptide at multiple sites by autoantibodies. *J. Biol. Chem.* **266**, 16128–16134.

Paul S., Sun M., Mody R., Tewary H.K., Stemmer P., Massey R.J., Gianferrara T., Mehrotra S., Dreyer T., Meldal M., and Tramontano A. (1992). Peptidolytic monoclonal antibody elicited by a neuropeptide. *J. Biol. Chem.* **267**, 13142–13145.

Pedersen J., Searle S., Henry A., and Rees A.R. (1992). Antibody modeling: beyond homology. *Immunomethods* **1**, 126–136.

Poljak R.J. (1994). An idiotope-anti-idiotope complex and the structural basis of molecular mimicking. *Proc. Natl. Acad. Sci. USA* **91**, 1599–1600.

Poljak R.J., Amzel L.M., Avey H.P., Chen B.L., Phizackerlay R.P., and Saul F. (1973). Three-dimensional structure of the Fab′ fragment of a human immunoglobulin at 2.8Å-resolution. *Proc. Natl. Acad. Sci. USA* **70**, 3305–3310.

Pollack S.J., Jacobs J.W., and Schultz P.G. (1986). Selective chemical catalysis by an antibody. *Science* **234**, 1570–1573.

Prasad L., Sharma S., Vandonselaar M., Quail J.W., Lee J.S., Waygood E.B., Wilson K.S., Dauter Z., and Delbaere L.T.J. (1993). Evaluation of mutagenesis for epitope mapping. Structure of an antibody-protein antigen complex. *J. Biol. Chem.* **268**, 10705–10708.

Reczko M., Martin A.C.R., Bohr H., and Suhai S. (1995). Prediction of hypervariable CDR-H3 loop structures in antibodies. *Protein Engin.* **8**, 389–395.

Rees A.R., Pedersen J.T., Searle S.M.J. Henry A.H., and Webster D.M. (1994). Antibody structure from x-ray crystallography and molecular modeling. In: *Immunochemistry* (Eds. van Oss C.J., and van Regenmortel M.H.V.). pp. 615–650. Marcel Dekker, New York.

Rini J.M., Stanfield R.L., Stura E.A., Salinas P.A., Profy A.T., and Wilson I.A. (1993). Crystal structure of a human immunodeficiency virus type 1 neutralizing antibody 50.1, in complex with its V3 loop peptide antigen. *Proc. Natl. Acad. Sci. USA* **90**, 6325–6329.

Rini J.M., Schulze-Gahmen U., and Wilson I.A. (1992). Structural evidence for induced fit as a mechanism for antibody-antigen recognition. *Science* **255**, 959–965.

Roberts V.A., and Getzoff E.D. (1995). Metalloantibody design. *FASEB J.* **9**, 94–100.

Roberts V.A., Iverson B.L., Iverson S.A., Benkovic S.J., Lerner R.A., Getzoff E.D., and Tainer J.A. (1990). Antibody remodeling: a general solution to the design of a metal-coordination site in an antibody binding pocket. *Proc. Natl. Acad. Sci. USA* **87**, 6654–6658.

Satow Y., Cohen G.H., Padlan E.A., and Davies D.R. (1986). Phosphocholine binding immunoglobulin Fab McPC603. An x-ray diffraction study at 2.7 Å. *J. Mol. Biol.* **190**, 593–604.

Saul F.A., and Alzari P.M. (1996). Crystallographic studies of antigen-antibody interactions. In: *Methods in Molecular Biology,* Vol. 66, *Epitope Maping Protocols* (Ed. Morris G.E.). pp. 11–23. Humana Press, Totowa, NJ.

Saul F.A., and Poljak R.J. (1992). Crystal structure of human immunoglobulin fragment Fab New refined at 2.0 Å resolution. *Proteins Struct. Funct. Genet.* **14**, 363–371.

Scherf T., Hiller R., and Anglister J. (1995). NMR observation of interactions in the combining site region of an antibody using a spin-labeled peptide antigen and NOESY difference spectroscopy. *FASEB J.* **9**, 120–126.

Schiweck W., and Skerra A. (1997). The rational construction of an antibody against cystatin: lessons from the crystal structure of an artificial Fab fragment. *J. Mol. Biol.* **268**, 934–951.

Schultz P.G., and Lerner R.A. (1995). From molecular diversity to catalysis: lessons from the immune system. *Science* **269**, 1835–1842.

Schulze-Gahmen U., Rini J., and Wilson I.A. (1993). Detailed analysis of the free and bound conformation of an antibody. X-ray structures of Fab 17/9 and three different Fab-peptide complexes. *J. Mol. Biol.* **234**, 1098–1118.

Segal D.M., Padlan E.A., Cohen G.H., Rudikoff S., Potter M., and Davies D.R. (1974). The three-dimensional structure of a phosphorylcholine-binding mouse immunoglobulin Fab and the nature of the antigen binding site. *Proc. Natl. Acad. Sci. USA* **71**, 4298–4302.

Searle S.J., Pedersen J.T., Henry A.H., Webster D.M., and Rees A.R. (1995). Antibody structure and function. In *Antibody Engineering* (Ed. Borrebaeck C.A.K.). pp. 3–51. Oxford University Press, New York.

Sheriff S. (1993a). Some methods for examining the interactions between two molecules. *Immunomethods* **3**, 191–196.

Sheriff S. (1993b). Antibody-protein complexes. *Immunomethods* **3**, 222–227.

Sheriff S., Chang C.Y., Jeffrey P.D., and Baiorath J. (1996). X-ray structure of the uncomplexed anti-tumor antibody BR96 and comparison with its antigen-bound form. *J. Mol. Biol.* **259**, 938–946.

Sheriff S., Silverton E.W., Padlan E.A., Cohen G.H., Smith-Gill S.J., Finzel B.C., and Davies D.R. (1987). Three-dimensional structure of an antibody-antigen complex. *Proc. Natl. Acad. Sci. USA* **84**, 8075–8079.

Shoham M., Proctor P., Hughes D., and Baldwin E.T. (1991). Crystal parameters and molecular replacement of an anticholera toxin peptide complex. *Proteins Struct. Funct. Genet.* **11**, 218–222.

Shuster A.M., Gololobov G.V., Kvashuk O.A., Bogomolova A.E., Smirnov I.V., and Gabibov A.G. (1992). DNA hydrolyzing antibodies. *Science* **256**, 665–667.

Spinelli S., and Alzari P.M. (1994). Structural implications of somatic mutations during the immune response to 2–phenyloxazolone. *Res. Immunol.* **145**, 41–44.

Spinelli S., Frenken L., Bourgeois D., de Ron L., Bos W., Verrips T., Anguille C., Cambillau C., and Tegoni M. (1996). The crystal structure of a llama heavy chain variable domain. *Nature Struct. Biol.* **3**, 752–757.

Stanfield R.L., Fieser T.M., Lerner R.A., and Wilson I.A. (1990). Crystal structures of an antibody to a peptide and its complex with peptide antigen at 2.8 Å. *Science* **248**, 712–719.

Stanfield R.L., and Wilson I.A. (1993). X-ray crystallographic studies of antibody-peptide complexes. *Immunomethods* **3**, 211–221.

Stanfield R.L., Takimoto-Kamimura M., Rini J.M., Profy A.T., and Wilson I.A. (1993). Major antigen-induced domain rearrangements in an antibody. *Structure* **1**, 83–93.

Stigler R.-D., Rüker F., Katinger D., Elliott G., Höhne W., Henklein P., Ho J.X., Keeling K., Carter D.C., Nugel E., Kramer A., Porstmann T., and Schneider-Mergener J. (1995). Interaction between a Fab fragment against gp41 of human immunodeficiency virus I and its peptide epitope: characterization using a peptide epitope library and molecular modeling. *EMBO J.* **15**, 471–479.

Stura E.A., Fieser G.G., and Wilson I.A. (1993). Crystallization of antibodies and antibody-antigen complexes. *Immunomethods* **3**, 164–179.

Tawfik D.S., Chap R., Green B.S., Sela M., and Eshhar Z. (1995). Unexpectedly high occurence of catalytic antibodies in MRL/*lpr* and SJL mice immunized with a transition state analog: is there a linkage to autoimmunity? *Proc. Natl. Acad. Sci. USA* **92**, 2145–2149.

Tawfik D.S., Green B.S., Chap R., Sela M., and Eshhar Z. (1993). catELISA: a facile general route to catalytic antibodies. *Proc. Natl. Acad. Sci. USA* **90**, 373–377.

Thanavala Y., and Pride M.W. (1994). Immunoglobulin idiotypes and anti-idiotypes. In: *Immunochemistry* (Eds. van Oss C.J., and van Regenmortel M.H.V.). pp. 69–91. Marcel Dekker, New York.

Thornton J.M. (1991). Modelling antibody combining sites: a review. In: *Catalytic Antibodies. Ciba Foundation Symposium* Vol. 159 (Eds. Chadwick D.J., and March J.). pp. 55–71. Wiley, Chichester, UK.

Tomlinson I.M., Cox J.P.L., Gherardi E., Lesk A.M., and Chothia C. (1995). The structural repertoire of the human V_κ domain. *EMBO J.* **14**, 4628–4638.

Tormo J., Blaas D., Parry N.R., Rowlands D., Stuart D., and Fita I. (1994). Crystal structure of a human rhinovirus neutralizing antibody complexed with a peptide derived from viral capsid protein VP2. *EMBO J.* **13**, 2247–2256.

Tormo J., Centeno N.B., Fontana E., Bubendorfer T., Fita I., and Blaas D. (1995). Docking of a human rhinovirus neutralizing antibody onto the viral capsid. *Proteins Struct. Funct. Genet.* **23**, 491–501.

Totrov M., and Abagyan R. (1994). Detailed *ab initio* prediction of lysozyme-antibody complex with 1.6 Å accuracy. *Nature Struct. Biol.* **1**, 259–263.

Tramontano A., Chothia C., and Lesk A.M. (1990). Framework residue 71 is a major determinant of the position and conformation of the second hypervariable region in the V_H domains of immunoglobulins. *J. Mol. Biol.* **215**, 175–182.

Tramontano A., Janda K.D., and Lerner R.A. (1986). Catalytic antibodies. *Science* **234**, 1566–1570.

Trinh C.H., Hemmington S.D., Verhoeyen M.E., and Phillips S.E.V. (1997). Antibody fragment Fv4155 bound to two closely related steroid hormones: the structural basis of fine specificity. *Structure* **5**, 937–948.

Tulip W.R., Varghese J.N., Laver W.G., Webster R.G., and Colman P.M. (1992a). Refined crystal structure of the influenza virus N9 neuraminidase–NC41 Fab complex. *J. Mol. Biol.* **227**, 122–148.

Tulip W.R., Varghese J.N., Webster R.G., Laver W.G., and Colman P.M. (1992b). Crystal structures of two mutant neuraminidase–antibody complexes with amino acid substitutions in the interface. *J. Mol. Biol.* **227**, 149–159.

Ulrich H.D., and Schultz P.G. (1998). Analysis of hapten binding and catalytic determinants in a family of catalytic antibodies. *J. Mol. Biol.* 275, 95–111.

Utsumi S., and Karush F. (1964). The subunits of purified rabbit antibody. *Biochemistry* **3**, 1329–1338.

Vargas-Madrazo E., Lara-Ochoa F., and Almagro J.C. (1995). Canonical structure repertoire of the antigen-binding site of immunoglobulins suggests strong geometrical restrictions associated to the mechanism of immune recognition. *J. Mol. Biol.* **254**, 497–504.

Viswanathan M., Anchin J.M., Droupadi P.R., Mandal C., Linthicum D.C., and Subramaniam S. (1995). Structural predictions of the binding site architecture for monoclonal antibody NC6.8 using computer-aided molecular modeling, ligand binding, and spectroscopy. *Biophys. J.* **69**, 741–753.

Wade W.S., Ashley J.A., Jahangiri G.K., McElhaney G., Janda K.D., and Lerner R.A. (1993). A highly specific metal-activated catalytic antibody. *J. Am. Chem. Soc.* **115**, 4906–4907.

Wade W.S and Scanlan T.S. (1997). The structural and functional basis of antibody catalysis. *Annu. Rev. Biphys. Biomol. Struct.* **26**, 461–493.

Webster D.M., and Rees A.R. (1995). Molecular modeling of antibody-combining sites. In: *Antibody Engineering Protocols* (Ed. Paul S.). pp. 17–49. Humana Press, Totowa, NJ.

Wedemayer G.J., Wang L.H., Patten P.A., Schultz P.G., and Stevens R.C. (1997a). Crystal structures of the free and liganded form an esterolytic catalytic antibody. *J. Mol. Biol.* **268**, 390–400.

Wedemayer G.J., Patten P.A., Wang L.H., Schultz P.G., and Stevens R.C. (1997b). Structural insights into the evolution of an antibody combining site. *Science* **276**, 1665–1669.

Wien M.W., Filman D.J., Stura E.A., Guillot S., Delpeyroux F., Crainic R., and Hogle J.M. (1995). Structure of the complex between the Fab fragment of a neutralizing antibody for type 1 poliovirus and its viral epitope. *Nature Struct. Biol.* **2**, 232–243.

Williams W.V., London S.D., Weiner D.B., Wadsworth S., Berzofsky J.A., Robey F., Rubin D.H., and Greene M.I. (1989). Immune response to a molecularly defined internal image idiotope. *J. Immunol.* **142**, 4392–4400.

Wilson I.A., and Stanfield R.L. (1993). Antibody-antigen interactions. *Curr. Opin. Struct. Biol.* **3**, 113–118.

Wu T.T., Johnson G., and Kabat E.A. (1993). Length distribution of CDRH3 in antibodies. *Proteins Struct. Funct. Genet.* **16**, 1–7.

Xiang J., Sha Y., Jia Z., Prasad L., and Delbaere L.T.J. (1995). Framework residues 71 and 93 of the chimeric B72.3 antibody are major determinants of the conformation of heavy-chain hypervariable loops. *J. Mol. Biol.* **253**, 385–390.

Ysern X., Fields B.A., Bhat T.N., Goldbaum F.A., Dall'Acqua W., Schwarz F.P., Poljak R.J., and Mariuzza R.A. (1994). Solvent rearrangement in an antigen-antibody interface introduced by site-directed mutagenesis of the antibody combining site. *J. Mol. Biol.* **238**, 496–500.

Yuhasz S.C., Parry C., Strand M., and Amzel L.M. (1995). Structural analysis of affinity maturation: the three-dimensional structures of complexes of an anti-nitrophenol antibody. *Mol. Immunol.* **32**, 1143–1155.

Zhou G.W., Guo J., Huang W., Fletterick R.J., and Scanlan T.S. (1994). Crystal structure of a catalytic antibody with a serine protease active site. *Science* **265**, 1059–1064.

CHAPTER 5

Antigen-Recognizing Molecules Other Than Antibodies

Members of the immunoglobulin superfamily in vertebrates are grouped in several families representing proteins with related functions and structure. The most significant of all proteins of the superfamily are cell surface-proteins. Their functions are associated with antigen recognition, cell adhesion, cell receptor function, and others.

Besides immunoglobulins, there are other molecules that participate in antigen recognition: T-cell antigen receptor (TCR), major histocompatibility complex (MHC) proteins (type I and II), and MHC-related C1 proteins. The structure of these molecules has been studied in detail and the data obtained has greatly facilitated our understanding of the vertebrate immune system (Bjorkman, 1997).

I. T-CELL ANTIGEN RECEPTOR

TCR complex is a group of cell membrane proteins responsible for recognition of antigenic peptides localized in the groove of MHC molecules by T cells. The activation of T cells by signal transducing is a result of the peptide–MHC recognition (Terhorst *et al.*, 1995). The antigen-specific part of the TCR is very similar to Fab fragments and represents a heterodimer composed of two peptide chains (α and β or γ and δ) (Fig. 66). Most $\alpha\beta$ heterodimers are linked by disulfide bridges but γ and δ chains are usually associated noncovalently. Each of the chains contains a variable domain, which is a product of rearranging *V(D)J* gene segments, and a constant domain (Lefranc, 1994; Prosser and Tonegawa, 1995; Bentley and Mariuzza, 1996). The most variable is the complementarity-determining region (CDR3) of both α and β chains; the other two CDR vary very little. The constant domains are anchored in the cell membrane by short transmembrane regions. T lymphocytes also secrete soluble antigen-specific products resembling peptide chains of the TCR but they are still poorly studied (Cone and Malley, 1996).

The three-dimensional structure of the extracellular portion of the $\alpha\beta$ TCR has been studied in detail by x-ray crystallography at high resolutions (Fields and Mariuzza, 1996). In these experiments two soluble TCR preparations were used: a fully glycosylated murine A6 TCR, which was expressed in Drosophila cells (Garcia *et al.*, 1996), and a nonglycosylated human 2C TCR expressed in bacteria. The structure of 2C TCR was solved in complex with the peptide–MHC complex (Garboczi *et al.*, 1996).

The general structure of three TCR domains (Vα, Vβ, and Cβ) is related to that of immunoglobulin domains, but the Cα and Cδ structures deviate con-

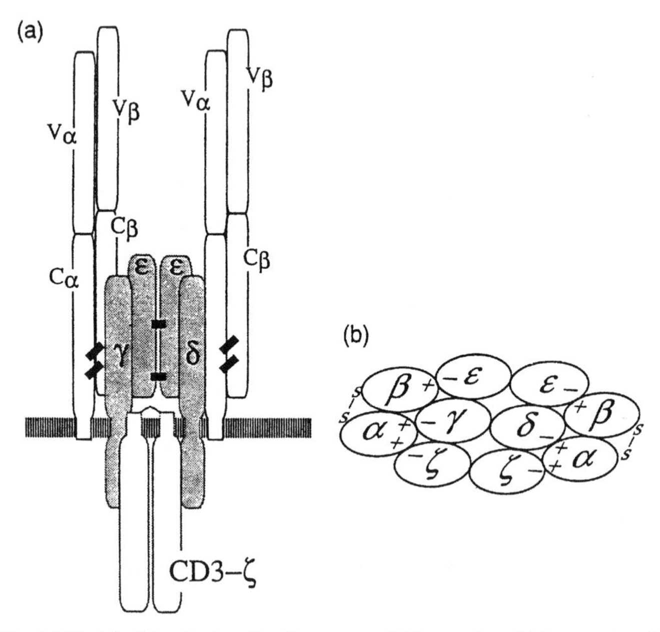

FIGURE 66 (a) Model of the divalent T-cell receptor–CD3 complex. (b) Interactions between the transmembrane regions of constituent peptide chains. (Terhorst *et al.*, 1995. Reprinted with permission from Oxford University Press.)

siderably from the common immunoglobulin fold. The Cβ and Cγ contain a large insertion that forms a protruding loop. The latter could be involved in contacts with CD3 proteins of the TCR–C3 complex. The V and C domains of the TCR contact closely and the surface areas of their interface are larger than the contact areas between VH and CH1 of Fabs. The Vβ–Cβ interaction area is highly polar but the Vα–Cα interface is composed mainly of van der Waals interactions. The dimensions of 2C TCR are 61 Å × 56 Å × 33 Å, which correspond well with that of the immunoglobulin Fab fragments. The elbow angle of human and murine TCR is 148 degrees, which is in the range of elbow angles characteristic for Fab. Of four visible carbohydrate units in 2CTCR, the largest one is located in Cα and is involved in Cα–Cβ interactions. The highly polar Cα–Cβ interface is like that of CH3–CH3.

The β-sheet framework of TCR domains is similar to those of immunoglobulins, with the exception of the Vα domain, in which the C″ strand is bonded to the D strand of the outer β sheet instead to the C′ strand as is usual in V domains (Fields *et al.*, 1995). The contact area between the Vα and Vβ domains of 2C TCR is smaller than in most Fabs (Garcia *et al.*, 1996). However, the Vα–Vβ contact area in A6 TCR is larger, about 780 Å^2 for each domain. The binding surface of murine and human TCRs is relatively flat, with one exception: a hydrophobic pocket exists between the third CDRs of both V domains. The overall dimension of the TCR-binding surface is 36 Å × 20 Å.

The T lymphocytes with the $\gamma\delta$ TCRs can recognize nonpeptide antigens; for example, antigens produced by mycobacteria (Tanaka *et al.*, 1995). There are no detailed structural data on the binding of such antigens to MHC proteins and the recognition of these complexes by T cells.

Other participants of the TCR–CD3 complex are three invariant polypeptide chains, γ, δ, and ε, which are related to each other by sequence. They belong to the immunoglobulin superfamily and have conserved intrachain disulfide bridges. These polypeptides are important in the assembly of the complex and are involved in transduction of the signal initiated by the formation of the complex between the TCR antigen-specific part and the peptide–MHC complex.

II. PROTEINS OF MAJOR HISTOCOMPATIBILITY COMPLEX

MHC proteins are widespread cell-surface glycoproteins that participate in binding antigen peptides inside the cell and presenting them to the antigen receptors of T cells (Klein, 1986; Bjorkman and Parham, 1990; Germain and Margulies, 1993; Browing and McMichael, 1996). The result of the formation of the complex between the TCRs and peptide–MHC is the stimulation of the T cells and activation of the immune response.

The variability of MHC proteins is one of the highest in vertebrates and a large number of allelic MHC variants were identified for human and mouse MHC loci. However, an individual expresses only several different MHC molecules (at most six of each class in humans) but each of them is able to bind a large array of different peptides (up to several thousand). Nevertheless, each allelic variant of the MHC proteins forms complexes with its own set of peptides and the peptide binding occurs with high affinity.

There are two main classes of MHC molecules, which have structural and functional differences. Class I molecules are expressed on virtually all nucleated cells, whereas class II molecules can be found only on a limited number of cells, such as dendritic cells, macrophages, and B lymphocytes (antigen-presenting cells). Class II molecules interact with peptides from extracellular antigens, whereas class I molecules bind peptides derived from intracellular processing of viral, bacterial, or endogenous proteins. Molecules of both classes are noncovalently linked heterodimers and composed of two types of chains, α and β. The α chain of class I is composed of three domains (totaly about 350 residues) and that of class II–of two domains (about 230 residues) encoded by separate exons. Each of them anchors to the cell membrane and has cytoplasmic tails. The β chain of class I represents only one domain (β_2-microglobulin) without a membrane part and the class II β chain consists of two domains (about 230 residues) with a membrane part and a cytoplasmic tail. The general dimensions of the extracellular part of the MHC molecules are equal to that of Fab but the structure of both types of molecules is different and MHC molecules are less compact than Fab. That MHC proteins are noncompact globules was confirmed by the spin-label method (Nezlin *et al.*, 1987). The values of the rotational correlation times (τ) determined for the isolated extracellular parts of MHC proteins class I and II are about two times lower (8 and 14 nsec, respectively) than those predicted for a rigid sphere with dimensions equal to these proteins (about 20 nsec).

Crystallographic studies performed in the past decade provided detailed information about the molecular organization of human and mice MHC proteins of both classes and the mechanism for peptide binding (Bjorkman *et al.*, 1987; Brown *et al.*, 1993). Several peptide–MHC complexes were studied at high resolution. Most of them contain peptides of defined structures and others contain a mixture of endogeneous peptides (Madden, 1995; Young *et al.*, 1995; Wang *et al.*, 1995).

Two extracellular membrane-proximal domains (α3- and β2-microglobulin of class I molecules and α2 and β2 domains of class II molecules) have β-sandwich structure resembling the C-type immunoglobulin fold. Their two sheets are composed of four and three β-strands. However, the interaction of membrane-proximal domains is quite different from that of the constant immunoglobulin domains. The proximal MHC domain pairs are not symmetrical: whereas $\alpha3$ of MHC class I interacts only with the adjacent $\beta2$-microglobulin, the latter has contacts with all three domains of the α chain and is located slightly higher from cell membrane (Fig. 67A).

The membrane-distal domains ($\alpha1$ and $\alpha2$ of the MHC class I molecules and $\alpha1$ and $\beta1$ of the MHC class II molecules) form a specific structure located on top of the membrane-proximal domains that is not related to the immunoglobulin fold. It looks like a platform composed of an eight-strand β-sheet and two long antiparallel α-helices that are placed diagonally across the sheet (Fig. 67B). Both distal domains participate equally in construction of the β-sheet and the α-helices.

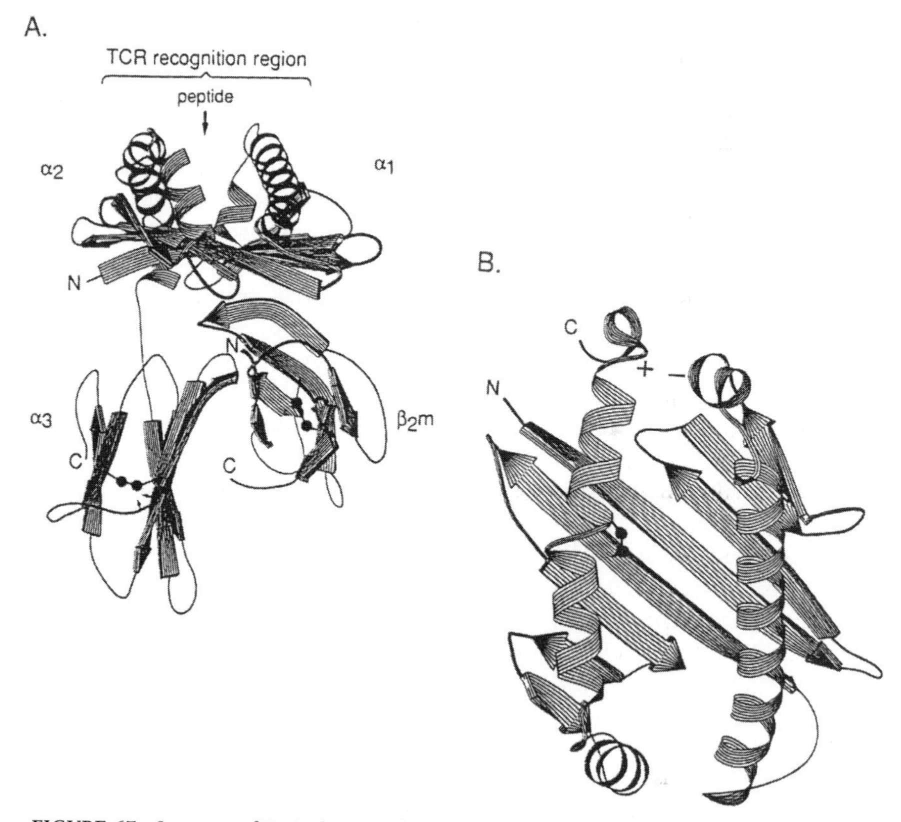

FIGURE 67 Structure of HLA class I molecule (HLA-A2). β-strands shown as thick arrows, α-helices as helical ribbons. N and C, termini of heavy and light chains. Disulfide bonds indicated as two connecting spheres. (A) View from the side. (B) $\alpha1$ and $\alpha2$ domains as viewed from the top of the molecule showing the peptide-binding site and surface contacted by a T-cell receptor (TCR). (Bjorkman and Parham, 1990. Reprinted with permission from Annual Reviews, Inc.)

Antigenic peptides are located in a groove formed by the β-sheet as a floor and the α-helices as sidewalls. The variable amino acid residues specific for allelic variants are spaced across the groove. Polymorphic as well as conserved residues participate in peptide binding. The groove of class I MHC molecules is blocked at both ends by side chains of conserved residues such as tyrosine and tryptophan. Therefore, the length of peptides bound to class I MHC molecules is limited, usually by 8–9 residues, which are buried completely in the groove (Rammensee *et al.*, 1993). There are extensive hydrogen-bonding interactions between peptide and the groove residues. Two peptide residues from a total of 8–9 are nonpolymorphic residues called anchors. Nearly all peptides that have specifically reacted with a given MHC allele contain such residues with preferred side-chain configuration at the same positions. These conserved residues have a primary importance for specific interactions with MHC molecules of both classes. The peptide-binding groove of the MHC class II molecules is open at both ends and longer peptides can be bound (about 12–25 residues) in extended conformation (Brown *et al.*, 1993). Both ends of the bound peptides are located out of the binding sites.

The interactions of peptides with residues of the binding site are different for the MHC class I and class II molecules (Adorini and Trembleau, 1994). In peptide binding by the class I molecules, most important are conserved residues (anchors) clustered at the ends of the groove and the bound peptides have little or no contacts in the middle. Therefore, the bound peptide has an arched shape (Guo *et al.*, 1992). In the peptide–class II molecule complexes, the bound peptides have a flatter conformation and the peptide main chain is close to the residues of the binding site along its entire length. The HLA-DR1 molecule binds an influenza virus peptide (13 residues long) as an extended chain with a pronounced twist (Fig. 68). About 35% of the peptide surface is accessible to solvent and potentially available for the reaction with a TCR. The peptide-binding site pockets adapt 5 of the 13 side chains of the peptide. These pockets determine the specificity of the HLA molecule for the peptide. Twelve of the hydrogen bonds between HLA-DR1 and the main chain of the peptide involve the residues conserved in most human and mouse MHC class II alleles. These bonds are probably universal for all peptide–MHC II complexes (Stern *et al.*, 1994). Several bound water molecules mediate contacts between HLA-DR1 and an endogenous peptide (Murthy and Stern, 1997). The buried water molecules may participate in accommodation of peptides with different conformations in the MHC-binding site.

The MHC molecule binding sites have an irregular surface that is different in various alleles of the MHC molecules. The participation of variable polymorphic residues in the construction of the binding site creates allele-specific peptide-binding motifs, which are responsible for the binding specificity of different MHC proteins. Structurally this specificity is provided by several allele-

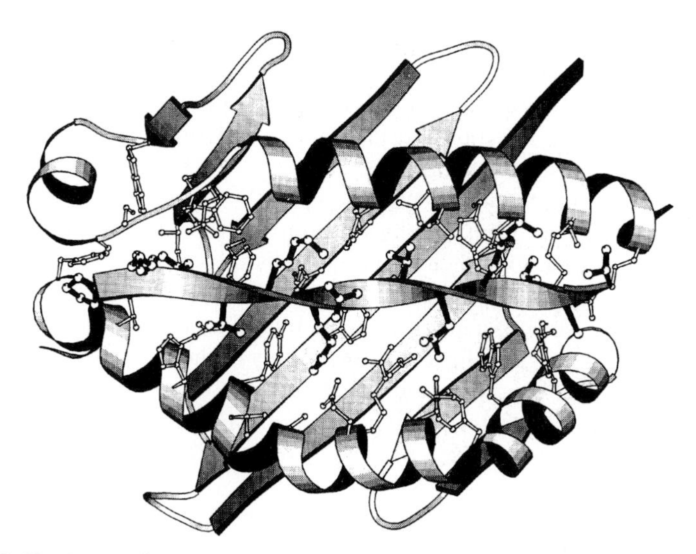

FIGURE 68 Contacts between the influenza virus peptide and the HLA-DR1 binding site. Top view of the peptide-binding site. HLA side chains contacting peptide or buried by interaction are shown by light bonds, peptide side chains by dark bonds (Stern *et al.*, 1994. Reprinted with permission from Macmillan Magazines Ltd.)

specific cavities or pockets for binding side chains of the bound peptide. Each of MHC proteins can bind different peptides with high affinity because the pockets can permit housing of various side chains and the binding is dependent also on interactions of the peptide backbone with nonpolymorphic residues.

III. COMPLEXES OF T-CELL RECEPTOR AND PEPTIDE–MHC

Crystal structure of two ternary TCR–peptide–MHC complexes was determined. One of them is composed of a human A6 $\alpha\beta$TCR specific for a Tax peptide of the human T-cell lymphothropic virus HTLV-1 and the Tax peptide (9 residues) located in the groove of a human HLA-A2 molecule (class I) (Garboczi *et al.*, 1996). Both these proteins were expressed in bacteria. The second complex was built from the murine 2C TCR molecule and the H-2K^b molecule (class I) with an octamer peptide dEV8 from a mitochondrial self-protein (Garcia *et al.*, 1996). In this case proteins were expressed in Drosophila cells and were glycosylated.

According to the 2.6 Å structure of the A6 TCR–Tax peptide–HLA complex, the TCR is located diagonally over the HLA peptide-binding site. The same orientation of TCR is also characteristic of the second complex. The HLA surface buried by the receptor is 671 Å^2, including parts of five polymorphic residues and 11 conserved residues. The total burried surface of the peptide–MHC by TCR is 998 Å^2. The TCR binding to the HLA polymorphic residues provides allelic specificity of the receptor and the common binding site is achieved by binding to the HLA conserved residues.

The recognition of peptide occurs through CDR loops of the TCR. Most of the Tax peptide surface buried by the receptor is covered by the CDR3 loop of the β chain and the CDR1 and CDR3 of the α chain each contribute about 25% (Fig. 69, see color plate) The total buried peptide surface is about 326 Å^2 and is smaller than the area of peptide antigens buried by anti-peptide antibodies (420–620 Å^2). The peptide is located more deeply in the HLA molecule than in the receptor. Of nine peptide residues, substantial contacts with receptors are made only by two residues and only a few atoms of the other residues are in contact with the receptor. These data point out the limited specificity of the TCR for the Tax peptide.

In the second complex, the CDR1 and CDR2 of the TCR Vα domain are positioned over the amino-terminal part of the bound peptide, the CDR1 and CDR2 of the Vβ domain are over the peptide carboxyl-terminal region and CDR3s of both receptor chains are placed over the central position of the peptide (Fig. 70, see color plate). In peptide recognition, the Vα and Vβ CDR3 loops of the TCR play the central role, whereas its CDR1s and CDR2s have contacts with both the peptide and the MHC α-helices. The long dimensions of the receptor and MHC–peptide-binding surfaces are not quite parallel, but tilted about 20–30 degrees toward the diagonal.

The limited specificity of the TCR could explain the origin of the T cell-mediated autoimmune diseases by cross-reactions between self and foreign peptides (Barnaba and Sinigaglia, 1997). Most of the MHC residues that contact the TCR are conserved among MHC alleles and one TCR can react with different MHC molecules or with one MHC molecule, which form complexes with different peptides (Jardetzky, 1997). Such cross-reactivity could explain how bacteria or viruses can activate autoimmune T-cell clones and cause an autoimmune attack due to molecular mimicry. For example, the stimulation of some clones of T cells, which are specific to myelin basic protein, by viral or bacterial peptides leads to an autoaggressive T-cell response (Wucherpfennig and Strominger, 1995).

IV. CD1 MOLECULES

CD1 proteins, a group of antigen-presenting molecules, are expressed on the cells of the thymus and on antigen-presenting cells of most of mammals, first of all on dendritic cells found in many tissues. (Porcelli, 1995). CD1 are nonpolymorphic molecules distantly related to MHC class I proteins. They are composed of an α chain and the β-microglobulin domain. Like that of MHC class I proteins, the CD1 α chain contains three domains. One of them, $\alpha1$, belongs to the IgC-domain family. Several isotypes of C1 were found in humans and rodents (Calabi *et al.*, 1989). In the human genome, there are five closely linked, nonpolymerphic CD1 genes not linked to the MHC complex. In the mouse genome, there are two, closely related CD1 genes.

The function of CD1 proteins is linked to the presentation of unusual antigens to a special subset of T cells. Most CD1-presenting antigens are lipids or glycolipids. For example, human CD1b and CD1c molecules can present mycolic acid from *Mycobacterium tuberculosis* and a glycolipid from *Mycobacterium leprae* (Sieling *et al.*, 1995). It was found that mouse CD1d molecules can bind long hydrophobic peptides (20–22 residues) with relatively high affinity. All these peptides have a definite sequence binding motif (Castaño *et al.*, 1995).

The crystal structure of mouse CD1d1 molecule has been determined (Zeng *et al.*, 1997). The overall structure of this molecule is more closely related to that of the MHC class I molecules. The CD1d1 binding groove is narrow (about 14 Å) and lined almost entirely by nonpolar and hydrophobic side chains. Its width is close to that of the FcRn receptor groove (10–13 Å), whereas the width of MHC I and II molecule grooves is larger (18–20 Å). The CD1d1 binding site is significanly larger, which is due to its increased depth, and also significantly more hydrophobic than the grooves of MHC molecules.

V. NATURAL KILLER CELL INHIBITORY RECEPTORS

Natural killer (NK) cells are cytolytic cells of the lymphocyte line that resemble cytotoxic T cells. They mediate cytolytic reactions, which do not require the presence of MHC molecules on the target cells (Lewis and McGee, 1992). Their target cells are tumor- or virus-infected cells that fail to express MHC class I molecules. The absence of even one of the MHC alleles activates the cytolytic activity of NK cells. NK cells possess inhibitory receptors specific for different groups of MHC class I allotypes. The interaction of the NK cell inhibitory receptors with class I MHC molecules of target cells leads to the inhibition of NK cell cytolytic activity.

The structure of the inhibitory receptors of NK cells is different from that of the TCRs. The NK receptors are nonpolymorphic and their genes do not undergo somatic rearrangements. One group of the mouse and human NK cell receptors belong to the C-type lectin superfamily and is expressed as a disulfide-bonded homodimer on most NK cells and on a subset of T cells (Lanier, 1997a,b). Some of these murine receptors (Ly49A) recognize different alleles of the H-2 complex; natural ligands of other murine and human receptors of this type have not been identified.

The inhibitory receptors of the second group are integral membrane glycoproteins that belong to the immunoglobulin superfamily (Colonna, 1996, 1997; Cantoni *et al.*, 1996). Human inhibitory receptors (KIR) block NK-cell cytotoxicity after binding to the $\alpha 1$ domain of different HLA class I molecules. The extracellular portion of KIR for HLA-C alleles (KIR-2D) is composed of two immunoglobulin-like C2 domains (Fig. 71). The inhibitory receptors for HLA-B alleles (KIR-3D) contain an extracellular portion composed of three C2-like domains (D'Andrea and Lanier, 1996). The receptors specific for HLA-A alleles are disulfide-bonded dimers of KIR-3D. These receptors posseses a long cytoplasmic tail containing two tyrosine-based inhibition motifs (ITIMs) with sequence Val/Ileu x Tyr x x Leu/Val. ITIM sequences are separated by 26 amino acids. The tyrosine residues of ITIM become phosphorilated after contacts of KIR with HLA-I molecules and bind to the protein tyrosine phosphotase SHP-1, which is an important participant of the negative signaling pathways. The specific HLA sequences recognized by KIR receptors are located in the carboxy-terminal half of the HLA $\alpha 1$-helix (between residues 77–83) and are distinct for different KIRs. The cytotoxic T cells recognize residues of wider areas belong to $\alpha 1$ as well as to $\alpha 2$ MHC domains (Gumperz and Parham, 1995). KIR binding to HLA ligands is probably not dependent on the presence of peptides in the peptide-binding groove of HLA molecules. The receptors with two extracellular domains but without ITIM-like motif in their truncated cytoplasmic tail are identified as stimulatory receptors. They probably stimulate NK cells upon binding to a specific HLA class I protein.

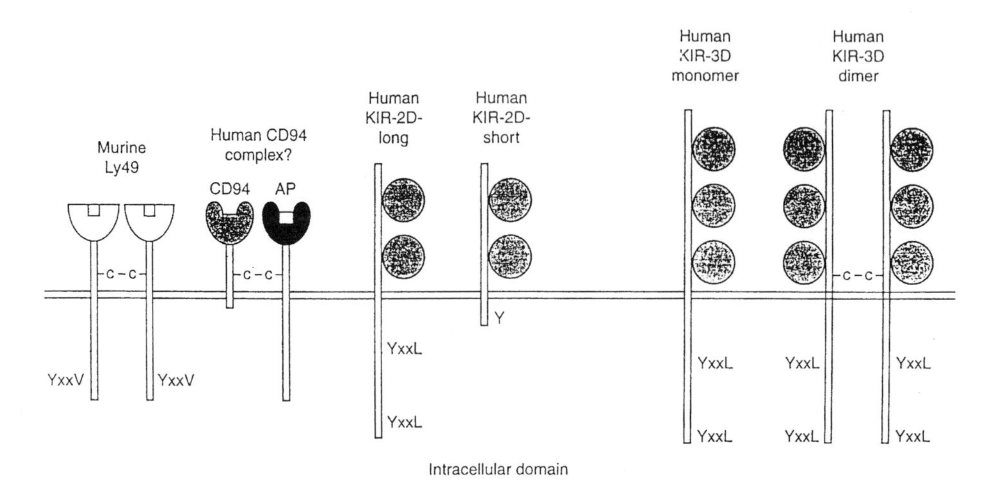

FIGURE 71 Natural killer cell receptors for MHC class I molecules. The Ly49 and CD94–AP molecules are type II proteins of the C-type lectin superfamily. The KIR glycoproteins are members of the immunoglobulin superfamily. YxxV and YxxL are immunoreceptor tyrosine-based inhibitory motifs (ITIM). (Lanier, 1997b. Reprinted with permission from Current Biology Ltd.)

VI. COMPARISON OF ANTIGEN BINDING BY VARIOUS ANTIGEN-RECOGNIZING MOLECULES

The main antigen-recognizing families of proteins, antibodies, TCRs, and MHC proteins, have different functions and accordingly different modes of interaction with antigen.

Antibody molecules, soluble or as a part of the B-cell antigen receptor complex, have the extreme variability of the antigen-combining site that originates during the joining of V(D)J gene segments with following somatic mutations. They are able to recognize an enormous range of various antigenic substances, either soluble or particulate, with quite different chemical natures. Variable amino acid residues of the combining site are responsible for contacts with antigen and antibody molecules are able to discriminate with great precision even very small structural differences between antigenic epitopes (Landsteiner, 1945; Pressman and Grossberg, 1968). Antibody affinity can reach very high values, up to $K = 10^9–10^{10}\ M^{-1}$ (Van Regenmortel and Azimzadeh, 1994). Thus, the main characteristic of the antigen-combining sites of antibodies is their universality, high specificity, and affinity.

MHC proteins have a different function than antibodies and a distinct strategy for reaction with antigens. All cells of an individual express the same set of several MHC alleles. Each allele can bind many peptides but all have a specific sequence motif that is adjusted to a given MHC allele (Adorini and Trembleau, 1994). The peptide-binding site of MHC proteins is composed of polymorphic, allele-specific amino acid residues, as well as from conserved residues. Such a combination permits a tight binding of many peptides to one binding site with a high affinity (equilibrium constant $K_a \sim 10^4–10^9 \cdot M^{-1}$).

The **TCR** binding site is much less variable than that of antibody molecules. The variable genes of the TCR are products of recombination of *V(D)J* gene segments like that of the antibody peptide chains. However, the CDR1 and CDR2 loops of the TCR variable regions vary very little, and the role of somatic mutations in the generation of the repertoire of TCRs is minimal if any. The main variations between different TCRs are concentrated in the CDR3, which varies in its primary structure and length. This type of variability can be explained by the specificity of the partners, with which the TCRs form complexes (i.e. peptide–MHC complex). According to x-ray structural studies, only about one-third of the peptide–MHC complex's total surface buried by the TCRs is related to the peptide surface. A significant part of the buried surface is composed of conserved, nonpolymorphic MHC residues. From a total of 20 hydrogen bonds formed between the TCR and Tax peptide–HLA complex, 8 are with conserved and only 3 with polymorphic HLA residues. The other nine hydrogen bonds are between the receptor and the peptide residues (Garboczi *et al.*, 1996).

The intrinsic affinities for TCR–peptide–MHC reactions vary with in a wide range, with the highest values about 10^7 M^{-1} (Sykulev *et al.*, 1994) but they do not match the highest intrinsic affinities found for antibody–antigen reactions (10^9–10^{10} M^{-1}). TCRs are membrane proteins with a high density and for such a multivalent system, high intrinsic affinities are not necessary for efficiency as they are for soluble antibody molecules (Eisen *et al.*, 1996). TCRs are able to recognize very small differences between peptides but at the same time a TCR can react with many different peptide–MHC complexes.

REFERENCES

Adorini *l.*, and Trembleau S. (1994). Peptide interactions with major histocompatibility complex class II molecules. In: *Immunochemistry.* (Eds. van Oss C.J., and van Regenmortel M.H.V.). pp. 159–180. Marcel Dekker, New York.

Barnaba V., and Sinigaglia F. (1997). Molecular mimicry and T cell-mediated autoimmune disease. *J. Exp. Med.* **185**, 1529–1531.

Bentley G.A., and Mariuzza R. (1996). The structure of the T cell receptor. *Annu. Rev. Immunol.* **14**, 563–590.

Bjorkman P.J. (1997). MHC restriction in three dimensions: a view of T cell receptor/ligand interactions. *Cell* **89**, 167–170.

Bjorkman P.J., and Parham P. (1990). Structure, function, and diversity of class I major histocompatibility complex molecules. *Annu. Rev. Biochem.* **59**, 253–288.

Bjorkman P.J., Saper M.A., Samraoui B., Bennett W.S., Strominger J.*l.*, and Wiley D.C. (1987). Structure of the human class I histocompatibility antigen HLA-A2. *Nature* **329**, 506–512.

Browing M., and McMichael A. (Eds.) (1996). *HLA and MHC: Genes, Molecules, and Function.* Bios Science Publishing, Oxford.

Brown J.H., Jardetzky T.S., Gorga J.C., Stern L.J., Urban R.G., Strominger J.*l.*, and Wiley D.C. (1993). Three-dimesional structure of the human class II histocompatibility antigen HLA-DR1. *Nature* **364**, 33–39.

Calabi F., Jarvis J.M., Martin *l.*, and Milstein C. (1989). Two classes of C1. *Eur. J. Immunol.* **19**, 285–292.

Cantoni C., Verdiani S., Falco M., Conte R., and Biassoni R. (1996). Molecular structures of HLA-specific human NK cell receptors. In: *Molecular Basis of NK Cell Recognition and Function.* (Ed. Moretta L.). pp. 88–103. Basel, Karger.

Castaño A.R., Tangri S., Miller J.E.W., Holcombe H.R., Jackson M.R., Huse W.D., Kronenberg M., and Peterson P. (1995). Peptide binding and presentation by mouse CD1. *Science* **269**, 223–226.

Colonna M. (1996). Natural killer cell receptors specific for MHC class I molecules. *Curr. Opin. Immunol.* **8**, 101–107.

Colonna M. (1997). Specificity and function of immunoglobulin superfamily NK cell inhibitory and stimulatory receptors. *Immunol. Rev.* **15**, 127–133.

Cone R.E., and Malley A. (1996). Soluble, antigen-specific T-cell proteins: T-cell-based humoral immunity? *Immunol. Today* **17**, 318–322.

D'Andrea A., and Lanier L.L. (1996). NKB1: a killer cell inhibitory receptor for class I HLA-B allotypes. In: *Molecular Basis of NK Cell Recognition and Function* (Ed. Moretta L.). pp. 104–115. Basel, Karger.

Eisen H.N., Sykulev Y., and Tsomides T.J. (1996). Antigen-specific T-cell receptors and their reactions with complexes formed by peptides with major histocompatibiity complex proteins. *Adv. Prot. Chem.* **49**, 1–56.

Fields B.A., and Mariuzza R.A. (1996). Structure and function of the T-cell receptor: insights from x-ray crysrallography. *Immunol. Today* **17**, 330–336.

Fields B.A., Ober B., Malchiodi E.*l*., Lebedeva M.I., Braden B.C., Ysern X., Kim J.-K., Shao X., Ward E.S., and Mariuzza R.A. (1995). Crystal structure of the V_α domain of a T cell antigen receptor. *Science* **270**, 1821–1824.

Garboczi D.N., Ghosh P., Utz U., Fan Q.R., Biddison W.E., and Wiley D.C. (1996). Structure of the complex between human T-cell receptor, viral peptiide and HLA-A2. *Nature* **384**, 134–141.

Garcia K.C., Degano M., Stanfield R.*l*., Brunmark A., Jackson M.R., Peterson P.A., Teyton *l*., and Wilson I.A. (1996). An $\alpha\beta$ T cell receptor structure at 2.5 Å and its orientation in the TCR–MHC complex. *Science* **274**, 209–219.

Germain R.N., and Margulies D.H. (1993). The biochemistry and cell biology of antigen processing and presentation. *Annu. Rev. Immunol.* **11**, 403–450.

Gumperz J.E., and Parham P. (1995). The enigma of the natural killer cell. *Nature* **378**, 245–248.

Gou H.-C., Jardetzky T.S., Garrett T.P.J., Lane W.S., Strominger J.*l*., and Wiley D.C. (1992). Different length peptides bind to HLA-Aw68 similarly at their ends but bulge out at the middle. *Nature* **360**, 364–366.

Jardetzky T. (1997). Not just another Fab: the crystal structure of a TcR–MHC–peptide complex. *Structure* **5**, 159–163.

Klein J. (1986). *Natural History of the Major Histocompatibility Complex.* Wiley, New York.

Landsteiner K. (1945). *The Specificity of Serological Reactions.* Harvard University Press. Cambridge, MA.

Lanier L.L. (1997a). Natutal killer cells: from no receptors to too many. *Immunity* **6**, 371–378.

Lanier L.L. (1997b). Natutal killer cell receptors and MHC class I interactions. *Curr. Opin. Immunol.* **9**, 126–131.

Lefranc M.-P. (1994). The T-cell receptor. In: *Immunochemistry.* (Eds. van Oss C.J., and van Regenmortel M.H.V.). pp. 129–157. Marcel Dekker, New York.

Lewis C.E., and McGee J.O'D. (Eds.) (1992). *The Natural Killer Cell.* IRL Press, Oxford.

Madden D.R. (1995). The three-dimensional structure of peptide–MHC complexes. *Annu. Rev. Immunol.* **13**, 587–622.

Murthy V.L., and Stern L.J. (1997). The class II MHC protein HLA-DR1 in complex with an endogeneous peptide: implications for the structural basis of the specificity of peptide binding. *Structure* **5**, 1385–1396.

Nezlin R., Pankratova E.V., Arutyunyan A.E., and Timofeev V.P. (1987). Extracellular portions of HLA antigens are not compact globule. *Mol. Immunol.* **24**, 803–806.

Porcelli S.A. (1995). The CD1 family: a third lineage of antigen-presenting molecules. *Adv. Immunol.* **58**, 1–98.

Pressman D., and Grossberg A.L. (1968). *The Structural Basis of Anrtibody Specificity.* W.A. Benjamin, New York.

Prosser H.M., and Tonegawa S. (1995). T cell receptor V(D)J recombinatuion: mechanisms and developmental regulation. In: *T cell Receptors* (Eds. Bell J.I., Owen M.J., and Simpson E.). pp. 326–351. Oxford University Press, Oxford.

Rammensee H.-G., Falk K., and Rötzschke O. (1993). Peptides naturally presented by MHC class I molecules. *Annu. Rev. Immunol.* **11**, 213–244.

Sieling P.A., Chatterjee D., Porcelli S.A., Prigozy T.I., Mazzaccaro R.J., Soriano T., Bloom B.R., Brenner M.B., Kronenberg M., Brennan P.J., and Modlin R.L. (1995). CD1-restricted T-cell recognition of microbial lipoglycan antigens. *Science* **269**, 227–230.

Stern L.J., Brown J.H., Jardetzky T.S., Gorga J.C., Urban R.G., Strominger J.l., and Wiley D.C. (1994). Crystal structure of the human class II MHC protein HLA-DR1 complexed with an influenza virus peptide. *Nature* **368**, 215–221.

Sykulev Y., Brunmark A., Jackson M., Cohen R.J., Peterson P.A., and Eisen H.N. (1994). Kinetics and affinity of reactions between an antigen-specific T cell receptor and peptide-MHC complexes. *Immunity* **1**, 15–22.

Tanaka Y., Morita C.T., Tanaka Y., Nieves E., Brenner M.B., and Bloom B.R. (1995). Natural and synthetic non-peptide antigens recognized by human $\gamma\delta$ T cells. *Nature* **375**, 155–158.

Terhorst C., Simpson S., Wang B., She J., Hall C., Huang M., Wileman T., Eichman K., Hollander G., Levelt C., and Exley M. (1995). Plasticity of the TCR–CD3 complex. In: *T cell Receptors* (Ed. Bell J.I., Owen M.J., and Simpson E.). pp. 369–402. Oxford University Press, Oxford.

Van Regenmortel M.H.V., and Azimzadeh A. (1994). Determination of antibody affinity. In: *Immunochemistry* (Eds. van Oss C.J., and van Regenmortel M.H.V.). pp. 805–828. Marcel Dekker, New York.

Wang C.-R., Castaño A.R., Peterson P.A., Slaughter C., Fisher Lindahl K., and Deisenhofer J. (1995). Nonclassical binding of formylated peptide in crystal structure of the MHC class Ib molecule H2–M3. *Cell* **82**, 655–664.

Wucherpfennig K.W., and Strominger J.L. (1995). Molecular mimicry in T cell-mediated autoimmunity: viral peptides activate human T cell clones specific for myelin basic protein. *Cell* **80**, 695–705.

Young A.C.M., Nathenson S.G., and Sacchettini J.C. (1995). Structural studies of class I major histocompatibility complex proteins: insights into antigen presentatiom. *FASEB J.* **9**, 26–36.

Zeng Z.-H., Castaño A.R., Segelke B.W., Stura E.A., Peterson P.A., and Wilson I.A. (1997). Crystal structure of mouse CD1: an MHC-like fold with a large hydrophobic binding groove. *Science* **277**, 339–345.

CHAPTER 6

Interactions Outside the Antigen-Combining Site

The antigen-binding capacity of the antibody molecules is an essential function of these molecules. The formation of the antigen–antibody complexes results in direct neutralization of infectious agents. Just as important, however, is the ability of immunoglobulins to react with other molecules at sites located outside the antigen-combining site (Table 19). Some of these interactions are related to the effector functions of antibodies, which are the most important part of the immune response. They include such well-known reactions as the activation of the complement cascade and the activation or the inhibition of cells after binding with the Fc receptors. These processes, which are stimulated by the formation of antigen–antibody complexes, induce inflammation and significantly enhance the response against infections. Other effector functions are related to the transportation of immunoglobulins through cell membranes and operate independently of antigen binding (Ward and Ghetie, 1995; Clark, 1997).

The significance of the interactions of immunoglobulins with many other ligands is not fully understood. In any case, the formation of the complexes are followed by modification of the surface of immunoglobulin molecules and as a consequence the closing of some functionally important sites or sites

TABLE 19 Reactive sites of immunoglobulin molecules

Sites	Localization
Antigen-combining site	Fv
Complement-binding sites	
C1q	$C_\gamma 2$ and $C_\mu 3$
C3b and C4b	Fab and Fc (C3b)
C3a, C4a, and C5a	Fab (C3a)
Fc receptor-binding sites	
$Fc_\gamma RI$ and $Fc_\gamma RII$	$C_\gamma 2$
$Fc_\varepsilon RI$ and $Fc_\varepsilon RII$	$C_\varepsilon 3$
Binding sites for	
Animal proteins:	
CD4 protein of T lymphocytes	Fv
Accesory chains of B cell antigen receptor (α, β, and γ)	Membrane part of heavy chains
Liver Fv protein	Fv_γ
Fibronectin	Fc_γ
Prolactin	Fd_γ
Clusterin	Fc (?)
Prostatic Fc binding proteins	Fc
Fc-binding peptides	Fc
Placental alkaline phosphotase	Fc
Heat shock protein HSP 70 (BiP)	Heavy chains
Animal lectins	Oligosachharide chains
Viral proteins	
Herpes simplex virus proteins gE and gl	Fc of Asian IgG3
HIV gp120	V_H3
Bacterial protein	
A	$C_\gamma 2$–$C_\gamma 3$ and Fv of V_H3
G	$C_\gamma 2$–$C_\gamma 3$ and $C_\gamma 1$
H	Fc
L	Light chains
Plant proteins	
Lectins	Oligosachharide chains

susceptible to proteolysis or the formation of a new site for immune attack. Biologically active substances (e.g., anaphylatoxins) after binding to immunoglobulins are eliminated from the circulation and inactivated. On cell membranes, there are receptors to different ligands. If two such ligands are in complex, their simultaneous reaction with two corresponding types of cell receptors can amplify signaling and change the cell response (the effect of co-ligation). The Fc receptors are widely expressed on different cells and they can be co-ligated with receptors to other ligands if these ligands are in complexes with immunoglobulins. For example, some leukemic lymphocytes are activated only by prolactin–IgG complexes but not by prolactin alone. It is quite probable that the polactin–IgG complex activity is due to the co-ligation of a prolactin receptor with the Fc receptor. Another example is the suppression of B-cell activity by the negative $Fc_{\gamma}IIB$ receptors, which is observed only if an antigen–antibody complex reacts simultaneously with $Fc_{\gamma}IIB$ by the Fc and with the B-cell antigen receptor by the antigen part of the immune complex.

Studies of immunoglobulin interactions with different ligands also have a practical aspect. Ligands with a high affinity and specificity for immunoglobulins are used for detection and isolation of immunoglobulin molecules and their fragments. The best example is the wide application of bacterial proteins A and G, which are highly specific reagents for most IgG. Gammaglobulin preparations, which usually contain practically only IgG, are widely used as a very effective pharmaceutical reagent and are introduced in large quantities for prophylaxis for therapy of patients with different illnesses, including infectious and autoimmune ones (Rosen, 1993; Lee *et al.*, 1997). Therefore, it is important to know whether these preparations also contain complexes of some biologically active substances with IgG that could be harmful to the body.

I. Fc RECEPTOR-BINDING SITES

Molecules of all immunoglobulin isotypes can interact through their Fc portions with cell Fc receptors (FcR), which are an important part of the immune system (Hulett and Hogarth, 1994; Fridman and Sautes, 1996; Raghavan and Bjorkman, 1996). FcR are expressed on the surface of various cells as transmembrane glycoproteins and are also present in soluble forms in biological fluids. The soluble FcR are generated by proteolytic cleavage of the FcR external domains or are due to splicing of the transmembrane exon of FcR membrane proteins (Fridman, 1991).

Two distinct types of the FcRs are known (Table 20). One type of FcR is responsible for transport of immunoglobulins across epithelial surfaces (the neonatal FcR, FcRn, and the polymeric immunoglobulin receptor, pIgR). Another type of FcR presents on effector cells and most of them can induce multiple biological processes. They include phagocytosis, antibody-dependent cell-mediated cytotoxity (ADDC), triggering the release of inflammatory mediators and cytokynes as well as signals for lymphocytes proliferation and differentiation (Burton and Woof, 1992; Ravetch, 1994; 1997). FcRs are of great importance in the destruction of pathogenic agents such as bacteria, viruses, and parasites and in the killing of cells expressing viral antigens. The significant role of FcRs in autoimmune pathology is well established (Deo *et al.*, 1997).

TABLE 20 Properties of Human Fc Cell Receptors

Receptor	Cell localization	Domain structure	Receptor binding site	Ig-binding site	Affinity for lg monomer K_a M^{-1}	Specificity
FcγRI	Neutrophils, monocytes, granulocytes, macrophages	3 V-like domains, TM, CT (72 kDa) γ chain	α1–3	$C_\gamma 2$	10^8-10^9	IgG1 = IgG3 >IgG4 > > > IgG2
FcγRII	Macrophages, neutrophils mast cells, basophils, platelets, Langerhans cells, B cells, placental cells	2 V-like domains, TM, CT (40 kDa)	α2	$C_\gamma 2$	$<10^7$	IgG1 = IgG3 >IgG4 > > > IgG2
FcγRIII[a]	Macrophages, monocytes, neutrophils, eosinophils NK cells, $\gamma\delta$ T cells	2 V-like domains, TM, CT (50–80 kDa) γ and ζ chains	α2	$C_\gamma 2$	10^5	IgG1 = IgG3 >IgG2 > > > IgG4
FcεRI	Mast cells, basophils, eosenophils, Langerhans cells	2 V-like, TM, CT, β and 2γ chains	α2	$C_\varepsilon 2$	10^{10}	IgE
FcεRII	T and B cells, monocytes, eosinophils, platelets, Langerhans cells, bone marrow and thymus epithelial cells	C-type lectin, TM, CT			10^7	IgE, CR2
FcαRI	Monocytes, macrophages, neutrophils, eosinophils	2 Ig-like domains, TM, CT (55–75 kDa), γ chain		$C_\alpha 2$–$C_\alpha 3$	10^7	IgA1, IgA2
FcRn	Intestinal epithelial cells, fetal yolk sac, placental cells, hepatocytes	MHC-like, β and γ chains	α1–α2 and β_2	$C_\gamma 2$–$C_\gamma 3$	10^7	IgG
pIgR	Glandular epithelial cells	5 V-like domains, TM, CT (100 kDa)	N-terminal domain	$C_\alpha 2$–$C_\alpha 3$	$10^8–10^9$	IgA, IgM

[a]FcγRIIIB is anchored to cell membrane by a glycosyl phosphatidylinositol moiety. TM, transmembrane segment; CT, cytoplasmic domain.

A. Binding Sites for Fc Receptors Involved in Effector Responses

There is a high structural homology between different members of the FcR family, which are located on effector cells and are able to trigger cell activation. They comprise an α subunit that determines isotype specificity and affinity of reaction with Fc and one or two accessory chains (β, γ and ζ) that define signaling properties and bear the tyrosine-containing activation motif (ITAM). The α subunit is composed of 2 or 3 immunoglobulin-like domains arranged in tandem. Several isoforms were found for human FcγRs, which differ in the structure of their α subunits, mode of the membrane anchorage, and function. The heteterogeneity of mouse FcγRs is less. The γ chain that is common for several FcRs is necessary for the expression and assembly of FcRs. The β chain enhances signaling through the γ chain.

1. Fcγ Receptors

The best-studied FcR are FcγRs specific for the Fc fragment of IgG and FcεRs specific for the Fc fragment of IgE. Most of them possess ITAMs and trigger cell activation after they are aggregated at the cell surface. The first group comprises three proteins in humans, FcγRI (CD64), FcγRII (CD32), and FcγRIII (CD16). Each of them are found in several isoforms (Hulett and Hogarth, 1994). FcγRI is composed of three immunoglobulin–like domains and two other receptors from two domains that are homologous to the first two N-terminal domains of FcγRI. Both FcγRI and FcγRIIIA are arranged in cell membrane with accessory chains. The human FcγRIIIB isoform is attached to the plasma membrane by a glycosylphosphatidylinositol (GPI) moiety.

The function of FcγRIIB is different from the other FcγRs. The cytoplasmic tail of the human and mouse FcγRIIB receptors contains an immunoreceptor tyrosine-based inhibitory motif (ITIM), a 13-residue-long sequence, which mediates inhibition of the B-cell receptor-triggered activation and anaphylactic responses (Gergely and Sármay, 1996; Ravetch, 1997). The FcγRIIB-deficient animals have elevated immunoglobulin levels in response to antigens and also demonstrate an enhanced passive cutaneous anaphylaxis reaction (Takai *et al.*, 1996). The FcγRIIB receptors suppress the B-cell activity only if they are coligated with the B-cell antigen receptors by antigen–antibody complexes. Therefore, FcγRIIB is used by B cells as a negative regulator of their response according to the amount of specific immune complexes (Doody *et al.*, 1996). This mechanism of the B-cell inhibition by the antigen–antibody complexes is important when the immune response is already strong and there is no need for the stimulation of additional B cells (DeFranco, 1996). The ITIM sequence is also found in some other negative coreceptors, such as some human inhibitory receptors (KIR) of natural killer (NK) cells and of a subset of T cells, CTLA-4 proteins of T cells, and CD22 proteins of B cells (Daëron, 1997).

The human FcγRI binds monomeric IgG1 and IgG3 with a high affinity (10^9 M^{-1}). The receptor also binds IgG4 but with a lower affinity and does not react with IgG2 at all. For high affinity binding all three FcγRI extracellular domains are important. The sites mediating the interaction of human IgG with FcγRI and FcγRII were identified using site-specific mutants (Lund *et al.*, 1991b; Chappel *et al.*, 1991; Morgan *et al.*, 1995). The residues 234–237 (Leu–Leu–Gly–Gly), presented on the N-terminal side of the human $C_\gamma2$ domain (low hinge region), are critical for IgG–FcγRI interaction. Mutations at position 235 dramatically change the FcγRI recognition. Similarly, mutations at position 234 and 237 decrease the capacity of IgG to react with FcγRII. The second important region for receptor binding is a hinge-proximal bend between two β strands of $C_\gamma2$ (particularly residues in position 331, proline in human IgG1 and IgG3 but serine in IgG4) (Canfield and Morrison, 1991). Thus, all Fcγ receptors including low affinity FcγRII and FcγRIII, appear to recognize nearly the same sites of $C_\gamma2$ (Sármay *et al.*, 1992). The sequence Leu–Leu–Gly–Gly in positions 234–237 is the optimal motif for these interactions. It is present in the human subclasses IgG1 and IgG3, mouse IgG2a, rat IgG2b, and rabbit IgG that are the most reactive with human FcγRI. By contrast, the less active human IgG4 and the nonreactive IgG2 have Phe–Leu–Gly–Gly and Val–Ala–deletion–Gly, respectively, at positions 234–237 of the γ chain. Two binding sites were identified on Cγ2 for mouse FcγRII (Lund *et al.*, 1991a). Three IgG mutants (Gly–237 $\rightarrow$ Ala, Asn–297 $\rightarrow$ Ala, and Glu–318 $\rightarrow$ Ala) fail to bind to FcγRII either in complexed or in monomeric form. An altered oligosaccharide moiety in $C_\gamma2$ affects the ability of IgG to react with $Fc_\gamma Rs$, probably by influencing the local structure of the binding site (Jefferis and Lund, 1997).

2. Fcε Receptor

The high affinity IgE receptor, FcεRI, is expressed on mast cells, basophils, skin Langerhans cells, and monocytes, whereas FcεRII, the low affinity IgE receptor, is expressed on lymphocytes and some other types of cells (Metzger and Kinet, 1988; Adamczewski and Kinet, 1994). Binding of IgE to FcεRI and aggregation of IgE–FcεRI complexes on mast cell and basophil surfaces, due to the multivalency of allergen molecules, results in signal transduction. This leads to cellular activation and release of pharmacologically active substances that mediate immediate hypersensitivity, thus triggering allergic reactions. Understanding the reaction between IgE and FcεRI at the molecular level will facilitate the design of peptides that can block IgE binding to this receptor, preventing allergic reactions.

Monomeric IgE reacts with FcεRI with a high affinity of about 10^{10} M^{-1} and with stoichiometry of one IgE molecule per one receptor. Human receptors bind not only human IgE but also rat and mouse IgE, while rodent FcεRIs bind only the rodent IgE molecules. IgE mutants, obtained by recombinant DNA technology, were used to determine which constant ε domain interacts with the FcεRI. Although deletion of the $C_{\varepsilon}4$ domain did not inhibit the FcεRI binding and cell activation capacities of IgE, altering the $C_{\varepsilon}2$ and $C_{\varepsilon}3$ domains does affect IgE–FcεRI binding. The interaction is located in the $C_{\varepsilon}3$ domain (Nissim *et al.*, 1991, 1993). The peptides that have the sequence Pro-343–Ser-353 (from human $C_{\varepsilon}3$) are able to recognize human FcεRI (Helm *et al.*, 1996). The importance of $C_{\varepsilon}2$ and $C_{\varepsilon}4$ is probably structural (e.g., to maintain an optimal binding conformation of the region). Disulfide bridges between the ε chains and oligosaccharides are not essential for IgE biological activity. According to resonance energy transfer experiments, the IgE molecules have a compact, bent structure in solution as well as after binding the cell FcεRI receptor (Baird *et al.*, 1993). Only one binding site form two of the IgE molecule is accessible for FcεRI. The model of the interaction of the bent IgE molecule with FcεRI in presented on Figure 72.

The low affinity FcεRII (CD23) receptor is not related to the immunoglobulin superfamily and belongs to the C-type (calcium-dependent) animal lectins (Delespesse *et al.*, 1991). The Fcε binding site for FcεRII is still not precisely located. The IgE–FcεRII interaction probably requires the presence of all three domains of Fcε. The reaction is not dependent on the IgE carbohydrate moiety as one could expect on the basis of the lectin nature of the FcεRII receptor. As a result of proteolysis, soluble fragments of the membrane FcεRII appear *in vivo.* These soluble FcεRIIs have various sizes and retain the ability to form complexes with IgE. One of FcεRII isoforms (FcεRIIa), expressed on B cells, participates in antigen presentation and the second one (FcεRIIb) is involved in cellular activation. Both membrane and soluble FcεRIIs participate also in the regulation of IgE biosynthesis.

3. Fcα Receptor

Receptors for the Fc region of IgA are located on different cell types, including eosinophils, neutrophils, monocytes, and phagocytes of mucosal tissues. The best-studied human FcαRI receptor (CD89) is a transmembrane glycoprotein expressed on myeloid cells. Two extracellular domains of the FcαRI are related to the immunoglobulin superfamily proteins (Morton *et al.*, 1996). The binding site of IgA for this receptor is confined to the boundary region between the second and third Cα domains (Carayannopoulos *et al.*, 1996). This site is different from the FcR recognition sites of other immunoglobulins, which are located in $C_{\gamma}2$ or the homologous $C_{\varepsilon}3$ domain, but it is analogous to the IgG site

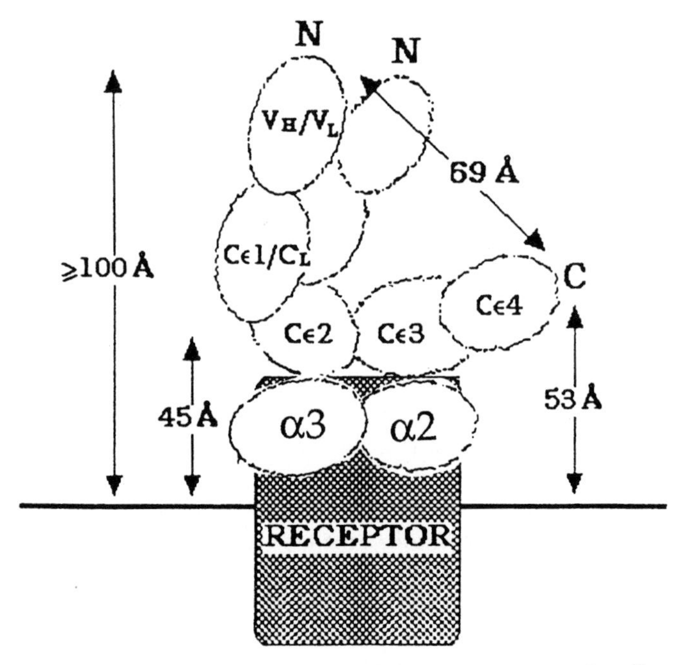

FIGURE 72 Complex of the bent IgE molecule and the FcεRI receptor on the cell membrane. N and C, the N-terminal and C-terminal ends of IgE peptide chains. The horizontal line represents the cell membrane. (Adapted from Zheng *et al.*, 1991.)

recognized by staphylococcus protein A and the neonatal intestinal receptor FcRn. The mutation preventing glycosylation of the $C_\alpha2$ domain abolishes the reaction between the Fcα portion and FcαR. The interactions between IgA and other cell membrane IgA-binding proteins including those on lymphocytes, have still not yet been studied on the molecular level.

4. Fcμ and Fcδ Receptors

The structure and functions of cell receptors for Fcμ and Fcδ expressed on lymphocytes are less well known than that of other FcRs (Hulett and Hogarth, 1994). On activated human B cells an FcμR (~60 kDa) was found. This protein is anchored to the cell surface through a glycosylphosphatidylinositol linkage (Ohno *et al.*, 1990). The important role of $C_\mu3$ and $C_\mu4$ domains for binding to the FcμR was shown in experiments with deletion mutants of IgM. The IgD receptor on murine T cells is a lectin specific for N-linked glycans located on $C_\delta1$ and $C_\delta3$ domains. (Amin *et al.*, 1991).

B. Binding Sites for Fc Receptors Involved in Transcytosis

Two Fc receptors belong to this group, a Fc neonatal receptor (FcRn) specific for IgG, and a receptor for the polymeric immunoglobulin molecules IgA and IgM (pIgR) (Raghavan and Bjorkman, 1996).

1. Neonatal Fc Receptor (FcRn)

The existence of a special receptor responsible for transmission and catabolism of IgG was predicted about 40 years ago (Junghans, 1997). FcRn was identified on the molecular level and its interaction with Fcγ was studied in detail. FcRn is expressed on intestinal epithelial cells of mammal sucklings. IgG from mother milk is bound to FcRn located on the apical part of these cells and transported to the basolateral cell surface where it is released into blood. FcRn is also expressed in the rodent fetal yolk sac and on placental cells. In the latter case, FcRn participates in transfer of the maternal IgG to the fetus, providing humoral immunity for the first days of newborns. The function of FcRn located on adult hepatocytes is to transfer IgG in complexes with antigen from bile to the parenchymal cells.

In all cases, IgG is bound by FcRn at pH 6.0–6.5 and released at pH 7.0–7.5. This pH gradient exists in the gut epithelial cells, and facilitates binding IgG at the apical side of these cells at acid pH and releasing IgG at the basolateral side at neutral pH. FcRn also serves as a catabolic receptor. IgG is degraded much faster in the absence of the receptor than in normal animals (Israel *et al.*, 1995; Ghetie *et al.*, 1996). It is proposed that IgG antibodies in antigen–antibody complexes are internalized into endosomes where at the low pH, antigen dissociates and IgG molecules are bound with FcRn. The IgG–FcRn complex is directed to cell surfaces where IgG separates from the receptor at the neutral pH and returns to circulation while antigen is degraded. Without the FcRn protection, IgG molecules are rapidly destroyed (Fig. 73).

The FcRn receptor is related to MHC class I proteins by sequence (Simister and Mostov, 1989), as well as by the general structure of the molecule (Burmeister *et al.*, 1994a). Similar to the MHC class I molecules, FcRns are composed of two peptide chains, an α chain built from three domains and β_2-microglobulin (Fig. 74). However, FcRn cannot bind peptides because its peptide-binding groove formed by $\alpha1$ and $\alpha2$ domains is closed and filled by amino acid side chains (Burmeister *et al.*, 1994b).

According to x-ray crystallographic studies, the Fc-binding site for FcRn is located at the interface between the $C_\gamma2$ and $C_\gamma3$ domains (Fig. 75) (Burmeister *et al.*, 1994b; Medesan *et al.*, 1997). The same IgG site is also involved in

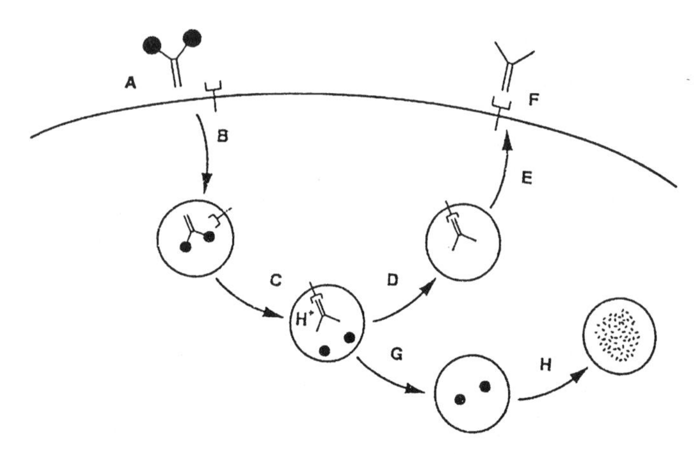

FIGURE 73 Catabolism model incorporating FcRn function. Monomeric IgG antibody molecule with antigen (A) is internalized into endosomes (B). In the acid pH of the endosome (C) the immune complex dissociates and IgG combines with FcRn. In this way IgG is protected from degradation, transported to the cell surface, and returned to circulation (D–F). Unbound antigen is degraded in lysosomes (Gγ, H). Without FcRn both antigen and antibody are catabolized. (Junghans and Anderson, 1996. Reprinted with permission.)

binding *staphylococcal* protein A and fragment B of protein A is able to inhibit the reaction between FcRn and IgG. The affinity of the FcRn–Fc reaction is dropped about two orders of magnitude after pH is raised from 6.0 to 7.0 (Raghavan *et al.*, 1995). It is known that the pKa of imidazole side chains of histidines decreases within the pH range of 6.0–7.0. This suggests that several his-

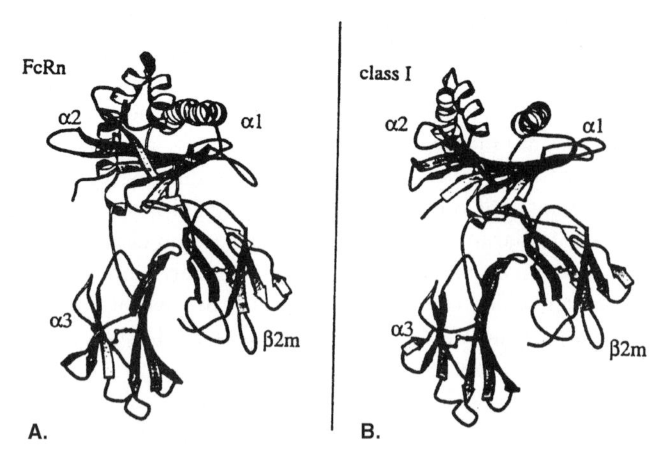

FIGURE 74 Ribbon diagrams of (A) FcRn and (B) murine MHC class I molecule H-2K^b. (Burmeister *et al.*, 1994a. Reprinted with permission from Macmillan Magazines Ltd.)

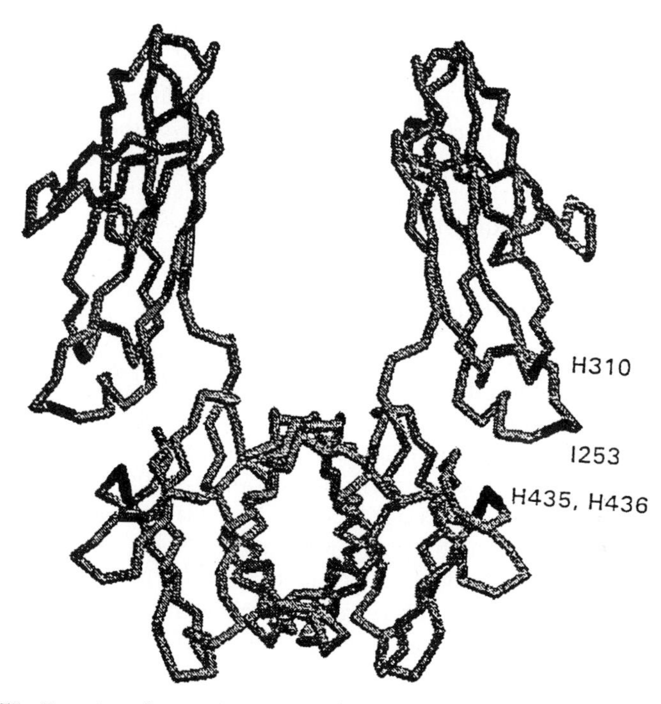

FIGURE 75 Fc region of IgG. The amino acids involved in transcytosis and catabolism are indicated. (Medesan *et al.*, 1997. Reprinted with permission from The American Association of Immunologists.)

tidine residues located at the $C_\gamma2$–$C_\gamma3$ interface may be responsible for the pH-dependent FcRn–IgG interaction (Raghavan *et al.*, 1994; 1995).

The results of studies with mutant Fcs support this view. The His → Ala mutations at the $C_\gamma2$–$C_\gamma3$ interface (particularly mutations of His-310, His-435, and, to a lesser extent, His-436) reduces the activity of the Fc fragment of murine IgG1 (Medesan *et al.*, 1997). The mutant Fcs are functionally deficient in transcytosis and have shot IgG serum half-life. All these data suggest that the histidine residues at the $C_\gamma2$–$C_\gamma3$ interface (Fig. 75) together with the exposed isoleucine residue Ileu-253 play a key role in mediating activities linked with FcRn receptor. The His-310 and Ileu-253 residues located at $C_\gamma2$ domain are highly conserved in all murine and human IgG molecules. Two other histidine residues (His-435 and His-436) are conserved to a lesser extent. But all residues, which are important for the interactions of FcRn with the Fc fragment of IgG, are absent in the molecules of other immunoglobulin classes (IgM, IgA, and IgE) that are unable to react with FcRn.

The serum persistence of the murine Fc was achieved by enhancing its binding affinity to FcRn. For that, the site-directed random mutagenesis of three Fc threonine residues (positions 252, 254, and 256) located in the vicinity of the Fc interaction site with FcRn was used (Ghetie *et al.*, 1997). An increased half-life of immunoglobulins in serum should contribute significantly to the effectiveness of immunotherapy by specific antibodies.

The FcRn dimer is found to have a higher binding affinity for IgG than the FcRn monomer (Raghavan *et al.*, 1995). It was proposed that in physiological conditions one IgG molecule is bound at the cell surface by one FcRn dimer (Vaughan *et al.*, 1997). However, according to other data stoichiometry for the IgG interaction with FcRn is 1:1 (Popov *et al.*, 1996).

2. Receptor for Polymeric Immunoglobulins (pIgR)

IgA dimers are found in mucosal secretions of respiratory and gastrointestinal systems where they form the first line of immunological defense against infection. Polymeric IgA is synthesized by submucosal plasma cells, transported in transcytosis vesicles across epithelial cells from their basolateral part to the apical part, and released into mucosal secretions (Mostov, 1994).

The receptor (pIgR) of serous type secretory epithelial cells, which participates in transepithelial translocation of polymeric immunoglobulins, IgA and IgM, is also related to the immunoglobulin superfamily. However, its structure is different from FcRn. The polyIgR is a membrane glycoprotein with a long extracellular part, a membrane-spanning segment, and a cytoplasmic tail. The extracellular part consists of five immunoglobulin-like domains (Mostov *et al.*, 1984). Four N-terminal domains have significant homology to the variable regions of immunoglobulin peptide chains and the fifth domain is more similar to C-type domain. The extracellular part of pIgR is cleaved-off on the basolateral part of cells by proteolytic enzymes as secretory component (SC) that is linked to IgA dimers.

The main site of the pIgR–SC interaction with IgA dimers (or with IgM) is the N-terminal domain of the pIgR extracellular part, which has a high affinity for noncovalent interaction with IgA ($K_a = 10^8\ M^{-1}$). The fragments of this domain were shown to be sufficient for noncovalent binding to IgA dimers (Frutiger *et al.*, 1986). The deletion of domains 2-5 of pIgR does not influence IgA dimer binding. Using a mutagenic approach, it was shown that each of the three CDR-like loops of the N-terminal domain of pIgR participates in the noncovalent binding of dimeric IgA (Coyne *et al.*, 1994). In some SC–IgA complexes, there is a disulfide bridge between $C\alpha 2$ and the fifth domain of SC. The exact location of the sites on Fc_α or Fc_μ for their interactions with pIgR as well the role of the J chain in such interactions are still unknown.

II. COMPLEMENT-BINDING SITES

The array of complement proteins plays a significant role in host defense against infections and in inflammation (Morgan, 1994; Reid, 1996). Interaction of immunoglobulins and immune complexes with the complement components is important for activation and function of the complement system. There are several aspects of the complement–immunoglobulin interactions (Miletic and Frank, 1995). First, IgG and IgM antibodies bind and activate C1 complement component, which initiates the classical complement pathway. Second, C3b and C4b components are able to form a covalent linkage with IgG, producing heterodimers with IgG. Third, immunoglobulin molecules can noncovalently bind anaphylatoxins C3a, C4a, and C5a, which are small fragments of C3, C4, and C5 complement components, respectively.

A. C1q Binding

Only two classes of immunoglobulins, IgG and IgM, are capable of activating the classical complement pathway. Despite the high sequence homology of IgG subclasses, their capacity to activate complement is different. In humans, the ability of IgG3 and IgG1 to activate complement is significantly higher than that of IgG2, while IgG4 cannot fix complement. In mice, the C1q binding hierarchy is IgG2a $<$ IgG2b $<$ IgG1 and in rats IgG2b $<$ IgG2c $<$ IgG1 $<$ IgG2a (Brüggemann *et al.*, 1989). Rat IgG2c is poor in complement-dependent hemolysis despite its ability to bind C1q.

Binding of the C1 complex is the first step in activation of complement. The C1q component of the complex has recognition functions and is able to interact with Fc by the reaction sites located on its six globular heads. Although the affinity for C1q of monomeric IgG is very low ($K_a = 5 \times 10^4\ M^{-1}$), the binding constant increases upon IgG aggregation ($K_a = 10^8\ M^{-1}$) due to polyvalent interactions. C1q binding to IgG is determined by the structure of $C_\gamma2$ domain (Fig. 76) (Tao *et al.*, 1991; Greenwood *et al.*, 1993). The complement-activating properties can be transferred with this domain by genetic manipulations from active to nonactive IgG molecules. The $C_\gamma2$ residues, Glu-318, Lys-320, and Lys-322, which are involved in murine IgG2b-C1q interaction and contain charged side chains, are conserved in all subclasses of human IgG and in most IgG subclasses of other mammals (Duncan and Winter, 1988). Mutation of any of these residues abolishes complement activity of murine IgG2b. By contrast, the mutation Lys-320 $\rightarrow$ Ala in human IgG1 has no effect on C1q binding or complement lysis. The substitution Leu-235 $\rightarrow$ Glu in human IgG1 abolishes human complement lysis, whereas the same mutation has no effect on the human C1q affinity of murine IgG2b (Morgan *et al.*, 1995). It seems likely that the C1q-binding sites of murine and human IgG have structural differences. The key murine C1q binding motif Glu-318–Lys-320–Lys-320 is also present in human IgG2, which has little complement activity, and in IgG4, which has no such activity. The C1q-binding

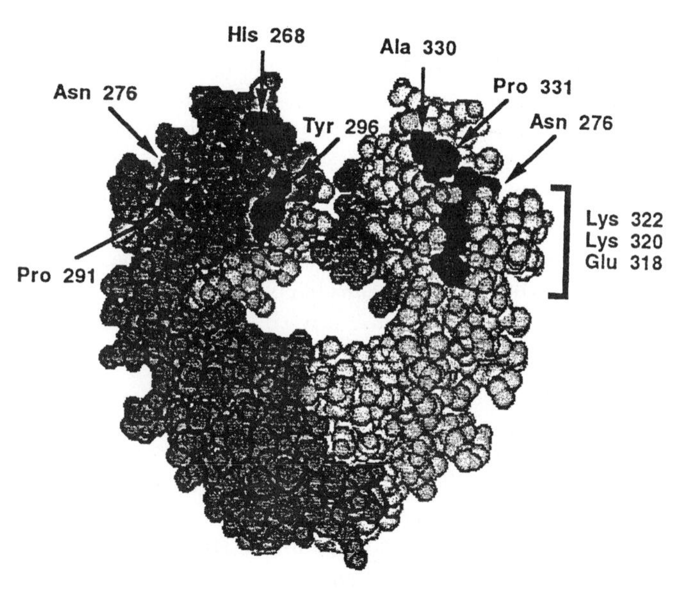

FIGURE 76 Polymorphic residues of human IgG1 Fc. The key binding motif for C1q is indicated by the square bracket on the right side. Residue Pro-331 is critical for determining inability of IgG4 to bind C1q. Residues 296 and 330 do not determine isotype-specific differences in complement activation. (Tao *et al.*, 1993. Reprinted with permission from The Rockefeller University Press.)

activity and complement lysis are probably dependent on not one but several $C_{\gamma}2$ sites and involve many amino acid residues (Morgan *et al.*, 1995).

A position 331 was identified as a critical element responsible for the distinct ability of the human IgG subclasses to activate complement (Tao *et al.*, 1993). Residues in this position are located near the key-binding motif composed of residues 318, 320, and 322 (Fig. 76). Human IgG4 has serine in position 331 but molecules of other subclasses have proline. Substitution of Ser-331 in IgG4 with proline partly restores the complement activity and the Pro-331 → Ser substitution decreases or even abolishes the capacity to activate complement of other IgG subclasses. Probably some other sites are also involved indirectly in C1q binding and complement activation as the removal of the $C_{\gamma}3$ domain reduces efficiency of complement activity (Utsumi *et al.*, 1985).

The main carbohydrate moiety of IgG located at Asn-297 in the $C_{\gamma}2$ domain has significant influence on various effector functions. The absence of the oligosaccharide or alteration of its structure changes the ability of immunoglobulin molecules to activate complement. Aglycosylated murine IgG2b molecule with the mutation Asn-297 → Ala has reduced C1 activation capacity (Duncan and Winter, 1988) and IgG molecules with truncated $C_{\gamma}2$ oligosaccharides are deficient in C1q binding, C1 activation, and complement consumption (Wright and Morrison, 1994).

IgG molecules with mutations in the hinge region were studied to test the role of the hinge for complement activity. The human IgG4 molecule, which has a shortened upper hinge region and a reduced flexibility, is not able to bind C1q. A mutant IgG4 with a long hinge from flexible human IgG3 also cannot bind C1q. Human IgG3 molecules without a hinge region are inactive, whereas IgG3 with the IgG4 hinge binds C1q efficiently (Tan *et al.*, 1990). The mutant IgG3 molecule, which has three alanine residues instead of the upper hinge but with cysteine residues that formed the interheavy disulfide bonds, has complement activity (Brekke *et al.*, 1993). Hence, a direct correlation between the length of the upper hinge and complement activation was not found.

The affinity for C1q for monomeric IgM molecules is low ($K_a = 2.5 \times 10^4$ to $5 \times 10^5\ M^{-1}$) and the IgM monomers cannot activate complement. However, when IgM molecules undergo conformational changes upon interaction with antigen, the affinity for C1q increases 10^3–10^4 fold. The $C_\mu3$ domain is involved in C1q binding. Because electrostatic interactions are predominant in the reaction with C1q, some conserved charged residues of this domain must participate in its contact with C1q. The mutations of three mouse $C_\mu3$ residues, Asn-432, Pro-434, and Pro-436, impair the complement–dependent cytolytic activity (Arya *et al.*, 1994). The last of these residues is homologous to Pro-331 of $C_\gamma2$, which is probably involved in C1q binding by IgG (Tao *et al.*, 1993), but the $C_\gamma2$ key binding motif for C1q binding linked with residues 318, 320, and 322 is absent in $C_\mu3$. It seems likely that the IgG and IgM sites for the C1q binding are not identical. Residues 430 and 432 are adjacent to the stretch connecting $C_\mu3$ to $C_\mu2$ and upon reaction of IgM with antigen, the disposition of the Fab_{μ_2} arms relative to the Fc_{μ_5} disc would make the proposed binding site accessible to the globular heads of C1q (Fig. 77) (Perkins *et al.*, 1991).

The interheavy disulfide bond Cys-414–Cys-414 is important for complement activity. The mutation Cys-414 → Ser abolishes complement binding due to changes in local conformation (Davis *et al.*, 1989). IgM molecules with abnormal glycosylation at $C_\mu3$ are defective in complement–dependent cytolysis. It is likely that the oligosaccharide moiety at residue Asp-402 participates indirectly in C1q binding by maintaining a $C_\mu3$ conformation optimal for IgM–C1q interaction (Wright *et al.*, 1990).

B. C3b and C4b Binding

C3 and C4 complement components become tightly bound to both antigen and antibody after activation by immune complexes (Campbell *et al.*, 1980; Gadd and Porter, 1981). The C3- and C4-binding sites are formed after

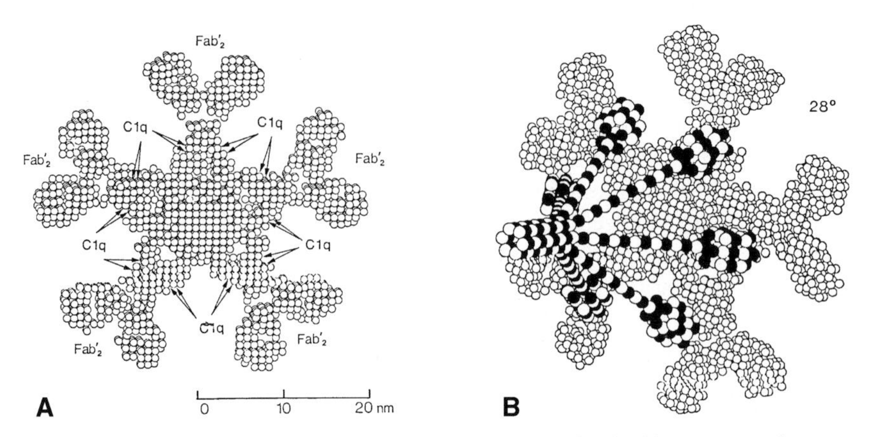

FIGURE 77 (A) Localization of C1q binding sites on the IgM molecule. (B) Interaction of six arms of one C1q molecule with the corresponding binding sites of the IgM molecule. (Perkins *et al.*, 1991. Reprinted with permission.)

enzymatic removal of the N-terminal parts of α chains (C3a or C4a) of C3 and C4 (Müller-Eberhardt, 1988). Due to conformational rearrangement, the intrachain thioester becomes reactive and its carbonyl group forms an ester bond with hydroxyl groups on receptive molecules. After incubation of C3 with heat-denaturated IgG, C3b is covalently linked to the $C_{\gamma}1$ domain of such artificially aggregated IgG in the region consisting of residues 123–156 (Shohet *et al.*, 1993). However, in experiments with immune complexes of ovalbumin and anti-ovalbumin rabbit antibodies, it was found that C3b binds not only to Fab but also to Fc and even with similar efficiency (Antón *et al.*, 1989; 1994). Deleting the $C_{\gamma}1$ domain alters neither the ability of IgG to bind C3 nor to activate the alternative pathway. Protein G inhibits the C3b covalent binding to the Fc region of human IgG1 molecule but does not interfere with the binding to the Fab fragment (Vivanco *et al.*, 1997).

C3b bound to IgG retains its capacity to react with the C3-cell receptors. Therefore, the immune complexes can react with cells not only through their Fc receptors but also through the C3 receptors, which facilitates binding immune complexes to cells.

C. C3a, C4a, and C5a Anaphylatoxin Binding

The C3a, C4a, and C5a anaphylatoxins are small proteolytic fragments (each with molecular weight of about 10 kDa), which are released from C3, C4 and C5 complement components, respectively, upon activation of the complement cascade. They exhibit various biological activities and are responsible for many inflammatory and anaphylactoid reactions, even at very low concentrations.

Anaphylatoxins are able to form noncovalent complexes with IgG at an approximately 1:1 molar ratio and can be separated from IgG in strong denaturing conditions, such as acidified 5M guanidinium chloride solution. The binding site for C3a is located in the Fab fragment (Nezlin *et al.*, 1993). IgG molecules can probably serve as scavengers of free anaphylatoxins, especially if the level of these highly active complement fragments increases significantly in serum due to complement activation. IgG–C3a complexes are found in commercial preparations of immunoglobulins (Nezlin, 1993). Adverse reactions to the injections of gammaglobulin preparations can be caused at least partly by anaphylatoxins bound to IgG molecules.

III. PROTEINS REACTING WITH THE Fc PORTION

A. Clusterin

Clusterin (Sp-40,40 or CLI) is a glycoprotein that was identified in blood and seminal plasma of several species. It has a molecular mass of 70 and consists of two 35-kDa subunits linked by a disulfide bond. Clusterin is an potent inhibitor of the terminal membrane attack complex of complement and can be isolated from this complex. It is able to react specifically with IgG and promote the formation of insoluble immune complexes (Wilson *et al.*, 1991).

B. Human Plasma Glycoprotein 60

Human plasma glycoprotein 60 (gp60) is a Fc binding protein that inhibits C1q binding; thus preventing complement activation (Ahmed *et al.*, 1989). It is in human hepatocytes, platelets, and a subpopulation of blood lymphocytes but it is distinct from known Fcγ receptors (Sandilands *et al.*, 1990).

C. Fibronectin

Fibronectin, a high molecular weight glycoprotein, is found in blood and other body fluids, in the extracellular matrix, and on the surface of many cells. It is built from two 220-kDa monomers linked by two disulfide bridges. Each monomer is composed of several copies of three different modules. Fibronectin binds specifically to normal human immunoglobulins of major classes, with the following hierarchy of affinity: IgG > IgM > IgA and IgG1 > IgG3 = IgG4 > IgG2 (Rostagno *et al.*, 1991). The binding site for fibronectin is located on Fc fragment of IgG and the affinity of the fibronectin binding to Fc is nearly identical

($K_d = 3.69 \times 10^9\ M^{-1}$) to that of its binding to the intact IgG molecule ($K_d = 3.77 \times 10^9 M^{-1}$). Fibronectin reacts with IgG at a site that is located within 93 amino acids in its N-terminus (residues 151–244) (Rostagno *et al.*, 1996). Fibronectin associates with circulating immune complexes and immunoglobulin aggregates that are present in the blood of patients with autoimmune, rheumatic, and myeloproliferative disorders. It may participate in the clearance of these complexes. Immune complexes can also react with fibronectin molecules present in the basement membranes of tissues, which may at least partially account for a deposition of immune complexes in different organs. Such localization of immune complexes could lead to the development of fibrotic diseases, such as pulmonary fibrosis and glomerulonephritis.

D. Fc-Binding Peptides

Fc-binding peptides can be separated from IgG of different species either by dilution of IgG solutions or by decreasing pH to below pH 6. The dissociation of the peptides from Fc is responsible for the concentration and pH dependence of the rotational relaxation time (ρ_h) of IgG molecules. The ρ_h value is dropped after acidification of the IgG solution or after its dilution below 2μM (Dudich *et al.*, 1978). The molecular weight of the peptides is about 2 kDa and they have a small ρ_h value. After separation from IgG molecules, the peptides rotate independently and decrease the experimentally obtained mean values of ρ_h of IgG solutions. It is proposed that the peptide-binding site is located in the $C_\gamma 2$ domain (Dudich and Dudich, 1983).

E. Seminal Proteins

Some soluble human seminal plasma proteins bind specifically to IgG. One of them is a protein that is able to react with the Fc fragment and antigenically resembles FcγRIII (Thaler *et al.*, 1989). In SDS-polyacrylamide electrophoresis, this protein shows two distinct bands at approximately 70 and 35 kDa. Another one is a prostatic secretory protein (β-microseminoprotein or β-inhibin) with a molecular weight of 27 kDa. It interacts with IgG of various species only after reduction, when its molecular weight is dropped to 16 kDa (Liang *et al.*, 1991a,b). This prostatic secretory protein also has common antigenic properties with FcγRIII. Some immunological reactions are affected by seminal plasma and immunoglobulin-binding seminal proteins could be partly responsible for this effect.

F. G-Actin

G-actin forms precipitates with immunoglobulins at low salt concentrations. Soluble complexes of IgG1, IgG2, or IgM with actin exist in 0.1 M KCl. (Fechheimer *et al.*, 1979). In physiological salt concentrations, actin reacts with antigen–IgG antibody complexes. However, it does not coprecipitate with complexes of antigen with the antibody $F(ab')_2$ fragment because the interaction requires the presence of Fc (Cebra *et al.*, 1977).

G. Placental Alkaline Phosphatase

Placental alkaline phosphatase (apparent molecular weight 64 kDa) is a membrane protein anchored by a glycan–phosphatidylinositol linkage. The extracellular domain of this protein binds to the Fc portion of IgG with $K_d = 3 \times 10^6 M^{-1}$ (Makiya and Stigbrand, 1992).

H. Herpes Simplex Virus Proteins

The cells infected by herpes simplex virus type 1 (HSV-1) express two viral proteins (gE and gI), which together form a receptor for the Fc portion of IgG. The molecules of human IgG1, IgG2, and IgG4 bind to this receptor. But the IgG3 binding to the HSV-1 receptor is dependent on the IgG3 phenotype. The HSV-1 Fcγ-binding activity has only IgG3 molecules in the Asian population with the Fc phenotype different from that of IgG3 in the Caucasian population (Johansson *et al.*, 1994). The Fcγ-binding site is located in the Cγ2–Cγ3 region like the binding site for staphylococcal protein A (Johansson *et al.*, 1988).

IV. PROTEINS REACTING WITH THE Fab PORTION

A. Prolactin

Prolactin is a polypeptide hormone with a molecular weight of 23 kDa that is synthesized in the anterior pituitary gland. Prolactin receptors are present on human B and T lymphocytes, monocytes, and NK cells. There is evidence that prolactin has an immunomodulatory role. Prolactin can form complexes with IgG molecules using the Fd part of all four human IgG subclasses as a place of attachment (Walker *et al.*, 1995). The isolated IgG–prolactin complex has about one mole of $N^{\varepsilon}(\gamma$-glutamyl)lysine crosslinks per mole of the complex, which points to the possible role of enzyme transglutaminase in formation of linkages between IgG and prolactin. About 0.8% of all IgG molecules in circulation is in complex with one or two molecules of prolactin. The molecular mass of such complexes is 192 or 219. The prolactin–IgG complexes can stimulate lymphocytes from some patients with chronic lymphocytic leukemia. However, prolactin alone is not active in this respect and the proliferating activity of the complex involves the engagement of both prolactin and IgG. This effect is probably due to the co-ligation of the receptors specific to both these molecules.

B. Protein Fv

Protein Fv, a sialoprotein, is synthesized in the liver and released into the digestive tract during viral hepatitis. It also presents in complexes with immunoglobulins in the stool of some normal persons (Bouvet *et al.*, 1993). Protein Fv is a homodimer with an apparent molecular weight of 175 kDa. It binds to V_H domains of most human and animal immunoglobulins. In feces, two types of protein Fv complexes with $F(ab')_2$ of secretory IgA were found with an apparent molecular weight of about 1800 and 800 kDa. The latter complex consists of six $F(ab')_2$ and one Fv dimer. Protein Fv and staphylococcal protein A can compete for the same binding site located on V_H3 regions of IgM (Silverman, 1997). Protein Fv binding does not interfere with the reaction of antibodies with antigen and it even increases the agglutinating properties of antibodies against viruses and *Salmonella typhi*. Protein Fv acts also as an activator of human basophils and mast cells by interacting with the V_H domain of IgE. Such reactions stimulate the release of chemical mediators like histamine and leukotriens from these cells (Patella *et al.*, 1993).

C. T-Cell Protein CD4

CD4 is a membrane glycoprotein (molecular weight 55 kDa) expressed on helper T lymphocytes (Brady and Barclay, 1996). It binds to MHC class II molecules, to the HIV gp120 protein and interacts also with immunoglobulin molecules of nearly all classes and subclasses. In the reaction with immunoglobulins, two sites on the first CD4 extracellular domain (residues 21–28 and 35–38) are involved. The same CD4 sites are also partly responsible for the human immunodeficiency virus (HIV)–gp120 binding (Lenert *et al.*, 1990, 1995). This fact explains the inhibition of the reaction of CD4 with immunoglobulins by gp120. One CD4 molecule is probably able to cross-link two immunoglobulin molecules. The immunoglobulin–V_H framework residues participate in reactions with soluble recombinant CD4. The interaction of CD4 with antibody molecules enhances antibody–antigen reactions, which could explain the antibody-mediated enhancement of the HIV infection.

D. HIV Protein gp120

The HIV protein gp120 is bound by immunoglobulins, the V_H domains of which coded preferentially by the V_H3 gene family, the largest family of the variable regions genes (Berberian *et al.*, 1993). The isolated gp120 reacts with a monoclonal IgM, which possesses the V_H3 region, with a high affinity ($K_d = 8.6 \times 10^9\ M^{-1}$) whereas no reaction was found with another monoclonal IgM that has the V_H1 region. The binding activity of IgG that has the V_H3 region for gp120 is significantly lower.

In acquired immunodeficiency syndrome (AIDS) patients, there is a clonal deficit of V_H3-expressing B cells, which is preceded by a stimulation and expansion of these B cells at earlier steps of the disease. Such cell dynamics could be accounted for by the fact that gp120 protein can react with the membrane-anchored (V_H3) immunoglobulins and activate B lymphocytes that bear V_H3-surface immunoglobulins (Berberian *et al.*, 1993). The V_H3-binding site for gp120 is probably determined by the structure of conserved, surface-exposed segments of the V_H3-framework regions.

Proteins that are able to react with the Fv part of immunoglobulin molecules in a nonantigen way can be termed **B-cell superantigens** by analogy with T-cell superantigens (Pascual and Capra, 1991; Silverman, 1997). The latter molecules can combine with MHC class II antigens and form ligands that stimulate T cells by reaction with a particular variant of V_β domain of T-cell receptors (Herman *et al.*, 1991). The Fv-reacting proteins can activate a group of B cells bearing a particular V-domain family of immunoglobulin receptors. Such interaction could lead to serious pathological events, especially if the stimulated B cells are producers of autoantibodies (Zouali, 1995). In addition to the proteins mentioned previously, some bacterial immunoglobulin-binding proteins, such as proteins A, G, and L, can be also assigned to the B-cell superantigens.

V. LECTINS

Lectins, sugar-binding proteins (Sharon, and Lis, 1989; Sharon, 1993) interact with immunoglobulins through their sugar residues (Table 21). The galactose-specific lectin from castor beans, *Ricinus communis,* reacts specifically with accessible terminal galactose residues of the Fc and Fab portions of IgG molecules. The interaction rate is higher after IgG aggregation, and the larger the mass of the IgG aggregates, the higher the rate of interaction. The lectin from the lentil, *Lens culinaris,* specific for α-mannose residues, binds to IgM molecules, and can be used for partial purification of monoclonal IgM antibodies from ascites fluid. Jacalin, a lectin isolated from seeds of a tropical plant, jackfruit, is specific for D-galactose (Roque-Bareira and Campos-Neto, 1985). This lectin binds to serum or secretory human IgA1 but not to molecules of the IgA2 subclass or other immunoglobulin classes. Jacalin also does not recognize IgA of rodents and other animals. The jacalin specificity for IgA1 is due to the presence of five galactose-containing *O*-linked oligosaccharides located in the IgA1 hinge. The IgA2 molecules have no glycans in this region due to a deletion of 13 amino acid residues. Jacalin is used effectively for the purification of human IgA1 and its separation from the second human IgA isotype, IgA2 (Haun *et al.*, 1989). A small fraction of rabbit IgG is also bound by jacalin, which is probably mediated through *O*-linked glycans present on the heavy chain of IgG (Kabir *et al.*, 1995).

Two dot blot assays for detection of agalactosylated IgG, in which lectins that specifically react with *N*-acetylglucosamine, are employed. In the first of these methods, a lectin isolated from *Bandeiraea simplicifolia* was used. Before the assay is performed, IgG must be denaturated by heat (Sumar *et al.*, 1990). More

TABLE 21 Plant lectins interacting with immunoglobulin oligosaccharides[a]

Source of lectin	Lectin specificity
Castor beans, *Ricinus communis* (RCA)	Galactose and *N*-acetylgalactosamine
Lens beans, *Lens culinaris* (LcH)	α-Mannose
Seeds of *Griffonia simplicifolia II* (GS II)	N-acetylglucosamine
Jack beans, *Canavalia ensiformis* (ConA)	α-Mannose
Snowdrop seeds, *Galanthus nivalis* (GNA)	α-Mannose
The bark of the elder tree, *Sambucus nigra* (SNA)	Sialic acid linked $\alpha 2 \rightarrow 6$ linked to galactose or N-acetylgalactosamine
Seeds of *Maackia amurensis* (MAA)	Sialic acid $\alpha 2 \rightarrow 3$ linked to galactose
Peanut, *Arachis hypogaea* (PNA)	Galactose $\beta 1 \rightarrow 3$ linked to *N*-acetylgalactosamine
Seeds of jimson weed, *Datura stramonium* (DSA)	Galactose $\beta 1 \rightarrow 4$ linked to *N*-acetylgalactosamine
Jackfriut *Artocarpus heterophyllus* (Jacalin)	Galactose $\beta 1 \rightarrow 3$ linked to *N*-galactosamine
Mushroom *Psathyrella velutina*	*N*-acetylglucosamine $\beta 1 \rightarrow 2$ linked to mannose

[a]Data taken from Roque-Bareira and Czampos-Neto (1985); Sumar *et al.* (1990); Mathov *et al.*, (1995).

useful is a lectin from the mushroom *Psathyrella velutina* that reacts preferentially with the *N*-acetylglucosamine$\beta 1 \rightarrow 2$Man group exposed in agalacto-IgG. For this assay, the immunoglobulin denaturation is not necessary (Tsuchiya *et al.*, 1993). Such simple methods can be used for the screening of IgG from a large number of patients.

Some animal lectins form complexes with immunoglobulins. For example, an IgE-binding factor Mac2/εBP is expressed on the surface as well as in the cytoplasm and nucleus of different kind of cells, including mast cells, macrophages, neutrophils, and eosinophils. This factor is a highly conserved multifunctional protein with β-galactoside specificity belonging to the galectin (S-type) lectin family (Liu, 1990; 1993). The receptors for Mac2/εBP on IgE are *N*-linked oligosaccharides and the interaction depends on sialylation of their terminal residue (Robertson and Liu, 1991). The serum lectin mannose-binding protein (MBP) that structurally resembles C1q complement component can bind to the terminal *N*-acetylglucosamine residues of the Fc galacto glycans. The binding of MBP, which can function as a surrogate C1 complement component, to agalacto-IgG results in complement activation (Malhotra *et al.*, 1995).

VI. MOLECULAR CHAPERONES

Several proteins inside the cell are able to interact with immunoglobulin peptide chains. They are related to molecular chaperones, which participate in folding, assembly, and transport of secreted and membrane-bound immunoglobulins (Hartl, 1996). The most studied of them is an immunoglobulin-binding protein or BiP (Haas and Wabl, 1983), a member of the heat shock protein Hsp70 family of molecular chaperones with a molecular weight of 70 kDa located in the endoplasmic reticulum. As with other chaperones, BiP is able to combine reversibly with hydrophobic segments of incompletely folded or unfolded peptide chains preventing formation of improper complexes with other unfolded peptides. One of the BiP physiological functions in antibody-forming lymphocytes is to protect hydrophobic patches of immunoglobulin chain folding intermediates until a proper association with the other chain partner can occur. The peptide-binding sites of BiP accept stretches of seven amino acid residues (Flynn *et al.*, 1991). The BiP binding areas of heavy chains are probably distributed within V_H and C_H domains and in the Fd portion they concentrate in sites responsible for contacts with light chains (Knarr *et al.*, 1995). After the formation of immunoglobulin peptide chains and their translocation into the lumen of the endoplasmic reticulum, BiP reacts with potential binding sites on both heavy and light chains. The reaction with incompletely folded C_H1 is especially important for the correct assembly of heavy and light chains (Kaloff and Haas, 1995). Following the termination of the folding and assembly processes, BiP releases and the completed immunoglobulin molecules are able to secrete from the antibody-forming cell (Fig. 78).

Two other chaperone-type proteins are also able to bind to immunoglobulin chains, GRP94 and calnexin (Melnick and Argon, 1995). The first of these proteins interacts with unassembled immunoglobulin chains probably after their binding to BiP. Calnexin binds many proteins and contributes particularly to proper assembly and transport of membrane immunoglobulins. It is required

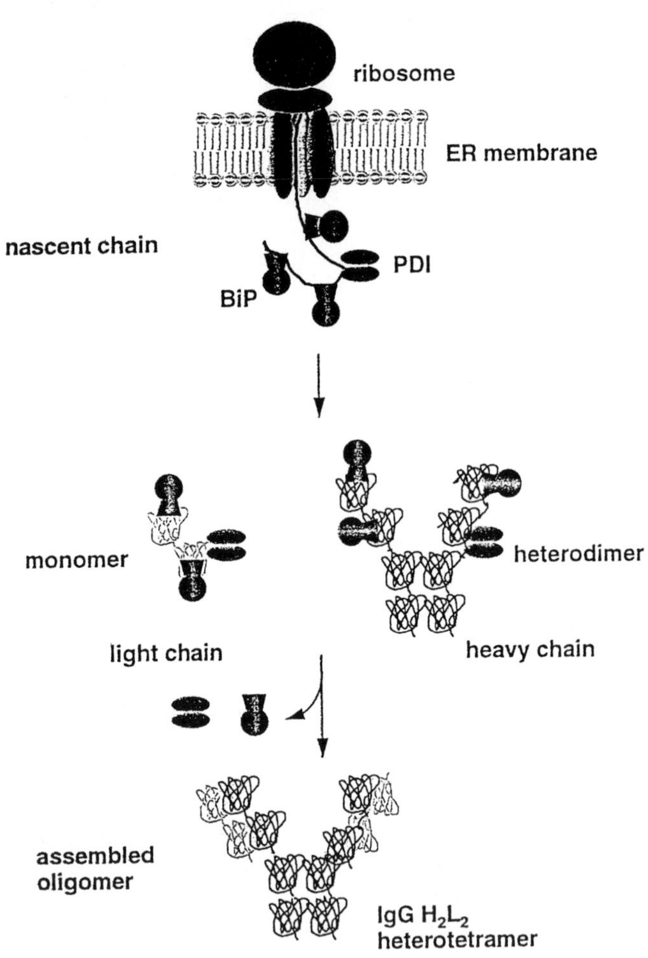

FIGURE 78 Function of the BiP molecular chaperone in antibody folding. BiP may interact with sites on the nascent chain and with additional sites at later stages of the folding to prevent aggregation and premature export. Protein disulfide isomerase (PDI) is involved in the formation of the correct inter- and intrachain disulfide bonds (Buchner, 1996. Reprinted with permission.)

for surface expression of membrane IgD molecules when Ig-$\alpha\beta$ chains of the B-cell receptor are absent (Wu *et al.*, 1997).

VII. BACTERIAL IMMUNOGLOBULIN-BINDING PROTEINS

Proteins capable of binding immunoglobulins in a nonimmune fashion are expressed on the surface of many microorganisms, such as staphylococci, streptococci, and peptostreptococci. They are probably involved in the process of infection and are able to weaken the immune response. The bacterial immunoglobulin-binding proteins are widely applied as powerful immunochemical reagents for isolation and quantitation of immunoglobulins and their fragments (Boyle and Metzger, 1994).

The first such protein discovered was protein A, present in the cell walls of *Staphylococcus aureus* (Forsgren and Sjöquist, 1966). Protein A binds to the γ heavy chains and some other bacterial proteins specifically bind to other heavy chains or to the light chains. A portion of bacterial proteins reacts exclusively with immunoglobulin molecules of one class, while the others have broader binding specificity. In a few cases a species specificity for binding bacterial proteins was observed (Boyle, 1990).

A. Staphylococcal Protein A

Staphylococcal protein A is a small cell wall protein with an extracellular part (27 kDa) composed of five highly homologous domains about 58 residues long (designated from the amino-terminal end E, D, A, B, and C). All domains bind the Fc fragment of IgG but only domains D and E bind the Fab fragments. The three-dimensional structures of B and E domains studied by nuclear magnetic resonance (NMR), spectroscopy are up–down three α-helical bundles (Tashiro and Montelione, 1995).

The structure of the binding site on the human Fcγ1 fragment for the B domain is known from the x-ray crystallography study of the Fc–B complex (Fig. 79, see color plate) (Deisenhofer, 1981; Derrick and Wrigley, 1993; Sauer-Eriksson *et al.*, 1995). The crystal structure of the B domain in the complex is a two-helical structure. The electron density of the third helical portion is very weak and is difficult to interpret. The fragment B binds to the region between the C_H2 and C_H3 domains. This contact is predominantly hydrophobic with few hydrogen bonds and involves 9 conserved residues in the Fc fragment and 11 in the B domain. The reactivity of animal immunoglobulins with protein A is summarized in Table 22 (Goodswaard *et al.*, 1978; Boyle and Metzger, 1994; Kerr and Thorpe, 1994).

TABLE 22 Immunoglobulin Reactivity with Protein A and Protein G[a]

Immunoglobulins		Protein A	Protein G
Human			
	IgG1	+++	+++
	IgG2	+++	+++
	IgG3	−	+++
	IgG4	+++	+++
Mouse			
	IgG1	+	+
	IgG2a	++	+++
	IgG2b	++	+++
	IgG3	+ +	+++
Rat			
	IgG1	+	+
	IgG2a	−	++
	IgG2b	+/−	+
	IgG2c	+++	+++
Cow			
	IgG1	−	++
	IgG2	++	+++
Sheep			
	IgG1	+	++
	IgG2	−	+++
Goat			
	IgG1	+/−	++
	IgG2	++	+++
Horse			
	IgG	+	+++
Dog			
	IgG	+++	++
Duck			
	IgY	+++	−
	IgY(ΔY)	+++	−
Pig			
	IgG	++	+
Chicken			
	IgG	−	−
Rabbit			
	IgG	+++	+++
Guinea pig			
	IgG 1 and 2	+++	+++

[a]Boyle and Metzger (1994) with additions.

The Fab-binding site for protein A is present in human immunoglobulins of different classes, the V_H regions of which are encoded by genes of the V_H3 family (Hillson *et al.*, 1993; Roben et al., 1995). The product of V_H3–23 gene has the highest affinity for protein A. The V_H3 residues of the framework regions 1 and 3 and of CDR2 are involved in the binding of protein A. All three of these regions are required for the activity, and the replacement of even one of them with the corresponding region from V_H of the nonbinding molecules results in loss of the binding activity (Potter *et al.*, 1996). The V_H3-binding site is conformationally dependent and its activity is destroyed after denaturation or if the light chains are absent. Variable regions of mouse immunoglobulins of different classes encoded by members of the S107 and J606 gene families are also recognized by protein A (Seppälä *et al.*, 1990).

B. Streptococcal Protein G

Streptococcal protein G is a cell wall protein with an extracellular part composed of two (or three) small domains that bind serum albumin (GA domains) and two (or three) immunoglobulin binding domains (B domains). According to NMR spectroscopy and x-ray crystallography studies, B domains are composed of a central α-helix packed against a four strand β-sheet (Gronenborn and Clore, 1993; Sauer-Eriksson *et al.*, 1995). The equilibrium constants of the reaction between protein G and human, rabbit, mouse, and goat IgG range between 10^{10} $M^{-1}$and 10^{11} M^{-1}, which is greater than the corresponding values for protein A binding (Åckerström and Björck, 1986). Protein G binding is optimal at pH 4–5, whereas the optimum for protein A binding is at pH 8.0.

All four human IgG subclasses interact with protein G with high affinity (Table 22). Protein G binds to the Fc as well as to the Fab portions. However, the interactions between protein G and Fab are much weaker than the interaction with Fc. Some immunoglobulins interact with protein G predominantly by Fab, such as mouse IgG1, while the others bind protein G primarily by Fc, such as human IgG (Lian *et al.*, 1994). Fab of human IgG2 is unable to react with protein G (Perosa *et al.*, 1997).

The interactions between protein G and the Fab and Fc regions were studied by NMR and by x-ray crystallography (Derrick and Wrigley, 1994; Sauer-Eriksson *et al.*, 1995; Lian *et al.*, 1994; Kato *et al.*, 1994). The Fc-binding site for B domain is located in the cleft between the C_H2 and C_H3 domains, similar to the binding site used by protein A (Fig. 79). Three residues of C_H2 and four residues of C_H3 are involved in the interfacial interactions. Eight residues of the immunoglobulin-binding B domain, including five residues from α-helix, contribute to the interface. Even though the Fc binding sites for protein G and protein A overlap extensively, modes of interactions are quite different in both cases. The protein G–Fc complex mainly involves charged and polar contacts: protein G has 12 polar or charged interactions with Fc and no hydrophobic contacts. Protein A and Fc contacts are stabilized mainly through nonspecific hydrophobic interactions. Five hydrophobic contacts but only six polar interactions occur between protein A and Fc.

The B domain of protein G uses different, nonoverlapping portions for binding to Fab and to Fc. The second β-strand of the B domain binds to Fab, forming an antiparallel β-sheet with the seventh β-strand from the C_H1 domain, including residues from Ser-209 to Lys-216 (Fig. 80). In addition, there is a second minor contact between the α-helix of B domain with the first β-strand in C_H1 (residues Pro-125 to Tyr-129). Both contacts form an extensive binding surface between the B domain and Fab, which is comparable in size with the antibody–antigen interaction sites. The contacting part of Fab is highly conserved between different subclasses and species and such binding strategy of protein G minimizes the effects of sequence variability on the immunoglobu-

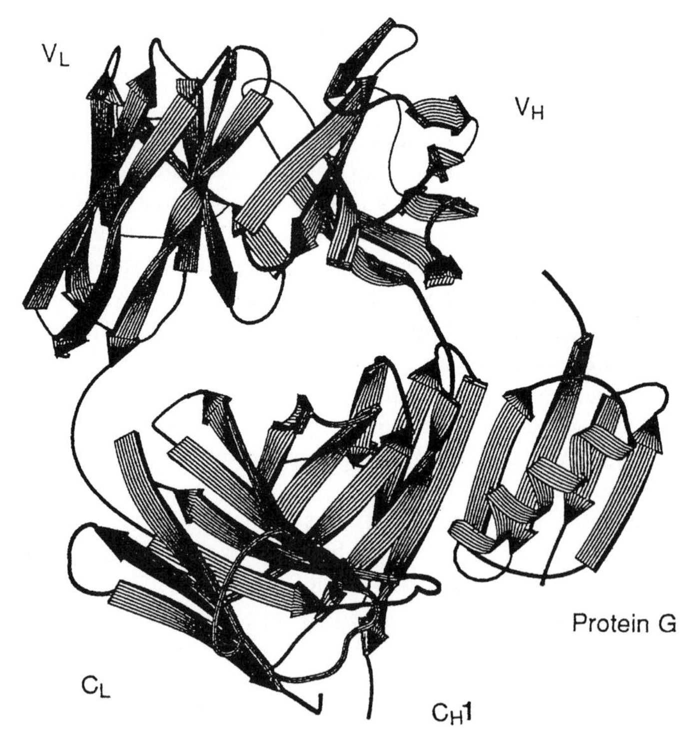

FIGURE 80 Interaction of the protein G domain III with the C_H1 domain of a mouse Fab fragment. (Derrick and Wrigley, 1992. Reprinted with permission.)

lin. No large changes in conformation were found in either protein on formation of the complex.

C. Protein H

Protein H is a polyreactive molecule, which is expressed at the surface of some strains of *Strepococcus pyogenes.* Its extracellular portion contains two IgG-binding domains followed by three albumin-binding domains. Protein H interacts with the Fc fragment of human IgG with a high affinity ($K_a = 1.6 \times 10^9 M^{-1}$). The Fc binding site is located at the $C\gamma2$–$C\gamma3$ interface nearly in the same region as the binding sites for proteins A and G (Frick *et al.*, 1994). There is a separate site on the N-terminal part of protein H for interaction with the fibronectin type III domain of the neural cell adhesion molecule (NCAM) (Frick *et al.*, 1995).

D. Protein L

Protein L, a multidomain protein, which is expressed by some strains of *Peptostreptococcus magnus,* binds exclusively to the κ light chains. The binding affinity is located at five small homologous domains in the N-terminal end of protein L. The three-dimensional structure of these domains is in general very similar to that of the protein G domains even though proteins G and L have no significant sequence homology. Both types of domains are composed of an α-helix on top of a four β-strand sheet (Wikström *et al.*, 1994). The interaction between human immunoglobulin light chains and protein L ($K_a = 10^9 M^{-1}$) involves only variable regions of three κ subgroups, $V_{\kappa I}$, $V_{\kappa III}$ and $V_{\kappa IV}$, but not of $V_{\kappa II}$ or V_λ subgroups (Nilson *et al.*, 1992). This interaction does not block antigen-binding activity of IgG antibodies because the binding site for protein L includes the framework residues. The protein L–κ chain complex probably has a similar structure to the protein G–C_H1 complex and involves β-sheet-like interactions (Wikström *et al.*, 1995).

REFERENCES

Åckerström B., and Björck L. (1986). A physicochemical study of protein G, a molecule with unique immunoglobulin G-binding properties. *J. Biol. Chem.* **261**, 10240–10247.

Adamczewski M, and Kinet J.-P. (1994). The high-affinity receptor for immunoglobulin E. In: *Lymphocyte Activation* (Ed. Samelson L.E.). pp.173–190. Karger, Basel.

Ahmed A.E.E., Bird P., Mkkay I.C., and Whaley K. (1989). The plasma protein which inhibits complement-mediated prevention of immune precipitation is an Fc binding protein. *Immunology* **66**, 20–25.

Amin A.R., Tamma S.M.L., Oppenheim J.D., Finkelman F.D., Kieda C., Coico R.F., and Thorbecke G.J. (1991). Specificity of murine IgD receptor on T cells is for N-linked glycans on IgD molecules. *Proc. Natl. Acad. Sci. USA* **88**, 9238–9242.

Antón L.C., Alcolea J.M., Sánchez-Corral P., Marqués G., Sánchez A., and Vivanco F. (1989). C3 binds covalently to the Cγ3 domain of IgG immune aggregates during complement activation by the alternative pathway. *Biochem. J.* **257**, 831–838.

Antón L.C., Ruiz S., Barrio E., Marqués G., Sánchez A., and Vivanco F. (1994). C3 binds with similar efficiency to Fab and Fc regions of IgG immune aggregates. *Eur. J. Immunol.* **24**, 599–604.

Arya S., Chen F., Spycher S., Isenman D.E., Shulman M.J., and Painter R.H. (1994). Mapping of amino acid residues in the Cμ3 domain of mouse IgM important in macromolecular assembly and complement-dependent cytolysis. *J. Immunol.* **152**, 1206–1212.

Baird B., Zheng Y., and Holowka D. (1993). Structural mapping of IgE-FcεRI, an immunoreceptor complex. *Acc. Chem. Res.* **26**, 428–434.

Berberian L., Goodglick L., Kipps T.J., and Braun J. (1993). Immunoglobulin V_H3 gene products: natural ligands for HIV gp120. *Science* **261**, 1588–1591.

Bouvet J.-P., Pirès R., Iscaki S., and Pillot J. (1993). Nonimmune macromolecular complexes of Ig in human gut lumen. Probable enhancement of antibody functions. *J. Immunol.* **151**, 2562–2571.

Boyle M.D.P. (Ed.) (1990). *Bacterial Immunoglobulin Binding Proteins.* Vols. I and II. Academic Press, San Diego.

Boyle M.D.P., and Metzger D.W. (1994). Antibody-binding bacterial proteins as immunoreagents. In: *Antibody Techniques.* pp. 177–209. Academic Press, San Diego.

Brady R.L., and Barclay A.N. (1996). The structure of CD4. In: *The CD4 Molecule* (Ed. Littman D.R.). pp. 1–18. Springer, Berlin.

Brekke O.H., Michaelsen T.E., Sandin R., and Sandlie I. (1993). Activation of complement by an IgG molecule without a genetic hinge. *Nature* **363**, 628–630; Correction *Nature* **383**, 103, (1996).

Brüggemann M., Teale C., Clark M., Bindon C., and Waldmann H. (1989). A matched set of rat/mouse chimeric antibodies. Identification and biological properties of rat H chain constant regions μ, γ1, γ2a, γ2b, γ2c, ε, and α. *J. Immunol.* **142**, 3145–3150.

Buchner J. (1996). Supervising the fold: functional principles of molecular chaperones. *FASEB J.* **10**, 10–19.

Burmeister W.P., Gastinel L.N., Simister N.E., Blum M.L., and Bjorkman P.J. (1994a). Crystal structure at 2.2 Å resolution of the MHC-related neonatal Fc receptor. *Nature* **372**, 336–343.

Burmeister W.P., Huber A., and Bjorkman P.J. (1994b). Crystal structure of the complex of rat neonatal Fc receptor with Fc. *Nature* **372**, 379–383.

Burton D.R., and Woof J.M. (1992). Human antibody effector function. *Adv. Immunol.* **51**, 1–84.

Campbell R.D., Dodds A.W., and Porter R.R. (1980). The binding of human complement component C4 to antibody-antigen aggregates. *Biochem. J.* **189**, 67–80.

Canfield S.M., and Morrison S.L. (1991). The binding affinity of human IgG for its high affinity Fc receptor is determined by multiple amino acids in the C_H2 domain and is modulated by the hinge region. *J. Exp. Med.* **173**, 1483–1491.

Carayannopoulos L., Hexham J.M., and Capra J.D. (1996). Localization of the binding site for the monocyte immunoglobulin (Ig) A-Fc receptor (CD89) to the domain boundary between Cα2 and Cα3 in human IgA1. *J. Exp. Med.* **183**, 1579–1586.

Cebra J., Brunhouse R., Cordle C., Daiss J., Fechheimer M., Ricardo M., Thunberg A., and Wolfe P.B. (1977). Isotypes of guinea pig antibodies: restricted expression and bases for interactions with other molecules. *Progr. Immunol.* **3**, 269–277.

Chappel M.S., Isenman D.E., Everett M., Xu Y.-Y., Dorrington K.J., and Klein M.H. (1991). Identification of the Fc_γ receptor class I binding site in human IgG through the use of recombinant IgG1/IgG2 hybrid and point-mutated antibodies. *Proc. Natl. Acad. Sci. USA* **88**, 9036–9040.

Clark M.R. (1997). IgG effector mechanisms. In: *Antibody Engineering.* (Ed. Capra J.D.). pp. 88–110. Karger, Basel.

Coyne R.S., Siebrecht M., Peitsch M.C., and Casanova J.E. (1994). Mutational analysis of polymeric immunoglobulin receptor/ligand interactions. Evidence for the involvement of multiple complementary determining region (CDR)-like loops in receptor domain I. *J. Biol. Chem.* **269**, 31620–31625.

Daëron M. (1997). ITIM-bearing negative coreceptors. *Immunologist* **5**, 79–86.

Davis A.C., Roux K.H., Pursey J., and Shulman M.J. (1989). Intermolecular disulfide bonding in IgM: effects of replacing cysteine residues in the μ heavy chain. *EMBO J.* **8**, 2519–2526.

DeFranco A.L. (1996). B-cell co-receptors: the two-headed antigen. *Curr. Biol.* **6**, 548–550.

Deisenhofer J. (1981). Crystallographic refinement and atomic models of a human Fc fragment and its complex with fragment B of protein A from *Staphylococcus aureus* at 2.9- and 2.8-Å resolution. *Biochemistry* **20**, 2361–2370.

Delespesse G., Suter U., Mossalayi D., Bettler B., Safrati M., Hofstetter H., Kilcherr E., Debre P., and Dalloul A. (1991). Expression, structure, and function of the CD23 antigen. *Adv. Immunol.* **49**, 149–191.

Deo Y.M., Graziano R.F., Repp R., and van de Winkel J.G.J. (1997). Clinical significance of IgG Fc receptors and FcγR-directed immunotherapies. *Immunol. Today* **18**, 127–135.

Derrick J.P., and Wrigley D.B. (1992). Crystal structure of a streptococcal protein G domain bound to an Fab fragment. *Nature* **359**, 752–754.

Derrick J.P. and Wrigley D.B. (1993). Analysis of bacterial immunoglobulin-binding proteins by X-ray crystallography. *Immunomethods* 2, 9–15.

Derrick J.P., and Wrigley D.B. (1994). The third IgG-binding domain from streptococcal protein G. An analysis by X-ray crystalography of the structure alone and in a complex with Fab. *J. Mol. Biol.* **243**, 906–918.

Doody G.M., Dempsey P.W., and Fearon D.T. (1996). Activation of B lymphocytes: intergarting signals from CD19, CD22 and FcγRIIb1. *Curr. Opin. Immunol.* **8**, 378–382.

Dudich E.I., and Dudich I.V. (1983). Polarization fluorescence, spin label and ultracentrifugal studies of specific interaction of low molecular weight proteins with the Fc fragment of human immunoglobulin G. *Mol. Immunol.* **20**, 1267–1272.

Dudich E.I., Nezlin R., and Franěk F. (1978). Fluorescence polarization analysis of various immunoglobulins. Dependence of rotational relaxation time on protein concentration and on ability to precipitate with antigen. *FEBS Lett.* **89**, 89–92.

Duncan A.R., and Winter G. (1988). The binding site for C1q on IgG. *Nature* **332**, 738–740.

Fechheimer M., Daiss J.L., and Cebra J. (1979). Intraction of immunoglobulin with actin. *Mol. Immunol.* **16**, 881–888.

Flynn G.C., Pohl J., Flocco M.T., and Rothman J.E. (1991). Peptide-binding specificity of the molecular chaperone BiP. *Nature* **353**, 728–731.

Forsgren A., and Sjöquist J. (1966). Protein A from *S. aureus*. I. Pseudoimmune reaction with human γ-globulin. *J. Immunol.* **97**, 822–827.

Frick I.-M., Åkesson P., Cooney J., Sjöbring U., Schmidt K.-H., Gomi H., Hattori S., Tagawa C., Kishimoto F., and Björck L. (1994). Protein H—a surface protein of *Streptococcus pyogenes* with separate binding sites for IgG and albumin. *Mol. Microbiol.* **12**, 143–151.

Frick I.-M., Crossin K.L., Edelman G.M., and Björck L. (1995). Protein H—a bacterial surface protein with affinity for both immunoglobulin and fibronectin type III domains. *EMBO J.* **14**, 1674–1679.

Fridman W.H. (1991). Fc receptors and imunoglobulin binding factors. *FASEB J.* **5**, 2684–2690.

Fridman W.H., and Sautes C. (1996). *Cell-mediated Effects of Immunoglobulins*. Springer, Berlin.

Frutiger S., Hughes G.J., Hanly W.C., Kingzette M., and Jaton J.-C. (1986). The amino-terminal domain of rabbit secretory component is responsible for noncovalent binding to immunoglobulin A dimers. *J. Biol. Chem.* **261**, 16673–16681.

Gadd K., and Porter R.R. (1981). The binding of complement component C3b to antibody–antigen aggregates after activation of the alternative pathway in human serum. *Biochem. J.* **195**, 471–480.

Gergely J., and Sármay G. (1996). FcγRII-mediated regulation of human B cells. *Scand. J. Immunol.* **44**, 1–10.

Ghetie V., Hubbard J.G., Kim J.-K., Tsen M.-F., Lee Y., and Ward E.S. (1996). Abnormally short serum half-lives of IgG in β_2–microglobulin-deficient mice. *Eur. J. Immunol.* **26**, 690–696.

Ghetie V., Popov S., Borvak J., Radu C., Matesoi D., Medesan C., Ober R.J., and Ward E.S. (1997). Increasing the serum persistence of an IgG fragment by random mutagenesis. *Nature Biotechnol.* **15**, 637–640.

Goodswaard J., van der Donk J.A., Noordzij A., van Dam R.H., and Vaerman J.P. (1978). Protein A reactivity of various mammalia immunoglobulins. *Scand. J. Immunol.* **8**, 21–28.

Greenwood J., Clark M., and Waldmann H. (1993). Structural motifs involved in human IgG antibody effector functions. *Eur. J. Immunol.* **23**, 1098–1104.

Gronenborn A.M., and Clore G.M. (1993). Structural studies of immunoglobulin-binding domains of streptococcal protein G. *Immunomethods* **2**, 3–8.

Haas I.G., and Wabl M. (1983). Immunoglobulin heavy chain binding protein. *Nature* **306**, 387–389.

Hartl F.U. (1996). Molecular chaperons in cellular protein folding. *Nature* **381**, 571–580.

Haun M., Incledon B., Alles P., and Wasi S. (1989). A rapid procedure for the purification of IgA_1 and IgA_2 subclasses from normal human serum using protein G and jackfruit lectin (jacalin) affinity chromatography. *Immunol. Lett.* **22**, 273–280.

Helm B.A., Sayers I., Higginbottom A., Machado D.C., Ling Y., Ahmad K., Padlan E., and Wilson A.P.M. (1996). Identification of the high affinity receptor binding region in human immunolgobulin E. *J. Biol. Chem.* **271**, 7994–7500.

Herman A., Kappler J.W., Marrack P., and Pullen A.M. (1991). Superantigens: mechanism of T-cell stimulation and role in immune response. *Annu. Rev. Immunol.* **9**, 745–772.

Hillson J.L., Karr N.S., Oppliger I.R., Mannik M., and Sasso E.H. (1993). The structural basis of germline-encoded V_H3 immunoglobulin binbding to Staphylococcal protein A. *J. Exp. Med.* **178**, 331–336.

Hulett M.D., and Hogarth P.M. (1994). Molecular basis of Fc receptor function. *Adv. Immunol.* **57**, 1–127.

Israel E.J., Patel V.K., Taylor S.F., Marshak-Rothstein A., and Simister N.E. (1995). Requirement for a β_2-microglobulin-associated Fc receptor for acquisition of maternal IgG by fetal and neonatal mice. *J. Immunol.* **154**, 6246–6251.

Jefferis R., and Lund J. (1997). Glycosylation of antibody molecules: structural and functional significance. In: *Antibody Engineering.* (Ed. Capra J.D.). pp. 111–128. Basel, Karger.

Johansson P.J.H., Nardella F.A., Sjöquist J., Schrÿder A.K., and Christensen P. (1988). Herpes simplex type 1–induced Fc receptor binds to the $C_\gamma2$–$C_\gamma3$ interface region of IgG in the area that binds staphylococcal protein A. *Immunology* **66**, 8–18.

Johansson P.J.H., Ota T., Tsuchiya N., Malone C.C., and Williams R.C. (1994). Studies of protein A and herpes simplex virus-1 induced Fcγ-binding specificities. Different binding patterns for IgG3 from Caucasian and Oriental subjects. *Immunology* **83**, 631–638.

Junghans R.P. (1997). The Brambell receptor (FcRB). *Immunol. Res.* **16**, 29–57.

Junghans R.P., and Anderson C.L. (1996). The protection receptor for IgG catabolism is the β_2-microglobulin-containing neonatal intestinal transport receptor. *Proc. Natl. Acad. Sci. USA* **93**, 5512–5516.

Kabir S., Ahmed I.S.A., and Daar A.S. (1995). The binding of jacalin with rabbit immunoglobulin G. *Immunol. Investig.* **24**, 725–735.

Kaloff C.R., and Haas I.G. (1995). Coordination of immunoglobulin chain folding and immunoglobulin chain assembly is essential for the formation of functional IgG. *Cell* **2**, 629–637.

Kato K., Lian L.-Y., Barsukov I.L., Derrick J.P., Kim H., Tanaka R., Yoshino A., Shiraishi M., Shimada I., Arata Y., and Roberts G.C.K. (1994). Model for the complex between protein G and an antibody Fc fragment in solution. *Structure* **3**, 79–85.

Kerr M.A., and Thorpe R., Eds. (1994). *Immunochemistry. LabFax.* Bios Scientific Publishers, Oxford, and Academic Press, San Diego.

Knarr G., Gething M.-J., Modrow S., and Buchner J. (1995). BiP binding sequences in antibodies. *J. Biol. Chem.* **46**, 27589–27594.

Lee M.L., Gale R.P., and Yap P.L. (1997). Use of intravenous immunoglobulin to prevent or treat infections in persons with immune deficiency. *Annu. Rev. Med.* **48**, 93–102.

Lenert P., Kroon D., Spiegelberg H., Golub E.S., and Zanetti M. (1990). Human CD4 binds immunoglobulins. *Science* **248**, 1639–1643.

Lenert P., Lenert G., and Zanetti M. (1995). Human recombinant CD4 and CD4–derived synthetic peptides agglutinate immunoglobulin-coated latex particles. Evidence that residues 25–28 and 35–38 of human CD4 form two separate immunoglobulin binding sites. *Mol. Immunol.* **32**, 1399–1404.

Lian L.-Y., Barsukov I.L., Derrick J.P., and Roberts G.C.K. (1994). Mapping the interactions between streptococcal protein G and the Fab fragment of IgG in solution. *Nature Struct. Biol.* **1**, 355–357.

Liang Z.G., Kamada M., and Koide S.S. (1991a). Structural identity of an immunoglobulin binding factor and prostatic secretory protein of human seminal plasma. *Bioch. Biophys. Res. Commun.* **180**, 356–359.

Liang Z.G., Kamada M., and Koide S.S. (1991b). Properties of an immunoglobulin binding factor from human seminal plasma. *Biochem. Intern.* **24**, 1003–1013.

Liu F.-T. (1990). Molecular biology of IgE-binding protein, IgE-binding factors and IgE receptors. *Crit. Rev. Immunol.* **10**, 289–306.

Liu F.-T. (1993). S-type mammalian lectins in allergic inflammation. *Immunol. Today* **14**, 486–490.

Lund J., Pound J.D., Jones P.T., Duncan A.R., Bentley T., Goodall M., Levine B.A., Jefferis R., and Winter G. (1991a). Multiple binding sites on the C_H2 domain of IgG for mouse FcγII. *Mol. Immunol.* **29**, 53–59.

Lund J., Winter G., Jones P.T., Pound J.D., Tanaka T., Walker M.R., Artymiuk P.J., Arata Y., Burton D.R., Jefferis R., and Woof J.M. (1991b). Human FcγRI and FcγRII interact with distinct but overlapping sites on human IgG. *J. Immunol.* **147**, 2657–2662.

Makiya R., and Stigbrand T. (1992). Placental alkaline phosphotase has a binding site for the human immunoglobulin-G Fc portion. *Eur. J. Biochem.* **205**, 341–345.

Malhotra R., Wormald M.R., Rudd P.M., Fischer P.B., Dwek R.A, and Sim R.B. (1995). Glycosylation changes of IgG associated with rheumatoid arthritis can activate complement via the mannose-binding protein. *Nature Med.* **1**, 237–243.

Mathov I., Plotkin L.I., Squiquera L., Fossati C.A., Margni R.A., and Leoni J. (1995). *N*-glycanase treatment of $F(ab')_2$ derived from asymmetric murine IgG_3 mAb determines the acquisition of precipitating activity. *Mol. Immunol.* **32**, 1123–1130.

Medesan C., Matesoi D., Radu C., Ghetie V., and Ward E.S. (1997). Deliniation of the amino acid residues involved in transcytosis and catabolism of mouse IgG1. *J. Immunol.* **158**, 2211–2217.

Melnick J., and Argon Y. (1995). Molecular chaperones and the biosynthesis of antigen receptors. *Immunol. Today* **16**, 243–250.

Metzger H., and Kinet J.-P. (1988). How antibodies work: focus on Fc receptors. *FASEB J.* **2**, 3–11.

Miletic V.D., and Frank M.M. (1995). Complement-immunoglobulin interactions. *Curr. Opin. Immunol.* **7**, 41–47.

Morgan B.P. (1994). Complement. In: *Immunochemistry* (Eds. van Oss C.J., and van Regenmortel M.H.V.). pp. 903–923. Marcel Dekker, New York.

Morgan A., Jones N.D., Nesbitt A.M., Chaplin L., Bodmer M.W., and Emtage J.S. (1995). The N-terminal end of the C_H2 domain of chimeric human IgG1 anti-HLA-DR is necessary for C1q, FcγRI and FcγRIII binding. *Immunology* **86**, 319–324.

Morton H.C., van Egmond M., and van de Winkel J.G.J. (1996). Structure and function of human IgA Fc receptor (FcaR). *Crit. Rev. Immunol.* **16**, 423–440.

Mostov K.E. (1994). Transepithelial transport of immunoglobulins. *Annu. Rev. Immunol.* **12**, 63–84.

Mostov K.E., Friedlander M., and Blobel G. (1984) The receptor for transepithelial transport of IgA and IgM contains muliple immunoglobulin-like domins. *Nature* **308**, 37–43.

Müller-Eberhardt H.J. (1988). Molecular organization and function of the complement system. *Annu. Rev. Biochem.* **57**, 321–347.

Nezlin R. (1993). Detection of C3a complement component in commercial gamma globulins by dot blotting. *J. Immunol. Meth.* **163**, 269–272.

Nezlin R., Freywald A., and Oppermann M. (1993). Proteins separated from human IgG molecules. *Mol. Immunol.* **30**, 935–940.

Nilson B.H.K., Solomon A., Björck L., and Åkerström B. (1992). Protein L from *Peptostreptococcus magnus* binds to the κ light chain variable domain. *J. Biol. Chem.* **267**, 2334–2239.

Nissim A., Jouvin M.-H., and Eshhar Z. (1991). Mapping of the high affinity Fcε receptor binding site to the third constant region domain of IgE. *EMBO J.* **10**, 101–107.

Nissim A., Schwarzbaum S., Siraganian R., and Eshhar Z. (1993). Fine specificity of the IgE interaction with the low and high affinity Fc receptor. *J. Immunol.* **150**, 1365–1374.

Ohno T., Kubagawa H., Sanders S.K., and Cooper M.D. (1990). Biochemical nature of an Fcμ receptor on human B-lineage cells. *J. Exp. Med.* **172**, 1165–1175.

Pascual V., and Capra J.D. (1991). B-cell superantigens? *Curr. Biol.* **1**, 315– 317.

Patella V., Bouvet J.-P., and Marone G. (1993). Protein Fv produced during viral hepatitis is a novel activator of human basophils and mast cells. *J. Immunol.* **151**, 5685–5698.

Perkins S.J., Nealis A.S., Sutton B.J., and Feinstein A. (1991). Solution structure of human and mouse immunoglobulin M by synchrotron x-ray scattering and molecular graphics modelling. A possible mechanism for complement activation. *J. Mol. Biol.* **221**, 1345–1366.

Perosa F., Luccarelli G., Neri M., and Dammacco F. (1997). The Fab region of IgG2 human myeloma proteins does not bear the streptococcal protein G-specific determinant. *J. Immunol. Meth.* **203**, 153–155.

Popov S., Hubbard J.G., Kim J.-K., Ober B., Ghetie V., and Ward E.S. (1996). The stoichiometry and affinity of the interaction of murine Fc fragments with the MHC class I-related receptor, FcRn. *Mol. Immunol.* **33**, 521–530.

Potter K.N., Li Y., and Capra J.D. (1996). Staphylococcal protein A simultaneously interacts with framework region 1, complementarity-determining region 2, and framework region 3 on human V_H3–encoded Igs. *J. Immunol.* **157**, 2982–2988.

Raghavan M., and Bjorkman P.J. (1996). Fc receptors and their interactions with immunoglobulins. *Annu. Rev. Cell Dev. Biol.* **12**, 181–220.

Raghavan M., Bonagura V.R., Morrison S.L., and Bjorkman P.J. (1995). Analysis of the pH dependence of the neonatal Fc receptor/immunolgobulin G interaction using antibody and receptor variants. *Biochemistry* **34**, 14649–14657.

Raghavan M., Chen M.Y., Gastinel L.N., and Bjorkman P.J. (1994). Investigation of the interaction between the class I MHC-related Fc receptor and its immunoglobulin G ligand. *Immunity* **1**, 303–315.

Ravetch J.V. (1994). Fc receptors: rubor redux. *Cell* **78**, 553–560.

Ravetch J.V. (1997). Fc receptors. *Curr. Opin. Immunol.* **9**, 121–125.

Reid K.B.M. (1996). The complement system. In: *Molecular Immunology* (Eds. Hames B.D., and Glover D.M.). pp. 326–381. IRL Press, Oxford.

Roben P.W., Salem A.N., and Silverman G.J. (1995). V_H3 family antibodies bind domain D of staphylococcal protein A. *J. Immunol.* **154**, 6437–6445.

Robertson M.W., and Liu F.-T. (1991). Heterogeneous IgE glycoforms charcterized by differential recognition of an endogeneous lectin (IgE-binding protein). *J. Immunol.* **147**, 3024–3030.

Roque-Bareira M.C., and Campos-Neto A. (1985). Jacalin: an IgA-binding lectin. *J. Immunol.* **134**, 1740–1743.

Rosen F.S. (1993). Putative mechansisms of the effect of intravenous γ-globulin. *Clin. Immunol. Immunopathol.* **67**, S41–S43.

Rostagno A., Frangione B., and Gold L.I. (1991). Biochemical studies of the interaction of fibronectin with Ig. *J. Immunol.* **146**, 2687–2693.

Rostagno A., Williams M., Frangione B., and Gold L.I. (1996). Biochemical analysis of the interaction of fibronectin with IgG and localization of the respective binding sites. *Mol. Immunol.* **33**, 561–572.

Sandilands G.P., Ahmed A.E.E., Griffiths M.R., and Whaley K. (1990). Immunochemical localization of a plasma protein (glycoprotein 60) which inhibits complement-mediated prevention of immune precipitation. *Immunology* **70**, 303–308.

Sármay G., Lund J., Rozsnyay Z., Gergely J., and Jefferis R. (1992). Mapping and comparison of the interaction sites on the Fc region of IgG responsible for triggering antibody dependent cellular cytotoxity (ADCC) through different types of human Fcγ receptor. *Mol. Immunol.* **29**, 633–639.

Sauer-Eriksson A.E., Kleywegt G.J., Uhlén M., and Jones T.A. (1995). Crystal structure of the C2 fragment of streptococcal protein G in complex with the Fc domain of human IgG. *Structure* **3**, 265–278.

Seppälä I., Kaartinen M., Ibrahim S., and Mäkelä O. (1990). Mouse Ig coded by V_H families S107 or J606 bind to protein A. *J. Immunol.* **145**, 2989–2993.

Sharon N. (1993). Lectin-carbohydrate complexes of plants and animals: an atomic view. *Trends Biochem. Res.* **18**, 221–226.

Sharon N., and Lis G. (1987). A century of lectin research (1888–1988). *Trends Biochem. Res.* **12**, 488–491.

Shohet J.M., Pemberton P., and Carroll M.C. (1993). Identification of a major binding site for complement C3 on the IgG_1 heavy chain. *J. Biol. Chem.* **268**, 5866–5871.

Silverman G.J. (1997). B-cell superantigens. *Immunol. Today* **18**, 379–386.

Simister N.E., and Mostov K.E. (1989). An Fc receptor structurally related to MHC class I antigens. *Nature* **337**, 184–187.

Sumar N., Bodman K.B., Rademacher T.W., Dwek R.A., Williams P., Parekh R.B., Edge J., Rook G.A.W., Isenberg D.A., Hay F.C., and Roitt I.M. (1990). Analysis of glycosylation changes in IgG using lectins. *J. Immunol. Meth.* **131**, 121–136.

Takai T., Ono M., Hikida M., Ohmori H., and Ravetch J.V. (1996). Augmented humoral and anaphylactic responses in FcγRII-deficient mice. *Nature* **379**, 346–349.

Tan L.K., Shopes R.J., Oi V.T., and Morrison S.L. (1990). Influence of the hinge region on complement activation, C1q binding, and segmental flexibility in chimeric human immunoglobulins. *Proc. Natl. Acad. Sci. USA* **87**, 162–166.

Tao M.-H., Canfield S.M., and Morrison S.L. (1991). The differential ability of human IgG1 and IgG4 to activate complement is determined by the COOH-terminal sequence of the C_H2 domain. *J. Exp. Med.* **173**, 1025–1028.

Tao M.-H., Smith R.I.F., and Morrison S.L. (1993). Structural features of human immunoglobulin G that determine isotype-specific diffrences in complement activation. *J. Exp. Med.* **178**, 661–667.

Tashiro M., and Montelione G.T. (1995). Structures of bacterial immunoglobulin-binding domains and their complexes with immunoglobulins. *Curr. Opin. Struct. Biol.* **5**, 471–481.

Thaler C.J., Faulk W.P., and McIntyre J.A. (1989). Soluble antigens of IgG receptor FcγRIII in human seminal plasma. *J. Immunol.* **143**, 1937–1942.

Tsuchiya N., Endo T., Matsuta K., Yoshinoya S., Takeuchi F., Nagano Y., Shiota M., Furukawa K., Kochibe N., Ito K., and Kobata A. (1993). Detection of glycosylation abnormality in rheumatoid IgG using N-acetylglucosamine-specific *Psathyrella velutina* lectin. *J. Immunol.* **151**, 1137–1146.

Utsumi S., Okada M., Udaka K., and Amano T. (1985). Preparation and biologic characterization of fragments containing dimeric and monomeric $C_\gamma2$ domain of rabbit IgG. *Mol. Immunol.* **22**, 811–819.

Vaughn D.E., Milburn C.M., Penny D.M., Martin W.L., Johnson J.L., and Bjorkman P.J. (1997). Identification of critical IgG binding epitopes on the neonatal Fc receptor. *J. Mol. Biol.* 274, 597–607.

Vivanco F., Muñoz E., Vidarte L., and Pastor C. (1997). Inhibition of C3 covalent binding to IgG immune aggregates by recombinant protein G (domain III). *Immunol. Lett.* **56**, 109.

Walker A.M., Montgomery D.W., Saraiya S., Ho T.W.C., Garewal H.S., Wilson J., and Lorand L. (1995). Prolactin-immunoglobulin G complexes from human serum act as costimulatory ligands causing proliferation of malignant B lymphocytes. *Proc. Natl. Acad. Sci USA* **92**, 3278–3282.

Ward E.S., and Ghetie V. (1995). The effector functions of immunoglobulins: implications for therapy. *Therapeutic Immunology* **2**, 77–94.

Wikström M., Sjöbring U., Drakenberg T., Forsén S., and Björck L. (1995). Mapping of the immunoglobulin light chain-binding site of protein L. *J. Mol. Biol.* **250**, 128–133.

Wikström M., Drakenberg T., Forsén S., Sjöbring U., and Björck L. (1994). Three-dimensional solution structure of an immunoglobulin light chain-binding domain of protein L. Comparison with the IgG-binding domains of protein G. *Biochemistry* **33**, 14011–14017.

Wilson M.R., Roeth P.J., and Easterbrook-Smith S.B. (1991). Clusterin enhances the formation of insoluble immune complexes. *Biochem. Biophys. Res. Comm.* **177**, 985–990.

Wright A., and Morrison S.L. (1994). Effect of altered C_H2–associated carbohydrate structure on the functional properties and in vivo fate of chimeric mouse-human immunoglobulin G1. *J. Exp. Med.* **180**, 1087–1096.

Wright J.F., Shulman M.J., Isenman D.E., and Painter R.H. (1990). C1 binding by mouse IgM. The effect of abnormal glycosylation at position 402 resulting from a serine to asparagine exchange at residue 406 of the μ chain. *J. Biol. Chem.* **265**, 10506–10513.

Wu Y., Pun C., and Hozumi N. (1997). Roles of calnexin and Ig-$\alpha\beta$ interactions with membrane Igs in the surface expression of the B cell antigen receptor of the IgM and IgD classes. *J. Immunol.* **158**, 2762–2770.

Zheng Y., Shopes B., Holowka D., and Baird B. (1991). Conformation of IgE bound to its receptor FcεRI and in solution. *Biochemistry* **30**, 9125–9132.

Zouali M. (1995). B-cell superantigens: implications for selection of the human antibody repertoire. *Immunol. Today* **16**, 399–405.

CHAPTER 7

Segmental Movements of Immunoglobulin Molecules

I. GENERAL ASPECTS

Immunoglobulin chains are composed of compact globules (domains) connected by short switch peptides. Fab and Fc portions composed of four domains each are tethered by the hinge region. The hinge region permits Fab and Fc to move relative to each other and switch peptides allow them to bend. The variety of internal movements common to all immunoglobulins are very important for the proper functioning of these proteins. The immunoglobulins are able not only to recognize antigens, but also to react with many other ligands and to participate in a number of other important reactions. The binding sites for antigens and other ligands are located in different parts of the immunoglobulin molecules, and the ability of their segments to rotate and to flex seems to be essential for optimal binding to several different ligands, often simultaneously.

The first proofs that immunoglobulins are flexible molecules were obtained more than 30 years ago by methods that are indirect for the study of molecular dynamics. With use of classical hydrodynamic methods, such as sedimentation analysis, determination of the frictional coefficient ratio, and viscosity measurements, it was found that both Fab and Fc behave as typical globular proteins, whereas the intact IgG molecule is a more extended molecule (Noelken *et al.*, 1965). The hypothesis was formulated that IgG molecules are built from three compact subunits resembling the papain Fab and Fc fragments and linked by flexible parts of the heavy chains.

This hypothesis was supported by electron microscopic studies of antibody–antigen complexes. After cross-linking of two molecules of antigen (ferritin), antibody molecules click open to varying degrees about a hinge point (Feinstein and Row, 1965). In the second study, it was shown that the angle between the Fab arms of anti-DNP antibodies varied from nearly 0 degrees in dimer–antibody complexes with a small bivalent hapten to about 180 degrees in circular complexes composed of several antibody molecules with the same bivalent hapten (Valentine and Green, 1967). These results form the basis for a flexible model of the IgG molecule, according to which the Fab arms could move more or less freely due to the existence of the flexible hinge located between Fab and Fc.

The measurements of the rotational relaxation time (τ) by relaxation methods (fluorescence polarization or spin-label technique) provide the direct evidence of segmental flexibility of protein molecules. If the experimental values of the τ are significantly lower than those calculated, assuming the rigidity of the macromolecule, the macromolecule is flexible and its separate parts are capable of more or less independent motion. Two variants of fluorescent depolarization measurements have been used. By steady-state fluorescence polarization measurements of aggregate-free IgG preparations, experimental τ values were found that were considerably lower (20–30 nsec) than those computed assuming the rigidity of IgG molecule (70 nsec) (Zagyansky *et al.*, 1969; Nezlin, 1990). Furthermore, it was found that the correlation time for the isolated Fab fragment is close to that determined for intact IgG (Nezlin *et al.*, 1970). Due to the short lifetime of the excited state of the fluorescent molecule used in this experiment (7.3 nsec), only Fab movements were practically registered, and not the tumbling of the whole molecule. By nanosecond pulse fluorescent polarization experiments (Yguerabide *et al.*, 1970; Hanson *et al.*, 1981), it has been shown that the Fab arms within the intact IgG molecule exhibit significant freedom of movement. These results were confirmed by other studies and now it is admitted that the Fab arms can rotate relatively freely between angles of 60–80 degrees and 180 degrees.

The valuable information on the dynamic aspects of the IgG structure was obtained by x-ray crystallography. The segmental flexibility impedes the formation of a crystal lattice and only a few IgG molecules have been crystallized. In crystals, the Fab of two intact IgG (Kol and Zie) molecules are well ordered but their Fc portion is disordered owing to the flexibility of the lower hinge. The Fc disorder is probably due to distribution of this portion of the molecule among several different sites of the crystal lattice. When anti-viral antibodies attach one viral particle by both combining sites, the Fc portions occupy random positions and no significant density that could correspond to Fc has been determined by cryoelectron microscopy and image reconstruction (Smith *et al.*, 1993; Thouvenin *et al.*, 1997).

In crystals of intact monoclonal mouse IgG2a (Mab231) all parts of the molecule, including Fc and the hinge region, are visible (Harris *et al.*, 1997). No evidence for the mobility of Fc is found in the IgG Dob and Mcg, which have a deletion of the hinge (Fett *et al.*, 1973; Steiner and Lopes, 1979). Their Fc portions are located tightly between Fab arms, thereby limiting their mobility (Rajan *et al.*, 1983). Such close contacts are not seen between subunits of the intact IgG.

The dynamic behavior of IgG molecules was studied by some other physicochemical methods and the results of these experiments are in agreement with a flexible model of the IgG molecule. According to measurements performed by the neutron spin-echo techniques, the Fab fragments can wobble within an angle of 50 degrees (Alpert *et al.*, 1985). The distances between the centers of antigen bound to monoclonal antibody molecules of three murine IgG subclasses (IgG1, IgG2a, and IgG2b) have been directly measured by neutron scattering (Sosnick *et al.*, 1992). The mean distances between two molecules of a small protein antigen, staphylococcal nuclease, located in the combining sites, are between 117 and 134 Å, with a large variance ($\approx$40 degrees). These data indicate a high degree of the dynamical flexibility of the studied IgG antibodies. A correlation between the subclass of the studied IgG antibodies and the degree of flexibility was not found in these experiments.

However, it was shown by nanosecond fluorescence polarization that the segmental flexibility correlates with the subclass differences in the length of the hinge region: the longer the upper hinge, the more flexible the IgG molecules were (Dangl *et al.*, 1988; Tan *et al.*, 1990). Distribution of the murine isotypes according to their flexibility was found as follows: IgG2b $>$ IgG2a $>$ IgG3 $>$ IgG1. For human isotypes the distribution is IgG3 $>$ IgG1 $>$ IgG4 $>$ IgG2. The same rank order of the hinge-mediated flexibility was found in electron microscopic studies (Roux *et al.*, 1997). The mean angle between Fab arms, as determined in these experiments, is equal to 136 degrees for IgG3, 128 degrees for IgG4, 127 degrees for IgG2, and 117 degrees for IgG1. The variation of this angle is also characteristic of igG subclasses: $\pm$53 degrees for IgG3, $\pm$43 degrees for IgG1, $\pm$39 degrees for IgG4, and $\pm$32 degrees for IgG2. The data on the extent of the IgG flexibility are represented on a diagram (Fig. 81.)

Several studies performed either by electron microscopy or by fluorescent polarization and spin-label methods yielded data supporting mouse and human IgA and IgM segmental flexibility (Cathou, 1978). The mobile parts of these molecules are most likely their Fab fragments. IgM molecules of lower vertebrates, like carp and shark, are less flexible than mammal IgMs. Probably the

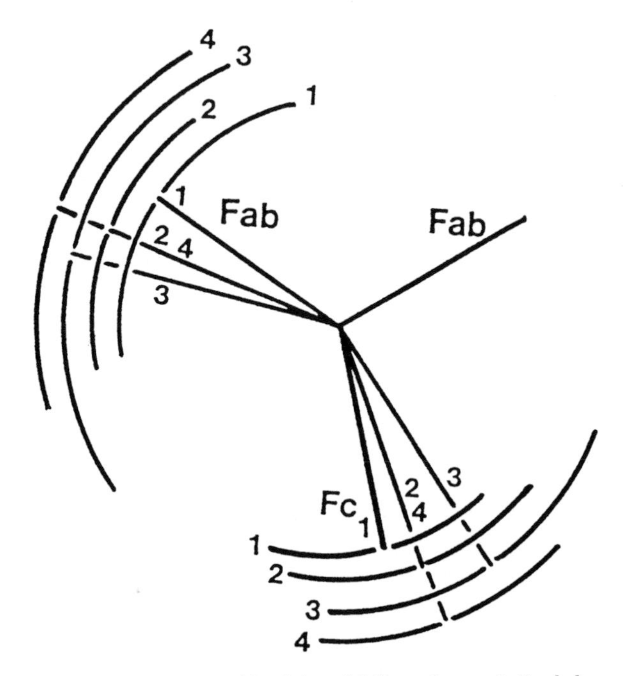

FIGURE 81 The hinge-bending mode of flexibility of different human IgG subclasses (as labeled, 1, 2, 3, and 4). The mean of Fab–Fab and Fab–Fc orientations are indicated by lines. The arcs represent the standard deviations of the mean measures of angles. The right Fab arm serves as a common reference. (Roux *et al.*, 1997. Reprinted with permission from The American Association of Immunologists.)

segmental flexibility of immunoglobulin molecules appeared relatively late in evolution (Zagyansky, 1975).

Human and murine IgE molecules are less flexible and have more compact configuration than IgG and IgM molecules (Nezlin *et al.*, 1973; Oi *et al.*, 1984; Slattery *et al.*, 1985). The measurement of the distance between the C-terminal of the IgE molecules and their antigen-binding sites was performed by fluorescence resonance energy transfer between a fluorescent dye at the end of the $C_\varepsilon3$ domain as a donor and a dye in the antigen-combining site of a IgE antibody molecule as an acceptor (Zheng *et al.*, 1991; 1992). These studies indicate that rather than existing in extended conformation, the IgE molecules are in bent conformation in solution and the axes of their Fab and Fc do not form a planar Y-shape (Fig. 72). After binding to the FcεRI receptors on the cell membrane the IgE molecules become even more rigid. However, the Fabs of IgE have some freedom of rotation, although it is less pronounced than that in IgG and IgM. By the method of resonance energy transfer it was found that the IgG molecules in solution are also bent; however, they posses significantly more segmental flexibility than IgE (Zheng *et al.*, 1992).

It is not an easy task to describe the precise character of Fab movements. The experimental values of the rotational relaxation times usually represent an average of several correlation times. Probably the Fab arms rotate more or less independently around the joints in the hinge region in one of the following ways: (1) angular waggling, (2) conlike wobbling, (3) twisting around the long axis, and (4) some movements around the short axes (Fig 82). $F(ab')_2$ rotation relative to Fc is due to the low hinge region. This kind of flexibility may increase the accessibility of Fc portions for interaction with effector ligands.

II. FUNCTIONAL ASPECTS OF SEGMENTAL FLEXIBILITY

For the formation of precipitate or agglutinate networks, antibody molecules must simultaneously bind to the epitopes of two antigen molecules or two cells. Epitope distribution varies greatly among antigens, and freedom of Fab movement is crucial for optimal antigen–antibody binding and formation of antibody bridges between antigens. The measurements of the flexibility of precipitating and nonprecipitating porcine IgG antibodies support this view. Fluorescence polarization and spin-label technique revealed that precipitating antibodies are significantly more flexible than nonprecipitating antibodies of the same specificity (Dudich *et al.*, 1978; Sykulev *et al.*, 1979).

When antibodies attach via both combining sites to antigens with repeating epitopes (bivalent attachment) such as bacterial polysaccharides, viruses, or synthetic antigens, their functional affinity enhances significantly, up to 100-fold (Karush, 1978). The arrangement of the repeating antigenic determinants varies among different bacteria and viruses. However, the gross conformation

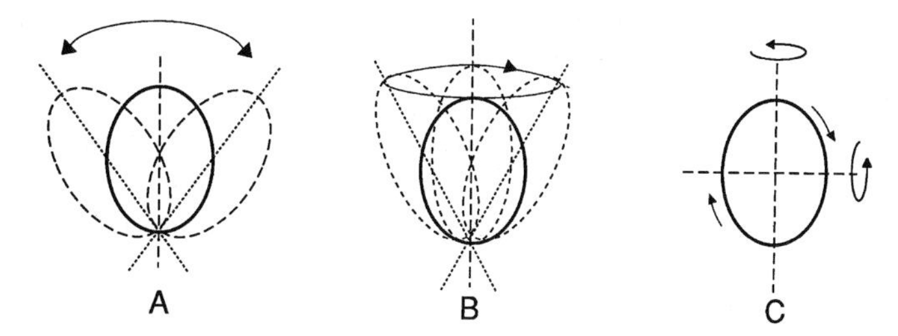

FIGURE 82 Possible models of rotation of the Fab in an intact IgG molecule. (A) Angular waggling. (B) Conelike wobbling. (C) Motions along short and long axes (Nezlin, 1990. Reprinted with permission.)

of antibody molecules of different specificities is the same. The ability of Fab portions bearing the antigen-combining sites to rotate relatively freely increases the probability that the antibody molecules can react with two epitopes simultaneously, resulting in a significant gain in binding strength provided by the IgG bivalency.

Bivalent ineractions of IgG antibodies with viral particles were studied by cryoelectron microscopy and three-dimensional reconstructon techniques. In the complexes of anti-viral antibodies with human rhinovirus (HRV2), the distance between two eitopes attached by the combining sites is as little as 60 Å (Hewat *et al.*, 1996) whereas in the complex with another variant of the same virus (HRV14), the spanning distance for bivalent antibody attachemnt is about two times larger (Smith *et al.*, 1993). The Fab elbow movements are probably most important for such bivalent IgG attachments. In the complexes of neutralizing antibody molecules with particles of rabbit calicivirus, another type of antibody flexibility (torsional flexibility) was noted (Thouvenin *et al.*, 1997). The bivalently bound antibody molecules in these complexes are not related by twofold symmetry axes and one Fab arm rotates by 60 degrees with respect to the other Fab arm.

Obviously, the pentameric structure of IgM molecules also enhances binding to antigens with many identical epitopes. In such instances, the Fab_{μ} arms can fold from the central Fc_5 disc to form a multilegged "table" on particulate antigens, such as bacterial flagella (Perkins *et al.*, 1991).

A direct correlation between the degree of segmental motion and their capacity to bind and activate complement was not observed (Tan *et al.*, 1990; Sandlie and Michaelsen, 1991). Although human IgG4 molecules exhibit restricted flexibility and are unable to activate complement, their aggregated Fc fragments can activate complement. Furthermore, a chimeric variant of IgG3, containing the hinge region of IgG4, activates the complement cascade very effectively. By contrast, the IgG4 chimeric molecule with the flexible hinge of IgG3 was unable to activate complement. Therefore, the sequence of the hinge region of IgG4 and the restricted flexibility alone cannot explain the inability of native IgG4 molecules to activate complement. The IgG3 molecule with a tripeptide instead of the long genetic hinge and with the interheavy disulfide bridge has the ability to initiate complement activation (Michaelsen *et al.*, 1994). Therefore, for complement activation, the presence of the normal hinge region is not necessary, but the connection of two heavy chain by a disulfide bridge is sufficient for effector functions.

The experiments with a chimeric mouse (V_H)–human ($C_{\gamma}1$–$L_{\varkappa}$) immunoglobulin molecule, which has a disulfide bridge between C_H1 domains, also suggest that the segmental flexibility is not important for complement activity of IgG (Shopes, 1993). Due to the additional disulfide linkage, the chimeric IgG has a restricted flexibility, which was confirmed by fluorescence polarization experiments. An average distance between its combining sites was significantly shorter than in normal IgG. Despite such restrictions, the chimeric molecule was able to initiate cytolysis of target cells by complement activation.

REFERENCES

Alpert Y., Cser L., Farago B., Franěk F., Mezei F., and Ostanevich Y.M. (1985). Segmental flexibility in pig IgG studied by neutron spin-echo technique. *Biopolymers* **24**, 1769–1784.

Cathou R.E. (1978). Solution conformation and segmental flexibility of immunoglobulins. *Comprehen. Immunol.* **5**, 37–83.

Dangl J.L., Wensel T.G., Morrison S.L., Stryer L., Herzenberg L.A., and Oi V.T. (1988). Segmental flexibility and complement fixation of genetically engineered chimeric human, rabbit and mouse antibodies. *EMBO J.* **7**, 1989–1994.

Dudich E.I., Nezlin R.S., and Franěk F. (1978). Fluorescence polarization analysis of various immunoglobulins. Dependence of rotational relaxation time on protein concentration and on ability to precipitate with antigen. *FEBS Lett.* **89**, 89–92.

Feinstein A., and Rowe A.J. (1965). Molecular mechanism of formation of an antigen-antibody complex. *Nature* **205**, 147–149.

Fett J.W., Deutsch H.F., and Smithies O. (1973). Hinge-region deletion localized in the IgG_1-globulin Mcg. *Immunochemistry* **10**, 115–118.

Hanson D.C., Yguerabide J., and Schumaker V.N. (1981). Segmental flexibility of IgG antibody molecules in solution: a new interpretation. *Biochemistry* **20**, 6842–6852.

Harris L.J., Larson S.B., Hasel K.W., and McPherson A. (1997). Refined structure of an intact IgG2a monoclonal antibody. *Biochemistry* **36**, 1581–1597.

Hewat E.A., and Blaas D. (1996). Structure of a neutralizing antibody bound bivalently to human rhinovirus 2. *EMBO J.* **15**, 1515–1523.

Karush F. (1978). The affinity of antibody range, variability, and the role of multivalence. *Comprehen. Immunol.* **5**, 85–116.

Michaelsen T.E., Brekke O.H., Aase A., Sandin R., Bremnes B., and Sandlie I. (1994). One disulfide bond in front of the second heavy chain constant region is necessary and sufficient for effector functions in human IgG3 without a genetic hinge. *Proc. Natl. Acad. Sci. USA* **91**, 9243–9247.

Nezlin R. (1990). Internal movements in immunoglobulin molecules. *Adv. Immunol.* **48**, 1–40.

Nezlin R., Zagyansky Y.A., and Tumerman L.A. (1970). Strong evidence for the freedom of rotation of IgG subunits. *J. Mol. Biol.* **50**, 569–572.

Nezlin R., Zagyansky Y.A., Käiväräinen A.I., and Stefani D.V. (1973). Properties of myeloma immunoglobulin E(Yu). Chemical, fluorescence polarization and spin-labeled studies. *Immunochemistry* **10**, 681–688

Noelken M.E., Nelson C.A., Buckley C.E., and Tanford C. (1965). Gross conformation of rabbit 7S γ immunoglobulin and its papain-cleaved fragments. *J. Biol. Chem.* **240**, 218–223.

Oi V.T., Vuong T.M., Hardy R., Reidler J., Dangl J., Herzenberg L.A., and Stryer L. (1984). Correlation between segmental flexibility and effector function of antibodies. *Nature* **307**, 136–140.

Perkins S.J., Nealis A.S., Sutton B.J., and Feinstein A. (1991). Solution structure of human and mouse immunoglobulin M by synchrotron x-ray scattering and molecular graphics modelling. A possible mechanism for complement activation. *J. Mol. Biol.* **221**, 1345–1366.

Rajan S.S., Ely K.R., Abola E.E., Wood M.K., Colman P.M., Athay R.J., and Edmundson A.B. (1983). Three-dimensional structure of the Mcg IgG1 immunoglobulin. *Mol. Immunol.* **20**, 787–799.

Roux K.H., Strelets *l.*, and Michaelsen T. E. (1997). Flexibility of human IgG subclasses. *J. Immunol.* **159**, 3372–3382.

Sandlie I., and Michaelsen T.E. (1991). Engineering monoclonal antibodies to determine the structural requirements for complement activation and complement mediated lysis. *Mol. Immunol.* **28**, 1361–1368.

Shopes B. (1993). A genetically engineerd human IgG with limited flexibility fully initiates cytolysis via complement. *Mol. Immunol.* **30**, 603–609.

Slattery J., Holowka D., and Baird B. (1985). Segmental flexibility of receptor-bound immunoglobulin E. *Biochemistry* **24**, 7810–7820.

Smith T.J., Olson N.H., Cheng R.H. Chase E.S., and Baker T.S. (1993). Structure of a human rhinovirus–bivalently bound antibody complex: implications for viral neutralization and antibody flexibility. *Proc. Natl. Acad. Sci. USA* **90**, 7015–7018.

Sosnick T.R., Benjamin D.C., Novotny J., Seeger P.A., and Trewhella J. (1992). Distances between the antigen-binding sites of three murine antibody subclasses measured using neutron and x-ray scattering. *Biochemistry* **31**, 1779–1786.

Steiner *l*., and Lopes A.D. (1979). The crystallizable human myeloma protein Dob has a hinge-region deletion. *Biochemistry* **18**, 4054–4067.

Sykulev Y.K., Timofeev V.P., Nezlin R., Misharin A., and Franěk F. (1979). Spin-label study of segmental flexibility of antihapten antibodies. Precipitating pig anti-DNP antibody is more flexible than non-precipitating. *FEBS Lett.* **101**, 27–30.

Tan L.K., Shopes R.J., Oi V.T., and Morrison S.L. (1990). Influence of the hinge region on complement activation, C1q binding, and segmental flexibility in chimeric human immunoglobulins. *Proc. Natl. Acad. Sci. USA* **87**, 162–166.

Thouvenin E., Laurent S., Madalaine M.-F., Rasschaert D., Vautherot J.-F., and Hewat E.A. (1997). Bivalent binding of a neutralizing antibody to a calcivirus involves the torsional flexibility of the antibody hinge. *J. Mol. Biol.* **270**, 238–246.

Valentine R.C., and Green N.M. (1967). Electron microscopy of an antibody–hapten complex. *J. Mol. Biol.* **27**, 615–617.

Yguerabide J., Epstein H.F., and Stryer L. (1970). Segmental flexibility in an antibody molecule. *J. Mol. Biol.* **51**, 573–590.

Zagyansky Y.A. (1975). Phylogenesis of the general structure of immunoglobulins. *Arch. Biochem. Biophys.* **166**, 371–381.

Zagyansky Y.A., Nezlin R.S., and Tumerman L.A. (1969). Flexibility of immunoglobulin G molecules as established by fluorescence polarization measurements. *Immunochemistry* **6**, 787–800.

Zheng Y., Shopes B., Holowka D., and Baird B. (1991). Conformations of IgE bound to its receptor FceRI and in solution. *Biochemistry* **30**, 9125–9132.

Zheng Y., Shopes B., Holowka D., and Baird B. (1992). Dynamic conformations compared for IgE and IgG1 in solution bound to receptors. *Biochemistry* **30**, 7446–7456.

INDEX

B

C

D

J

L

M

N

P

R

S

T